Classical and Multilinear Harmonic Analysis

This two-volume text in harmonic analysis introduces a wealth of analytical results and techniques. It is largely self-contained and is intended for graduates and researchers in pure and applied analysis. Numerous exercises and problems make the text suitable for self-study and the classroom alike.

The first volume starts with classical one-dimensional topics: Fourier series; harmonic functions; Hilbert transforms. Then the higher-dimensional Calderón–Zygmund and Littlewood–Paley theories are developed. Probabilistic methods and their applications are discussed, as are applications of harmonic analysis to partial differential equations. The volume concludes with an introduction to the Weyl calculus.

This second volume goes beyond the classical to the highly contemporary and focuses on multilinear aspects of harmonic analysis: the bilinear Hilbert transform; Coifman–Meyer theory; Carleson's resolution of the Lusin conjecture; Calderón's commutators and the Cauchy integral on Lipschitz curves. The material in this volume has not been collected previously in book form.

Camil Muscalu is Associate Professor in the Department of Mathematics at Cornell University.

Wilhelm Schlag is Professor in the Department of Mathematics at the University of Chicago.

CAMBRIDGE STUDIES IN ADVANCED MATHEMATICS

All the titles listed below can be obtained from good booksellers or from Cambridge University Press. For a complete series listing visit: www.cambridge.org/mathematics.

Already published

93 D. Applebaum *Lévy processes and stochastic calculus (1st Edition)*
94 B. Conrad *Modular forms and the Ramanujan conjecture*
95 M. Schechter *An introduction to nonlinear analysis*
96 R. Carter *Lie algebras of finite and affine type*
97 H. L. Montgomery & R. C. Vaughan *Multiplicative number theory, I*
98 I. Chavel *Riemannian geometry (2nd Edition)*
99 D. Goldfeld *Automorphic forms and L-functions for the group GL(n,R)*
100 M. B. Marcus & J. Rosen *Markov processes, Gaussian processes, and local times*
101 P. Gille & T. Szamuely *Central simple algebras and Galois cohomology*
102 J. Bertoin *Random fragmentation and coagulation processes*
103 E. Frenkel *Langlands correspondence for loop groups*
104 A. Ambrosetti & A. Malchiodi *Nonlinear analysis and semilinear elliptic problems*
105 T. Tao & V. H. Vu *Additive combinatorics*
106 E. B. Davies *Linear operators and their spectra*
107 K. Kodaira *Complex analysis*
108 T. Ceccherini-Silberstein, F. Scarabotti & F. Tolli *Harmonic analysis on finite groups*
109 H. Geiges *An introduction to contact topology*
110 J. Faraut *Analysis on Lie groups: An Introduction*
111 E. Park *Complex topological K-theory*
112 D. W. Stroock *Partial differential equations for probabilists*
113 A. Kirillov, Jr *An introduction to Lie groups and Lie algebras*
114 F. Gesztesy *et al. Soliton equations and their algebro-geometric solutions, II*
115 E. de Faria & W. de Melo *Mathematical tools for one-dimensional dynamics*
116 D. Applebaum *Lévy processes and stochastic calculus (2nd Edition)*
117 T. Szamuely *Galois groups and fundamental groups*
118 G. W. Anderson, A. Guionnet & O. Zeitouni *An introduction to random matrices*
119 C. Perez-Garcia & W. H. Schikhof *Locally convex spaces over non-Archimedean valued fields*
120 P. K. Friz & N. B. Victoir *Multidimensional stochastic processes as rough paths*
121 T. Ceccherini-Silberstein, F. Scarabotti & F. Tolli *Representation theory of the symmetric groups*
122 S. Kalikow & R. McCutcheon *An outline of ergodic theory*
123 G. F. Lawler & V. Limic *Random walk: A modern introduction*
124 K. Lux & H. Pahlings *Representations of groups*
125 K. S. Kedlaya *p-adic differential equations*
126 R. Beals & R. Wong *Special functions*
127 E. de Faria & W. de Melo *Mathematical aspects of quantum field theory*
128 A. Terras *Zeta functions of graphs*
129 D. Goldfeld & J. Hundley *Automorphic representations and L-functions for the general linear group, I*
130 D. Goldfeld & J. Hundley *Automorphic representations and L-functions for the general linear group, II*
131 D. A. Craven *The theory of fusion systems*
132 J. Väänänen *Models and games*
133 G. Malle & D. Testerman *Linear algebraic groups and finite groups of Lie type*
134 P. Li *Geometric analysis*
135 F. Maggi *Sets of finite perimeter and geometric variational problems*
136 M. Brodmann & R. Y. Sharp *Local cohomology (2nd Edition)*
137 C. Muscalu & W. Schlag *Classical and multilinear harmonic analysis, I*
138 C. Muscalu & W. Schlag *Classical and multilinear harmonic analysis, II*
139 B. Helffer *Spectral theory and its applications*

Classical and Multilinear Harmonic Analysis

Volume II

CAMIL MUSCALU
Cornell University

WILHELM SCHLAG
University of Chicago

CAMBRIDGE
UNIVERSITY PRESS

University Printing House, Cambridge CB2 8BS, United Kingdom

One Liberty Plaza, 20th Floor, New York, NY 10006, USA

477 Williamstown Road, Port Melbourne, VIC 3207, Australia

4843/24, 2nd Floor, Ansari Road, Daryaganj, Delhi - 110002, India

79 Anson Road, #06-04/06, Singapore 079906

Cambridge University Press is part of the University of Cambridge.

It furthers the University's mission by disseminating knowledge in the pursuit of education, learning and research at the highest international levels of excellence.

www.cambridge.org
Information on this title: www.cambridge.org/9781107031821

First published 2013
Reprinted 2015

A catalogue record for this publication is available from the British Library

Library of Congress Cataloging in Publication data
Muscalu, C. (Camil), author.
Classical and multilinear harmonic analysis / C. Muscalu and W. Schlag.
volumes cm. – (Cambridge studies in advanced mathematics ; 138–)
Includes bibliographical references.
ISBN 978-0-521-88245-3 (v. 1 : hardback)
1. Harmonic analysis. I. Schlag, Wilhelm, 1969– author. II. Title.
QA403.M87 2013
515′.2422 – dc23 2012024828

ISBN 978-1-107-03182-1 Hardback

Contents

Preface

This is the second volume of our textbook devoted to harmonic analysis. The first volume commenced with the one-dimensional theory of Fourier series, harmonic functions, their conjugates, the Hilbert transform and its boundedness properties. It then moved on to the higher-dimensional theory of singular integrals, the Calderón–Zygmund and Littlewood–Paley theorems, and the restriction theory of the Fourier transform, as well as a brief introduction to pseudodifferential operators via the Weyl calculus. As Vol. I aims for breadth it also includes some basic probability theory and demonstrates how this relates to Fourier analysis. In addition, some application to PDEs are described. For example, in Chapter I.10 we discuss the uncertainty principle and how it allows for a simple proof of the Malgrange–Ehrenpreis theorem on the local solvability of constant-coefficient PDEs. In Chapter 11, which is devoted to restriction phenomenon for the Fourier transform, applications to dispersive evolution equations such as the wave and Schrödinger equations appear in the form of Strichartz estimates.

This second volume is more specialized in the sense that it is entirely devoted to multilinear aspects of singular integrals and pseudodifferential operators. However, at the same time it covers a wide range of topics within that area. By design, each topic is presented within the framework of a few overarching principles. Amongst these the most fundamental notion is that of a *paraproduct*, and we devote the first three chapters of the present volume to the introduction, motivation, and development of this basic idea. The immediate aim of these three chapters is a systematic and unifying treatment of fractional Leibnitz rules. In the later chapters, which analyze more difficult operators such as the Carleson maximal function, the bilinear Hilbert transform, and Calderón commutators, it becomes necessary to understand the combinatorics involved in handling many paraproducts simultaneously.

While paraproducts made an appearance in the first volume of our book, in the context of Haar functions and their use, in the proof of the $T(1)$ theorem, in this present volume we delve more deeply into the structure of these objects.

The core of Vol. II consists of three main strands, which are very much interwoven:

(1) Calderón commutators and the Cauchy integral on Lipschitz curves;
(2) the bilinear Hilbert transform;
(3) Carleson's theorem on the almost everywhere convergence of L^2 Fourier series.

While the relation between topics (1) and (2) was observed many years ago by Calderón, who also introduced them into harmonic analysis, the close relation between (2) and (3) is a more recent discovery.

Let us now give a brief synopsis of the history of each of these topics. In his thesis of 1915, Lusin conjectured that the Fourier series of any L^2 function converges almost everywhere. The question of whether this is true proved to be difficult to answer and, indeed, out of the reach of classical methods. In 1922 Kolmogorov famously constructed an example of an L^1 function for which the associated Fourier series diverges almost everywhere. In an attempt to shed some light on the L^2 case, Paley and Zygmund studied random Fourier series and established a theorem stating that an L^2 series with random coefficients converges almost surely almost everywhere; see Chapter 6 of Vol. I for both Kolmogorov's example and random Fourier series.

It was not until 1966 that Carleson established Lusin's conjecture as correct. The L^p version of Carleson's theorem for $1 < p < \infty$ was obtained shortly thereafter by Hunt. Carleson's theorem is based on a deep phase-space analysis of L^2 functions.

If convergence in L^2 of Fourier series is equivalent to the L^2-boundedness of each partial-sum operator $S_N f(x)$, independently of N (which is an easy consequence of Plancherel's theorem), it is natural to expect that proving the almost everywhere convergence requires the handling of infinitely many such (localized) partial Fourier sums simultaneously. This is a very difficult task since in any collection the partial sums may overlap with each other both in space and in frequency. To prove his theorem, Carleson invented an intricate combinatorial and analytical way of organizing these partial sums, based on geometric intuition coming from the phase-space picture and the Heisenberg principle, and which rendered these carefully selected partial sums almost orthogonal.

A few years after Carleson's breakthrough, Fefferman gave a new proof of the Carleson–Hunt theorem by building upon, as well as simplifying, some of these ideas. In addition, he also introduced the modern language of tiles and trees, which has been used in the field ever since. A proof of the Carleson–Hunt theorem is given in Chapter 7 of this volume.

In the 1960s, Calderón introduced commutators and the Cauchy integral on Lipschitz curves as part of his program in the study of PDEs. These interesting and natural singular integrals are no longer of the convolution type and cannot be treated by means of the earlier Calderón–Zygmund theory. By using a combination of methods from complex and harmonic analysis, Calderón proved the L^p-boundedness of his first commutator in 1965, but his method could not handle the second commutator let alone the higher ones. In spite of serious efforts by the harmonic analysis community, these problems resisted solution until 1975 when Coifman and Meyer proved the desired bounds for the second commutator and shortly afterwards for all Calderón commutators. Their proof builds on some of Calderón's ideas, but it was based entirely on harmonic analysis techniques. These authors were the first to realize in a profound way that the commutators are in fact multilinear operators, and their method of proof used this observation crucially. Around the same time they proved what is now called the Coifman–Meyer theorem on paraproducts. It is also interesting to note that all these multilinear operators came out of studies of linear PDE problems. A few years later, however, Bony realized that paraproducts play an important role in nonlinear PDEs, and they continue to do so to this day.

After all these developments, the Cauchy integral on Lipschitz curves (which initially appeared in complex analysis) was the last operator that remained to be understood. The Cauchy integral possesses the remarkable feature of being naturally decomposable as an infinite sum of all Calderón commutators, in which the simplest (linear) term equals the classical Hilbert transform. In order to prove estimates for it one therefore needed to prove sufficiently good bounds for the operator norms of all the commutators, so as to be able to sum their contributions. The initial proof by Coifman and Meyer was very complicated and it was not clear what type of bounds it yielded for the commutators. Calderón was the first to obtain exponential bounds, thus proving L^p estimates for the Cauchy integral under the assumption that the Lipschitz constant is small. It had also been observed that polynomial bounds for the commutators would allow for a complete understanding of the Cauchy integral on Lipschitz curves and many of its natural extensions. The final breakthrough was achieved in 1982 by Coifman, McIntosh, and Meyer, who established the desired polynomial bounds for Calderón commutators. Chapter 4 of this volume is devoted to the proofs of these results. The proofs we give there differ from those in the original literature.

Prior to the resolution of his conjectures on the commutators, Calderón proposed the study of the bilinear Hilbert transform. He viewed this as a step towards the commutators and eventually to the Cauchy integral on Lipschitz curves. It had been observed that the first commutator was equal to an average of such bilinear operators. While the entire Calderón program was settled without the use of the bilinear Hilbert transform, the question whether this operator satisfies any L^p estimates remained open. It was finally answered affirmatively by Lacey and Thiele in two papers, in 1997 and 1999. As it turned out, the analysis of the bilinear Hilbert transform is very closely related to the analysis of the Carleson maximal operator, which appeared implicitly in the proof of the almost everywhere convergence of Fourier series. A brief explanation for this is as follows. When viewed as a bilinear multiplier, the symbol of the bilinear Hilbert transform is singular along a one-dimensional line. This line regulates the one-dimensional modulation symmetry of the bilinear Hilbert transform, which is identical to the modulation symmetry of the Carleson operator. In particular, both these objects have precisely the same symmetries: translation, dilation, and modulation invariances. By comparison, paraproducts have classical Marcinkiewicz–Mikhlin–Hörmander symbols and these are smooth away from the origin; in other words they have a zero-dimensional singularity. They have only translation and dilation invariance, the usual symmetries of the classical Calderón–Zygmund convolution operators. A proof of the boundedness of the bilinear Hilbert transform appears in Chapter 6 of this volume.

Let us conclude with a few words about Chapter 5, which describes the more recent theory of iterated Fourier series and integrals. This chapter is almost entirely devoted to motivation and is somewhat speculative in character. To be more specific, we describe what appears to be a very natural "physical" problem where both the Carleson maximal operator and the bilinear Hilbert transform appear simultaneously. In fact, they are the simplest operators in an infinite series that determines the solutions of a certain ODE. This problem goes beyond *multilinear harmonic analysis* and may be said to belong to *nonlinear harmonic analysis*, since the study of its multilinear building blocks alone is insufficient for its complete understanding. It is towards such a theory of nonlinear harmonic analysis that Chapter 5 aspires.

How to use this volume. There are several options. For instance, after completing Chapters 7 and 8 of Vol. I on the classical theory of singular integrals, it is natural to move on to some of the more advanced topics presented in this volume, such as the theory of Calderón commutators and the Cauchy integral on Lipschitz curves. We would like to emphasize that this is indeed possible since Chapter 4 of the present volume, where these topics are covered, assumes familiarity with only basic harmonic analysis such as maximal functions, Calderón–Zygmund operators, and Littlewood–Paley square functions. It does not rely on any other material from Vol. I such as, for example, the $T(1)$ theorem, bounded mean oscillation space (BMO), or Carleson measures.

However, for an audience that is more inclined towards learning techniques useful for PDEs, it might be advisable to focus on paraproducts and Leibnitz rules. In that case, Chapters 1, 2, 3, and 8 would be the natural order in which to proceed.

A more mature reader wishing to study the almost everywhere convergence of Fourier series can in principle start with Chapter 7. However, that chapter does rely on some technical results from the preceding chapter, where the bilinear Hilbert transform is presented. We therefore recommend that Chapter 7 should be attempted only after mastering Chapter 6, which can be read independently by anyone familiar with paraproducts and the John–Nirenberg inequality (which are covered in Chapter 2).

Finally, we would like to stress that this volume was designed for the specific purpose of fitting seemingly disparate objects into a unifying framework. The clarity and transparency that we hope to have achieved by doing so will only be appreciated by the patient reader willing to take the journey from beginning to end.

A few words are in order concerning interpolation. It is used frequently either in the more standard linear form of the Riesz–Thorin and Marcinkiewicz theorems (both in the Banach and the quasi-Banach context) or in the multilinear setting described in detail in the appendix. For statements of the standard interpolation theorems, see for example Chapter 1 of Vol. I.

Feedback. The authors welcome comments on this book and ask that they be sent to harmonic@math.uchicago.edu.

Acknowledgements

Wilhelm Schlag expresses his gratitude to Rowan Killip for detailed comments on his old harmonic analysis notes from 2000, from which the first volume of this book eventually emerged. Furthermore, he thanks Serguei Denissov, Charles Epstein, Burak Erdogan, Patrick Gérard, David Jerison, Carlos Kenig, Andrew Lawrie, Gerd Mockenhaupt, Paul Müller, Casey Rodriguez, Barry Simon, Chris Sogge, Wolfgang Staubach, Eli Stein, and Bobby Wilson for many helpful suggestions and comments on a preliminary version of Vol. I. Finally, he thanks the many students and listeners who attended his lectures and classes at Princeton University, the California Institute of Technology, the University of Chicago, and the Erwin Schrödinger Institute in Vienna over the past ten years. Their patience, interest, and helpful comments have led to numerous improvements and important corrections.

The second volume of the book is partly based on two graduate courses given by Camil Muscalu at Şcoala Normală Superioară, Bucureşti, in the summer of 2004 and at Cornell University in the fall of 2007. First and foremost he would like to thank Wilhelm Schlag for the idea of writing this book together. Then, he would like to thank all the participants of those classes for their passion for analysis and for their questions and remarks. In addition, he would like to thank his graduate students Cristina Benea, Joeun Jung, and Pok Wai Fong for their careful reading of the manuscript and for making various corrections and suggestions and Pierre Germain and Raphaël Côte for their meticulous comments. He would also like to thank his collaborators Terry Tao and Christoph Thiele. Many ideas that came out of this collaboration are scattered through the pages of the second volume of the book. Last but not least, he would like to express his gratitude to Nicolae Popa from the Institute of Mathematics of the Romanian Academy for introducing him to the world of harmonic analysis and for his unconditional support and friendship over the years.

Many thanks go to our long-suffering editors at Cambridge University Press, Roger Astley and David Tranah, who continued to believe in this project and support it even when it might have been more logical not to do so. Their cheerful patience and confidence is gratefully acknowledged. Barry Simon at the California Institute of Technology deserves much credit for first suggesting to David Tranah roughly ten years ago that Wilhelm Schlag's harmonic analysis notes should be turned into a book.

The authors were partly supported by the National Science Foundation during the preparation of this book.

1

Leibnitz rules and the generalized Korteweg–de Vries equation

A primary role of this first chapter is motivational. We aim to convince the reader that *paraproducts* are important objects, which appear in analysis in a natural way. Paraproducts were discussed in Section 9.4 of Vol. I (i.e., Section I.9.4), in the context of the $T(1)$ theorem. Our goal here is in some sense complementary, since now we want to describe some of their connections to the theory of differential equations. We plan to do this in two steps. In the present chapter we explain the appearance of the Leibnitz rules, which play an important role in nonlinear PDEs and in Chapter 2 we show why paraproducts are the correct objects to use in understanding these estimates.

The Leibnitz rules are inequalities of the type

$$\|D^\alpha(fg)\|_r \lesssim \|D^\alpha f\|_{p_1}\|g\|_{q_1} + \|f\|_{p_2}\|D^\alpha g\|_{q_2}{}^1. \tag{1.1}$$

which hold as long as $1 < p_i, q_i \le \infty$, $1/r = 1/p_i + 1/q_i$ for $i = 1, 2$ and $1/(1+\alpha) < r < \infty$. The fractional derivative $D^\alpha h$ is defined for every $\alpha > 0$ by, as usual, $\widehat{D^\alpha h}(\xi) = (2\pi|\xi|)^\alpha \widehat{h}(\xi)$, and all the functions involved are defined on the real line. Such inequalities are valid in higher dimensions and also for an arbitrary number of functions, but for simplicity we restrict ourselves to this particular *bilinear* one-dimensional case. However, the method that we will develop to understand them works equally well in the general case.

There are many natural questions that the reader may have about these inequalities. Such questions will be addressed in detail in Chapter 2. For now, let us just point out that if instead of $D^\alpha(fg)$ one considers the simpler expressions fg or $(fg)'$ then the corresponding estimates follow easily from Leibnitz's formula and Hölder's inequality.

To motivate the inequalties (1.1), we shall describe some natural dispersive estimates for a certain generalized Korteweg–de Vries (gKdV) equation , which

[1] The sign $\lesssim$ means "less than or equal to within a multiplicative constant".

rely (among other things) on such inequalities. However, there are other things that the reader will find in this chapter. In particular, in order to understand the so called *Airy function*, which will be defined in a natural way later on, we need to introduce *wave packets* and *phase-space portraits* concepts that will play a fundamental role throughout the rest of the book.

Let us start by considering the following initial-value problem (IVP) for the gKdV equation on the real line $\mathbb{R}$:

$$\begin{cases} \partial_t u + \partial_x^3 u + \partial_x F(u) = 0, \\ u(0, x) = g(x), \end{cases} \tag{1.2}$$

where the solution $u(t, x)$ is a real-valued function of two real variables and the given function $g(x)$ is its initial profile. This equation models weakly nonlinear shallow-water waves. The above function F is continuous and satisfies $F(0) = 0$. In the particular case $F(x) = x^2/2$ we obtain the classical KdV equation. The following notations are standard.

For every time t, one naturally defines the function $u(t)$ by $u(t)(x) := u(t, x)$. Then, if B is an arbitrary Banach space, $C(\mathbb{R}, B)$ denotes the space of all B-valued continuous functions while $C_w(\mathbb{R}, B)$ denotes the space of all B-valued weakly continuous functions.

We will also rely on the following classical facts about the gKdV equation, which will be taken for granted hereafter.

First, if $F \in C^2$ and if $g \in H^1$ with $\|g\|_{H^1}$ small then the IVP (1.2) has a solution $u \in C(\mathbb{R}, L^2) \cap C_w(\mathbb{R}, H^1)$; here, we denote by H^1 the classical Sobolev space with one derivative in L^2. This is a classical theorem of Kato. Second, for $F(u) = |u|^s$ the gKdV equation has solitary-wave solutions of the form $u(t, x) = w(x - ct)$, called *solitons*. As one can see, these solutions do not change their shape and travel with speed c.

1.1. Conserved quantities

The interesting fact about the gKdV equation is that it has infinitely many conserved quantities (i.e., expressions involving the solution that remain constant in time). We list the first three of them:

(1) $\int_{\mathbb{R}} u dx$, *the conservation of mass;*
(2) $\int_{\mathbb{R}} u^2 dx$, *conservation of the* L^2 *norm;*
(3) $\int_{\mathbb{R}} (\frac{1}{2}u_x^2 - V(u))\, dx$, *conservation of the Hamiltonian, where* V *is an integral of* F.

Here the function u is assumed to be smooth enough that these formulae are all well defined. Later, when we are going to use them, we will see that this will always be the case. The proofs below use the Fourier transform. The more standard approach based on integration by parts is left to the reader as one of the problems at the end of the chapter.

Proof of (1) Taking the Fourier transform (with respect to the x variable) of the equation, we obtain

$$\widehat{\partial_t u}(t,\xi) + (2\pi i\xi)^3\widehat{u}(t,\xi) + (2\pi i\xi)\widehat{F(u)}(t,\xi) = 0, \tag{1.3}$$

from which we get $\widehat{\partial_t u}(t,0) = 0$. This implies that

$$\frac{d}{dt}\int_{\mathbb{R}} u(t,x)\,dx = 0$$

or, in other words, that $\int_{\mathbb{R}} u(t,x)\,dx$ is constant in time. □

Proof of (2) Here we assume for simplicity that $F(u) = u^5$. In fact, later on we will present dispersive estimates in this particular case.

In general, if f and g are real-valued functions then one has

$$\int_{\mathbb{R}} f(x)g(x)\,dx = \int_{\mathbb{R}} \widehat{f}(\xi)\bar{\widehat{g}}(\xi)\,d\xi = \int_{\mathbb{R}} \widehat{f}(\xi)\widehat{\bar{g}}(-\xi)\,d\xi$$
$$= \int_{\mathbb{R}} \widehat{f}(\xi)\widehat{g}(-\xi)\,d\xi = \int_{\xi_1+\xi_2=0} \widehat{f}(\xi_1)\widehat{g}(\xi_2)\,d\xi_1 d\xi_2,$$

where an overbar indicates the complex conjugate. In particular, we can write

$$\int_{\mathbb{R}} u^2\,dx = \int_{\xi_1+\xi_2=0} \widehat{u}(\xi_1)\widehat{u}(\xi_2)\,d\xi_1 d\xi_2$$

and so

$$\frac{d}{dt}\int_{\mathbb{R}} u^2 dx$$
$$= 2\int_{\xi_2+\xi_2=0} \partial_t\widehat{u}(\xi_1)\,\widehat{u}(\xi_2)\,d\xi_1 d\xi_2. \tag{1.4}$$

From (1.3) we know that

$$\partial_t\widehat{u}(\xi) = -(2\pi i\xi)^3\widehat{u}(\xi) - (2\pi i\xi)\widehat{F(u)}(\xi);$$

then (1.4) becomes

$$16\pi^3 i\int_{\xi_1+\xi_2=0} \xi_1^3\widehat{u}(\xi_1)\widehat{u}(\xi_2)\,d\xi_1 d\xi_2 - 4\pi i\int_{\xi_1+\xi_2=0} \xi_1\widehat{F(u)}(\xi_1)\widehat{u}(\xi_2)\,d\xi_1 d\xi_2$$
$$= I + II.$$

By symmetry, the first term, I, is equal to

$$8\pi^3 i \int\limits_{\xi_1+\xi_2=0} (\xi_1^3+\xi_2^3)\,\widehat{u}(\xi_1)\widehat{u}(\xi_2)\,d\xi_1 d\xi_2,$$

which is clearly identically equal to zero. The second term, II, becomes

$$\begin{aligned}
&-4\pi i \int\limits_{\xi_1+\xi_2=0} \xi_1 \widehat{u^5}(\xi_1)\widehat{u}(\xi_2)\,d\xi_1 d\xi_2 \\
&\quad = -4\pi i \int\limits_{\xi_1+\xi_2=0} \xi_1 \left(\int\limits_{\lambda_1+\cdots+\lambda_5=\xi_1} \widehat{u}(\lambda_1)\cdots\widehat{u}(\lambda_5)d\lambda_1\cdots d\lambda_5 \right) \widehat{u}(\xi_2)\,d\xi_1 d\xi_2 \\
&\quad = 4\pi i \int\limits_{\lambda_1+\cdots+\lambda_5+\xi_2=0} \xi_2 \widehat{u}(\lambda_1)\cdots\widehat{u}(\lambda_5)\widehat{u}(\xi_2)\,d\lambda_1\cdots d\lambda_5 d\xi_2 \\
&\quad = \frac{4\pi}{6} i \int\limits_{\lambda_1+\cdots+\lambda_5+\xi_2=0} (\lambda_1+\cdots+\lambda_5+\xi_2)\widehat{u}(\lambda_1)\cdots\widehat{u}(\lambda_5)\widehat{u}(\xi_2)\,d\lambda_1\cdots d\lambda_5 d\xi_2 \\
&\quad = 0.
\end{aligned}$$

This proves that $\int_{\mathbb{R}} u^2(t,x)\,dx$ indeed is independent of time. □

Proof of (3) As before, it is enough to show that the derivative with respect to time of the expression

$$\int\limits_{\mathbb{R}} \left(\frac{1}{2} u_x^2 - V(u) \right) dx$$

is zero. Now one can write

$$\begin{aligned}
&\frac{d}{dt} \int_{\mathbb{R}} \left(\frac{1}{2} u_x^2 - V(u) \right) dx \\
&\quad = \frac{d}{dt} \left(\frac{1}{2} \int\limits_{\xi_1+\xi_2=0} \widehat{u}_x(\xi_1)\widehat{u}_x(\xi_2)\,d\xi_1 d\xi_2 - \int_{\mathbb{R}} V(u)\,dx \right) \\
&\quad = \frac{d}{dt} \left(-2\pi^2 \int\limits_{\xi_1+\xi_2=0} \xi_1\xi_2 \widehat{u}(\xi_1)\widehat{u}(\xi_2)\,d\xi_1 d\xi_2 - \int_{\mathbb{R}} V(u)\,dx \right) = A + B.
\end{aligned}$$

By using the equality (1.3) we see that A equals

$$
\begin{aligned}
-4\pi^2 & \int_{\xi_1+\xi_2=0} \xi_1\xi_2\partial_t\widehat{u}(\xi_1)\widehat{u}(\xi_2)\, d\xi_1 d\xi_2 \\
&= -32\pi^5 i \int_{\xi_1+\xi_2=0} \xi_1\xi_2\xi_1^3\widehat{u}(\xi_1)\widehat{u}(\xi_2)\, d\xi_1 d\xi_2 \\
&\quad + 8\pi^3 i \int_{\xi_1+\xi_2=0} \xi_1\xi_2\xi_1\widehat{F(u)}(\xi_1)\widehat{u}(\xi_2)\, d\xi_1 d\xi_2 \\
&= 32\pi^5 i \int_{\xi_1+\xi_2=0} \xi_1^5\widehat{u}(\xi_1)\widehat{u}(\xi_2)\, d\xi_1 d\xi_2 \\
&\quad + 8\pi^3 i \int_{\xi_1+\xi_2=0} \xi_1^2\xi_2\widehat{F(u)}(\xi_1)\widehat{u}(\xi_2)\, d\xi_1 d\xi_2 \\
&= -8\pi^3 i \int_{\xi_1+\xi_2=0} \xi_1^3\widehat{F(u)}(\xi_1)\widehat{u}(\xi_2)\, d\xi_1 d\xi_2. \qquad (1.5)
\end{aligned}
$$

The last line follows since, as before, the first integral is zero by symmetry. Given that V is an integral of F, the second term B can be written as

$$
\begin{aligned}
-\int_{\mathbb{R}} F(u)\partial_t u\, dx &= -\int_{\xi_1+\xi_2=0} \widehat{F(u)}(\xi_1)\partial_t\widehat{u}(\xi_2)\, d\xi_1 d\xi_2 \\
&= -8\pi^3 i \int_{\xi_1+\xi_2=0} \widehat{F(u)}(\xi_1)\xi_2^3\widehat{u}(\xi_2)\, d\xi_1 d\xi_2 \\
&\quad + 2\pi i \int_{\xi_1+\xi_2=0} \xi_2\widehat{F(u)}(\xi_1)\widehat{F(u)}(\xi_2)\, d\xi_1 d\xi_2 \\
&= -8\pi^3 i \int_{\xi_1+\xi_2=0} \widehat{F(u)}(\xi_1)\xi_2^3\widehat{u}(\xi_2)\, d\xi_1 d\xi_2. \qquad (1.6)
\end{aligned}
$$

Now one observes that the sum of (1.5) and (1.6) is zero. □

1.2. Dispersive estimates for the linear equation

Let us consider now the *linear part* of the initial-value problem (IVP) (1.2), which is given by

$$
\begin{cases} \partial_t u + \partial_x^3 u = 0, \\ u(0) = g. \end{cases} \qquad (1.7)
$$

We will see that one can calculate the solution explicitly in this case. By taking the Fourier transform with respect to the x variable of the first equation, we obtain

$$\partial_t \widehat{u}(t,\xi) = 8\pi^3 i\xi^3 \widehat{u}(t,\xi),$$

while the second equation gives

$$\widehat{u}(0,\xi) = \widehat{g}(\xi).$$

By combining these two we obtain immediately

$$\widehat{u}(t,\xi) = \widehat{g}(\xi) e^{8\pi^3 it\xi^3}$$

or, equivalently, $u(t,x) = (g * K_t)(x) := S(t)g$, where

$$\widehat{K_t}(\xi) = e^{8\pi^3 it\xi^3}.$$

It is easy to observe that

$$K_t(x) = \frac{1}{(4\pi^2 t)^{1/3}} Ai\left(\frac{x}{(4\pi^2 t)^{1/3}}\right),$$

where Ai is the *Airy function* whose Fourier transform is $e^{2\pi i\xi^3}$. Of course, the functions K_t are defined a priori as distributions, but later we will see that they are in fact functions, whose asymptotic behavior will be studied in detail. The reader should also recall the related topics described in Chapter I.4.[2] The following lemma will play an important role.

Lemma 1.1 *One has the following:*
(i) *$Ai(x)$ is a bounded function and is $O(|x|^{-1/4})$ as $|x| \to \infty$;*
(ii) *$D^\alpha(Ai)$ is bounded for any $\alpha \in \left[0, \frac{1}{2}\right]$.*

The proof will be postponed to the end of the chapter. Using this lemma one can easily prove the following *dispersive estimates* for the solutions of the linear equation (1.7).

Lemma 1.2 *Let $g \in L^1$. Then*
(i) $\|S(t)g\|_\infty \lesssim t^{-1/3}\|g\|_1$.
(ii) $\|S(t)g\|_p \lesssim t^{(-1/3)(1-1/p)}\|g\|_1$ *for every $p > 4$.*

Proof To prove the first statement one can write

$$(S(t)g)(x) = (g * K_t)(x) = \int_{\mathbb{R}} K_t(x-y)g(y)\,dy,$$

[2] Chapter I.4 refers to Chapter 4 in Vol. I of the book.

from which one obtains

$$\|S(t)g\|_\infty \lesssim \left\| \int_{\mathbb{R}} |K_t(\cdot - y)||g(y)|\,dy \right\|_\infty$$
$$\lesssim t^{-1/3}\|g\|_1,$$

using Lemma 1.1(i).

Similarly, to prove the second statement one can write

$$\|S(t)g\|_p \lesssim \int_{\mathbb{R}} \|K_t(\cdot - y)\|_p |g(y)|\,dy \leq \|K_t\|_p\|g\|_1,$$

and it is easy to see, again using Lemma 1.1(i) and the fact that $p > 4$, that

$$\|K_t\|_p \lesssim t^{(-1/3)(1-1/p)}.$$ □

The next three lemmas will also be useful later.

Lemma 1.3 *Let $g \in L^1$. Then*

$$\|D^{1/2}(S(t)g)\|_\infty \lesssim t^{-1/2}\|g\|_1.$$

Proof One has on the one hand

$$D^{1/2}(S(t)g) = D^{1/2}(K_t * g) = (D^{1/2}K_t) * g.$$

On the other hand,

$$\left|D^{1/2}K_t(x)\right| \lesssim \left| \frac{1}{t^{1/3}}(D^{1/2}Ai)\left(\frac{x}{(4\pi^2)t^{1/3}}\right)\frac{1}{t^{1/6}} \right| \lesssim \frac{1}{t^{1/2}},$$

using Lemma 1.1(ii). As a consequence,

$$\|D^{1/2}(S(t)g)\|_\infty \lesssim \|D^{1/2}K_t\|_\infty\|g\|_1 \lesssim t^{-1/2}\|g\|_1,$$

as desired. □

Lemma 1.4 *Let $p \geq 2$ and $1/p + 1/p' = 1$. Then*

$$\left\|D^{1/2-1/p}(S(t)g)\right\|_p \lesssim t^{-1/2+1/p}\|g\|_{p'}.$$

Proof There are two distinct cases, which we have met already. For $p = 2$ the inequality follows immediately from Plancherel's theorem, while for $p = \infty$ it was the object of Lemma 1.3. The general case then follows from Stein's complex interpolation theorem, since one can clearly extend the definition of $D^\alpha h$ to complex exponents α. □

Lemma 1.5 *Let $p \in \left(1, \frac{4}{3}\right)$. Then there exist $\beta > 0$, $\gamma > 0$ such that*

$$\left\| D^{\beta}(S(t)g) \right\|_{\infty} \lesssim t^{-\gamma} \, \|g\|_{p} \tag{1.8}$$

with $\beta \to \frac{1}{2}$ and $\gamma \to \frac{1}{2}$ as $p \to 1$.

Proof From Lemma 1.2 we know that the linear operator $S(t)$ maps L^1 into L^p for $p > 4$ with an operatorial bound of the type $O\left(t^{(-1/3)(1-1/p)}\right)$. As a consequence, by duality, since $S(t)$ is a convolution operator, it also maps $L^{p'}$ to L^{∞} with the same bound, which can be rewritten as $O\left(t^{-1/3p'}\right)$. Also, since $p > 4$ it follows that $1 < p' < \frac{4}{3}$. In other words, changing the notation a little, we have shown that

$$\|S(t)g\|_{\infty} \lesssim t^{-1/3p} \|g\|_{p}$$

for every $1 < p < \frac{4}{3}$. However, from Lemma 1.3 we know that

$$\left\| D^{1/2}(S(t)g) \right\|_{\infty} \lesssim t^{-1/2} \, \|g\|_{1}.$$

The desired conclusion (1.8) follows by complex interpolation between the two estimates above. □

It is also natural to ask what happens if one keeps the nonlinearity and instead drops the linear term in equation (1.2). In the case of the KdV equation (i.e. $F(x) = x^2/2$) the initial-value problem becomes

$$\begin{cases} \partial_t u + u \partial_x u = 0, \\ u(0, x) = g(x), \end{cases} \tag{1.9}$$

and it is a well-known fact that the solutions of (1.9) may develop *shocks*. For instance, one can check directly that for $g(x) = -x$ the solution is given by $u(t, x) = -x/(1 - t)$. While this is well defined for t strictly between 0 and 1, the solution breaks down and a shock is developed at time $t = 1$.

What we can conclude from this is that there is always a competition between the good influence of the linear term and the bad influence of the nonlinear term, in the IVP (1.2). It is remarkable that there is sometimes a perfect balance between the two, as one can see from the existence of solitons.

Exercise 1.1 Fill in the complex interpolation details in the proofs of Lemmas 1.4 and 1.5.

1.3. Dispersive estimates for the nonlinear equation

Given all these dispersive estimates that the solutions of the linear IVP (1.7) satisfy, it is natural to ask whether we have dispersion in the nonlinear case (1.2) as well. Clearly, assuming that $F(u) = |u|^s$, in general there is no dispersion because of the presence of solitons. However, it is natural to believe that if one starts with a small initial datum g, and if s is big enough, the dispersion should exist since then the nonlinear PDE is close to its linear counterpart.

More precisely, the goal of the section is to prove the following theorem.

Theorem 1.6 *Consider* (1.2) *with* $F(u) = u^5$. *Then there exists* $\varepsilon_0 > 0$ *such that if* g *satisfies*

$$\|g\|_1 + \|g\|_{H^1} < \varepsilon_0,$$

the solution to the corresponding gKdV equation is dispersive; more precisely, it satisfies

$$\sup_{t \in \mathbb{R}} \langle t \rangle^{1/3} \|u(t)\|_\infty < \infty. \tag{1.10}$$

In general by $\langle t \rangle$ one denotes the so-called *Japanese bracket* given by $\langle t \rangle = (1 + |t|^2)^{1/2}$. Let us also remark that it is crucial that the exponent of the nonlinearity be large enough ($s = 5$ in our case). For generic functions of the type $F(u) = |u|^s$, such a theorem does not hold for $1 < s < 3$, for example. To see this, let us recall the existence of solitons in this particular case. If

$$(t, x) \mapsto w(x - t)$$

is such a solution then a straightforward calculation shows that

$$(t, x) \mapsto \lambda^{1/(s-1)} w\left(\lambda^{1/2} x - \lambda^{3/2} t\right) \tag{1.11}$$

is also a solution, for every $\lambda > 0$. At time $t = 0$ this solution becomes $x \mapsto \lambda^{1/(s-1)} w\left(\lambda^{1/2} x\right) = w_\lambda(x)$. Then one has

$$\begin{aligned}
\|w_\lambda\|_1 + \|w_\lambda\|_{H^1} &\simeq \|w_\lambda\|_1 + \|w_\lambda\|_2 + \|w_\lambda'\|_2 \\
&= \lambda^{1/(s-1)-1/2} \|w\|_1 + \lambda^{1/(s-1)-1/4} \|w\|_2 + \lambda^{1/(s-1)+1/4} \|w'\|_2 \\
&= \lambda^{(3-s)/(2(s-1))} \|w\|_1 + \lambda^{(5-s)/(4(s-1))} \|w\|_2 \\
&\quad + \lambda^{(s+3)/(4(s-1))} \|w'\|_2.
\end{aligned}$$

Thus, if $1 < s < 3$ and λ is small enough, one can make $\|w_\lambda\|_1 + \|w_\lambda\|_{H^1}$ smaller than ε_0 while it is clear that the solution (1.11) is not dispersive.

To prove Theorem 1.6, we use a method of Christ and Weinstein. This method can be extended to cover more general nonlinearities, such as $F(u) = |u|^s$ for $s > 4$. However, since our goal here is mostly motivational we will describe it in the case $F(u) = u^5$, when many of its technicalities become easier. In fact, this particular case is part of an earlier result of Ponce and Vega.

Let us start by stating the following lemmas.

Lemma 1.7 *If g and u are as in Theorem 1.6 then, for any time t, one has*

$$\|u(t)\|_2 + \|\partial_x u(t)\|_2 \lesssim \|g\|_{H^1}.$$

Proof Let us remark that the inequality is completely trivial in the linear case. In the nonlinear case one observes that $\|u(t)\|_2 = \|u(0)\|_2 = \|g\|_2 \le \|g\|_{H^1}$, by the conservation of energy. It is therefore enough to prove the corresponding estimates for the term $\|\partial_x u(t)\|_2$.

We will first show that

$$\|\partial_x u(t)\|_2 \lesssim 1$$

for every t.

From the conservation of the Hamiltonian, we know that

$$\int_{\mathbb{R}} \left(\tfrac{1}{2}u_x^2 - V(u)\right) dx = \text{constant}$$

which implies that

$$\int_{\mathbb{R}} \left(u_x^2 - 2V(u)\right) dx = \text{constant}.$$

Then

$$\begin{aligned}
\|\partial_x u(t)\|_2^2 &= \int_{\mathbb{R}} u_x^2(t)\, dx = \int_{\mathbb{R}} \left(u_x^2(t) - 2V(u)(t)\right) dx + 2\int_{\mathbb{R}} V(u)(t)\, dx \\
&= \int_{\mathbb{R}} \left(u_x^2(0) - 2V(u)(0)\right) dx + 2\int_{\mathbb{R}} V(u)(t)\, dx \\
&= \int_{\mathbb{R}} u_x^2(0)\, dx + 2\left(\int_{\mathbb{R}} (V(u)(t) - V(u)(0))\, dx\right) \\
&\le \varepsilon_0^2 + 2\left|\int_{\mathbb{R}} V(u)(0)\, dx\right| + 2\left|\int_{\mathbb{R}} V(u)(t)\, dx\right|.
\end{aligned}$$

Also, since $V(u) = \frac{1}{6}u^6$ in our case, we have

$$\begin{aligned}
\left|\int_{\mathbb{R}} V(u)(0)\, dx\right| &\lesssim \int_{\mathbb{R}} |u^6(0)|\, dx \le \|u(0)\|_\infty^4\, \|u(0)\|_2^2 \\
&\lesssim \|u(0)\|_{H^1}^4\, \|u(0)\|_{H^1}^2 = \|u(0)\|_{H^1}^6 \lesssim \varepsilon_0^6.
\end{aligned}$$

Similarly,

$$\begin{aligned}\left|\int_{\mathbb{R}} V(u)(t)dx\right| &\lesssim \int_{\mathbb{R}} u^6(t)\,dx \lesssim \|u(t)\|_2^2\,\|u(t)\|_\infty^4 \\ &= \|u(0)\|_2^2\,\|u(t)\|_\infty^4 \lesssim \varepsilon_0^2\,\|u(t)\|_\infty^4 \\ &\lesssim {\varepsilon_0}^2\|u(t)\|_{H^1}^4 \lesssim {\varepsilon_0}^2\,(\|u(t)\|_2 + \|\partial_x u(t)\|_2)^4 \\ &= \varepsilon_0^2\,(\|u(0)\|_2 + \|\partial_x u(t)\|_2)^4 \\ &\lesssim \varepsilon_0^2\,(\varepsilon_0 + \|\partial_x u(t)\|_2)^4 .\end{aligned}$$

In the above inequalities we used Sobolev embedding several times. See Lemma I.4.11 for a detailed proof. Putting everything together, we conclude that

$$\|\partial_x u(t)\|_2^2 \lesssim \varepsilon_1 + \varepsilon_2\,(\varepsilon_3 + \|\partial_x u(t)\|_2)^4$$

for some small numbers $\varepsilon_1, \varepsilon_2, \varepsilon_3$ depending on ε_0. In particular, there exists a small number ε such that

$$\|\partial_x u(t)\|_2^2 \lesssim \varepsilon\big(1 + \|\partial_x u(t)\|_2^4\big) .$$

Now, it is easy to see that an inequality of the type

$$x^2 \le \varepsilon\big(1 + x^4\big) \tag{1.12}$$

holds either when x is close to zero or when x is large enough. Using Kato's theorem, mentioned just before Section 1.1, we have $u \in C_w\left(\mathbb{R}, H^1\right)$ and, since at time $t = 0$ $\|\partial_x u(0)\|_2$ is small and satisfies (1.12), the only option for the norm $\|\partial_x u(t)\|_2$ is to satisfy the inequality (1.12) for all times t, with $\|\partial_x u(t)\|_2$ lying in the connected component containing the origin. This shows that $\|\partial_x u(t)\|_2$ is indeed uniformly bounded.

A careful inspection of the above argument shows that we have also proved that

$$\|\partial_x u(t)\|_2^2 \lesssim \|g\|_{H^1}^2 + \|g\|_{H^1}^6 + \|g\|_{H^1}^2\big(\|g\|_{H^1}^4 + \|\partial_x u(t)\|_2^4\big) ,$$

which implies that

$$\|\partial_x u(t)\|_2^2 \lesssim \|g\|_{H^1}^2\,\big(1 + \|\partial_x u(t)\|_2^4\big) .$$

Furthermore, this, together with the inequality $\|\partial_x u(t)\|_2 \lesssim 1$ mentioned at the start of the proof, yields that

$$\|\partial_x u(t)\|_2 \lesssim \|g\|_{H^1},$$

as desired. □

Now if $a \in \mathbb{R}$ is a real number, we denote by $a+$ a generic number arbitrarily close to a but strictly large than a. One defines $a-$ similarly.

Lemma 1.8 *Let f be a function in H^1. Then*

$$\|D^{1/2+} f\|_{2+} \lesssim \|f\|_2 + \|D^1 f\|_2. \tag{1.13}$$

Proof To prove (1.13) we recall the standard Littlewood–Paley decomposition from the first volume of the book and write f as

$$f = \sum_{k\in\mathbb{Z}} f * \Psi_k,$$

where the Ψ_k are as usual, smooth L^1-normalized Schwartz functions whose Fourier supports are intervals of the type $[2^{k-1}, 2^{k+1}]$. Now, we split f further as

$$f = f^{\text{high}} + f^{\text{low}},$$

where

$$f^{\text{high}} := \sum_{k\geq 0} f * \Psi_k$$

and

$$f^{\text{low}} := \sum_{k<0} f * \Psi_k.$$

To estimate $D^{1/2+}(f^{\text{high}})$, one writes

$$\begin{aligned} D^{1/2+}(f^{\text{high}}) &= \sum_{k\geq 0} D^{1/2+}(f * \Psi_k) \\ &= \sum_{k\geq 0} D^{-1/2-} D^1(f * \Psi_k) = \sum_{k\geq 0} D^{-1/2-}(D^1 f * \Psi_k) \\ &=: \sum_{k\geq 0} 2^{-\alpha k}(D^1 f * \widetilde{\Psi}_k), \end{aligned}$$

with $\alpha \sim 1/2$ and $\widetilde{\Psi}_k$ defined naturally. As a consequence, we have

$$\|D^{1/2+}(f^{\text{high}})\|_{2+} \lesssim \sum_{k\geq 0} 2^{-\alpha k}\|D^1 f * \widetilde{\Psi}_k\|_{2+}. \tag{1.14}$$

We now claim that, for any function g, one has

$$\|g * \widetilde{\Psi}_k\|_{2+} \lesssim 2^{\varepsilon k}\|g\|_2$$

for a certain small $\varepsilon > 0$. The reader may remember this from Lemma I.4.13.[3] It is a particular case of the so-called Bernstein inequality.

[3] That is, Lemma 4.13 in Vol. I.

Proving this special case is easy since one has

$$\|g * \widetilde{\Psi}_k\|_p \lesssim \|g\|_p$$

for every $1 < p < \infty$ and also

$$\|g * \widetilde{\Psi}_k\|_\infty \lesssim 2^k \|g\|_1.$$

By interpolating carefully between these two inequalities we obtain our claim.

Using it, (1.14) can be estimated by

$$\sum_{k \geq 0} 2^{-\alpha k} 2^{\varepsilon k} \|D^1 f\|_2 \lesssim \|D^1 f\|_2.$$

Similarly, to estimate $D^{1/2+}(f^{\text{low}})$ one writes

$$\begin{aligned} D^{1/2+}(f^{\text{low}}) &= \sum_{k<0} D^{1/2+}(f * \Psi_k) \\ &= \sum_{k<0} \left(f * D^{1/2+}\Psi_k\right) =: \sum_{k<0} 2^{\beta k} f * \widetilde{\Psi}_k, \end{aligned}$$

where as before $\beta \sim 1/2$. As a consequence,

$$\begin{aligned} \|D^{1/2+}(f^{\text{low}})\|_{2+} &\lesssim \sum_{k<0} 2^{\beta k} \|f * \widetilde{\Psi}_k\|_{2+} \\ &\lesssim \sum_{k<0} 2^{\beta k} 2^{\varepsilon k} \|f\|_2 \lesssim \|f\|_2, \end{aligned}$$

as desired. □

Exercise 1.2 Show the following extension of the above estimate. If $0 < \alpha < \beta$ and $1 < p < \infty$ then one has

$$\|D^\alpha f\|_{p+} \lesssim \|f\|_p + \|D^\beta f\|_p.$$

Finally, we will also need the Leibnitz rule (1.1). After all these preparations, we can start the proof of Theorem 1.6.

Proof of Theorem 1.6 Write

$$M_\infty(t) := \sup_{|\tau| \leq t} \langle \tau \rangle^{1/3} \|u(\tau)\|_\infty,$$

$$M_6(t) := \sup_{|\tau| \leq t} \langle \tau \rangle^{5/18} \|u(\tau)\|_6,$$

and

$$\|g\| := \|g\|_1 + \|g\|_{H^1}.$$

We will show that (for $s = 5$)

$$M_\infty(t) \le C\|g\| + \widetilde{C} M_6(t)^3 M_\infty(t)^{s-4}, \tag{1.15}$$

$$M_6(t) \le C\|g\| + \widetilde{C} M_6(t)^3 M_\infty(t)^{s-11/3}, \tag{1.16}$$

where $\widetilde{C} = \widetilde{C}(\|g\|) \to 0$ as $\|g\| \to 0$.

Let us assume (1.15) and (1.16) for a moment and conclude the proof. Write $M(t) := M_\infty(t) + M_6(t)$ and observe that as a consequence of (1.15) and (1.16) one has

$$M(t) \le \varepsilon\big(1 + M(t)^\alpha + M(t)^\beta\big),$$

where α, β are both larger than 1 and ε is small. As before, if one considers the inequality

$$x \le \varepsilon\left(1 + x^\alpha + x^\beta\right)$$

for $x \ge 0$, one sees that it holds either if x is sufficiently small or x is sufficiently larger. Again by Kato's theorem, since at time $t = 0$ the function $M(t)$ lies in the connected component containing the origin, it has to remain there for all future times t; this means that $M(t)$ is bounded and in particular that $M_\infty(t)$ is bounded, as the theorem claims.

We are therefore left with proving (1.15) and (1.16). We will prove (1.15) only and leave (1.16) as an exercise for the reader.

To establish (1.15), we need the following identity of Duhamel:

$$u = S(t)g - \int_0^t S(t-\tau)\left(\partial_x F(u)(\tau)\right) d\tau. \tag{1.17}$$

To prove (1.17) one has from (1.3) that

$$\partial_\tau \widehat{u}(\tau,\xi) - 8\pi^3 i\xi^3 \widehat{u}(\tau,\xi) + \widehat{\partial_x F(u)}(\tau,\xi) = 0.$$

In particular, this implies that

$$\partial_\tau \widehat{u}(\tau,\xi)\, e^{-8\pi^3 i\tau\xi^3} - 8\pi^3 i\xi^3 e^{-8\pi^3 i\tau\xi^3} \widehat{u}(\tau,\xi) + e^{-8\pi^3 i\tau\xi^3} \widehat{\partial_x F(u)}(\tau,\xi) = 0$$

and so

$$\frac{\partial}{\partial\tau}\big(e^{-8\pi^3 i\tau\xi^3} \widehat{u}(\tau,\xi)\big) + e^{-8\pi^3 i\tau\xi^3} \widehat{\partial_x F(u)}(\tau,\xi) = 0.$$

Integrating the above equality, one obtains

$$e^{-8\pi^3 it\xi^3} \widehat{u}(t,\xi) - \widehat{u}(0,\xi) = -\int_0^t e^{-8\pi^3 i\tau\xi^3} \widehat{\partial_x F(u)}(\tau,\xi)\, d\tau$$

which implies that

$$\widehat{u}(t,\xi) = e^{8\pi^3 it\xi^3}\widehat{g}(\xi) - \int_0^t e^{8\pi^3 i(t-\tau)\xi^3}\,\widehat{\partial_x F(u)}(\tau,\xi)\,d\tau.$$

On taking the inverse Fourier transform of the above identity, one obtains (1.17). Using it one can write

$$\begin{aligned}\|u(t)\|_\infty &\le \|S(t)g\|_\infty + \int_0^t \|S(t-\tau)\partial_x F(u)(\tau)\|_\infty\,d\tau\\ &\le C\langle t\rangle^{-1/3}\|g\| + C\int_0^t \|D^{1/2-}S(t-\tau)\big[\big(\partial_x D^{-1}\big)D^{1/2+}F(u)(\tau)\big]\|_\infty\,d\tau\\ &\le C\langle t\rangle^{-1/3}\|g\| + C\int_0^t |t-\tau|^{-1/2+}\|D^{1/2+}F(u(\tau))\|_{1+}\,d\tau,\end{aligned}$$

by Lemmas 1.2 and 1.5 and the usual boundedness properties of the Hilbert transform. Now employing the Leibnitz rule (1.1), the above expression can be estimated by

$$C\langle t\rangle^{-1/3}\|g\| + C\int_0^t |t-\tau|^{-1/2+}\|D^{1/2+}u(\tau)\|_{2+}\,\|u(\tau)\|_6^3\,\|u(\tau)\|_\infty\,d\tau.$$

Using Lemma 1.8 followed by Lemma 1.7 this can be further estimated by

$$C\langle t\rangle^{-1/3}\|g\| + \widetilde{C}\int_0^t |t-\tau|^{-1/2+}\|u(\tau)\|_6^3\,\|u(\tau)\|_\infty\,d\tau.$$

In particular, one can write

$$\begin{aligned}\langle t\rangle^{1/3}\|u(t)\|_\infty \le C\|g\| + \widetilde{C}\langle t\rangle^{1/3}\int_0^t &|t-\tau|^{-1/2+}\langle\tau\rangle^{-15/18}\big(\langle\tau\rangle^{15/18}\|u(\tau)\|_6^3\big)\\ &\times\langle\tau\rangle^{-(s-4)/3}\big(\langle\tau\rangle^{-(s-4)/3}\|u(\tau)\|_\infty^{s-4}\big)\,d\tau,\end{aligned}$$

from which one can immediately deduce the desired inequality

$$M_\infty(t) \le C\|g\| + \widetilde{C}M_6(t)^3 M_\infty(t)^{s-4}$$

as long as one has proved that

$$\langle t\rangle^{1/3}\int_0^t \frac{1}{|t-\tau|^{\frac{1}{2}-}}\langle\tau\rangle^{-15/18-(s-4)/3}d\tau \tag{1.18}$$

is uniformly bounded in t.

There are two cases to consider, to verify (1.18).

Case 1: t is small. Clearly the expression is now bounded, since the inner function is integrable near the origin.

Case 2: t larger. Now $\langle t\rangle \simeq t$ and, by changing to the variable $\tau = t\alpha$, one obtains that on the one hand (1.18) is smaller than

$$t t^{1/3} t^{-1/2+} \int_0^1 \frac{1}{|1-\alpha|^{1/2-}} \langle t\alpha\rangle^{-15/18-(s-4)/3}\, d\alpha.$$

On the other hand, $\langle t\alpha\rangle \simeq 1 + |t\alpha| = t(1/t + |\alpha|)$, which allows us to further estimate the above expression by

$$\begin{aligned} & t^{(1/2+)+1/3} t^{-15/18-(s-4)/3} \int_0^1 \frac{1}{|1-\alpha|^{1/2-}} \frac{1}{|1/t+\alpha|^{15/18+(s-4)/3}}\, d\alpha \\ & = t^{\varepsilon-(s-4)/3} \left(\int_0^{1/2} \frac{1}{|1-\alpha|^{1/2-}} \frac{1}{|1/t+\alpha|^{15/18+(s-4)/3}}\, d\alpha \right. \\ & \qquad \left. + \int_{1/2}^1 \frac{1}{|1-\alpha|^{1/2-}} \frac{1}{|1/t+\alpha|^{15/18+(s-4)/3}}\, d\alpha \right). \end{aligned} \tag{1.19}$$

The second term on the right-hand side is clearly $O(1)$ and this gives a contribution of the type $O(t^{-1/3+\epsilon})$, which is satisfactory, since t is now large. The first term can be calculated precisely:

$$\begin{aligned} & \int_0^{1/2} \frac{1}{|1-\alpha|^{1/2-}} \frac{1}{|1/t+\alpha|^{15/18+(s-4)/3}}\, d\alpha \\ & \lesssim \int_0^{1/2} \frac{1}{|1/t+\alpha|^{15/18+(s-4)/3}}\, d\alpha =: \int_0^{1/2} \frac{1}{|1/t+\alpha|^{\mu}}\, d\alpha = \int_{1/t}^{1/2+1/t} \frac{1}{y^{\mu}}\, dy \\ & = \frac{1}{-\mu+1} \left(\left(\frac{1}{2}+\frac{1}{t}\right)^{-\mu+1} - \left(\frac{1}{t}\right)^{-\mu+1} \right) \end{aligned}$$

and, as a consequence, its contribution to (1.19) is

$$O\left(t^{\varepsilon-1/3}t^{\mu-1}\right) = O(t^{\varepsilon-1/6})$$

given that $s = 5$ in our case; this is again acceptable since t is large. Thus, our proof of the uniform boundedness of (1.18) and of Theorem 1.6 is complete. □

1.4. Wave packets and phase-space portraits

We are left with the proof of Lemma 1.1. Our goal is first to describe the so-called *phase-space portraits* of the functions $e^{2\pi i x^2}$ and $e^{2\pi i x^3}$. As we will see, once we understand them the proof of Lemma 1.1 will become an easy exercise. We start with some definitions, which will play an important role in the later chapters of the book as well.

Definition 1.9 Let $J \subseteq \mathbb{R}$ be an arbitrary interval. A smooth function φ is said to be *adapted* to J if and only if one has

$$\left|\varphi^{(\ell)}(x)\right| \leq C_\ell C_M \frac{1}{|J|^\ell} \frac{1}{(1 + |x - x_J|/|J|)^M}$$

for sufficiently many derivatives ℓ, where x_J denotes the center of J and M is a large number.

The intuition here is that φ is *essentially supported* inside J, in the sense that φ and many of its derivatives decay rapidly away from the interval J at scale $|J|$.

An arbitrary dyadic rectangle of area 1 in the plane is called a *Heisenberg box* (see Figure 1.1). A smooth function φ is said to be a *wave packet* associated with the Heisenberg box $[0, 1] \times [0, 1]$ if and only if both φ and $\widehat{\varphi}$ are adapted to the interval $[0, 1]$, in the sense of the above definition.

For example, the function e^{-x^2} is a wave packet associated to with Heisenberg box $[0, 1] \times [0, 1]$.

Given any Heisenberg box $I \times \omega = [n2^k, (n+1)2^k] \times [m2^{-k}, (m+1)2^{-k}]$ and a wave packet φ associated with $[0, 1] \times [0, 1]$, one can easily modify the latter to define a wave packet associated with $I \times \omega$ as follows. First, one dilates the function so that

$$x \mapsto \varphi\left(\frac{x}{2^k}\right),$$

thus obtaining a wave packet associated with $[0, 2^k] \times [0, 2^{-k}]$; then one translates the new wave packet in such a way that

$$x \mapsto \varphi\left(\frac{x - n2^k}{2^k}\right),$$

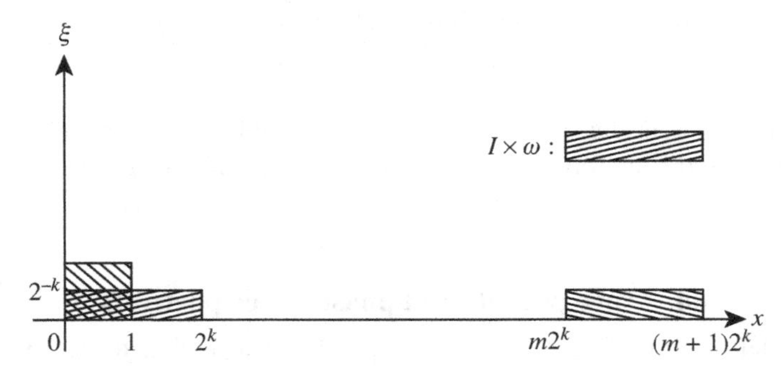

Figure 1.1. Heisenberg boxes.

obtaining a wave packet associated with $[n2^k, (n+1)2^k] \times [0, 2^{-k}]$. Finally one modulates this wave packet according to

$$x \mapsto \varphi\left(\frac{x - n2^k}{2^k}\right) e^{2\pi i m 2^{-k} x},$$

obtaining a wave packet one associated with the Heisenberg box $I \times \omega$. We will see later on that sometimes it is convenient to assume that the wave packet associated with $I \times \omega$ satisfies $\operatorname{supp} \widehat{\varphi} \subseteq \omega$. The intervals I are usually called time intervals or space intervals while the intervals ω are called frequency intervals. For $1 \leq p \leq \infty$, if φ is a wave packet associated with $I \times \omega$, we say that $|I|^{-1/p}\varphi$ is an L^p-normalized wave packet associated with $I \times \omega$.

> **Exercise 1.3** Show that φ is an L^p-normalized wave packet associated with the Heisenberg box $I \times \omega$ if and only if $\widehat{\varphi}$ in an $L^{p'}$-normalized wave packet associated with the Heisenberg box $\omega \times I$, where $1/p + 1/p' = 1$.

Suppose now that $P = I_P \times \omega_P$ and $Q = I_Q \times \omega_Q$ are two distinct Heisenberg boxes in the phase plane; assume that $P \cap Q = \emptyset$. Consider two L^∞-normalized wave packets φ_P and φ_Q adapted to them and having the property that $\operatorname{supp} \widehat{\varphi_P} \subseteq \omega_P$ and $\operatorname{supp} \widehat{\varphi_Q} \subseteq \omega_Q$. We observe here that if P and Q are sufficiently far from each other then φ_P and φ_Q are *almost orthogonal*. There are clearly two possibilities. Either $\omega_P \cap \omega_Q = \emptyset$ or $I_P \cap I_Q = \emptyset$. In the first case we have perfect orthogonality, $\langle \varphi_P, \varphi_Q \rangle = 0$, by Plancherel's theorem, while in the second (assuming that, say $|I_Q| \leq |I_P|$) we have

$$|\langle \varphi_P, \varphi_Q \rangle| \lesssim \left(1 + \frac{\operatorname{dist}(I_P, I_Q)}{|I_P|}\right)^{-100} |I_Q|, \tag{1.20}$$

which is a way of quantifying the almost orthogonality mentioned above. Clearly, since there is no way to localize both a function and its Fourier transform to compact intervals simultaneously, one can never obtain *perfect orthogonality* in general.

Exercise 1.4 Prove (1.20).

There is also a very useful and suggestive Fourier integral representation, which we describe next. Suppose that φ is a smooth function with Heisenberg box $[0, 1] \times [0, 1]$ and supported on the interval $[0, 1]$. Pick another smooth function $\widetilde{\varphi}$, which is equal to 1 on $[0, 1]$ and is supported inside a certain larger interval. Then, one can write $\varphi(x)$ as follows:

$$\varphi(x) = \int_{\mathbb{R}} \widetilde{\varphi}(x)\widehat{\varphi}(\xi)e^{2\pi i x\xi}\, d\xi =: \int_{\mathbb{R}} a_\varphi(x, \xi)e^{2\pi i x\xi}\, d\xi. \tag{1.21}$$

Similarly, any smooth function φ that has Heisenberg box $I \times \omega = [n2^k, (n+1)2^k] \times [m2^{-k}, (m+1)2^{-k}]$ admits a representation of the type

$$\varphi(x) = \int_{\mathbb{R}} b_\varphi(x2^{-k} - n, 2^k\xi - m)e^{2\pi i x\xi}\, d\xi. \tag{1.22}$$

Exercise 1.5 Calculate explicitly the function b_φ in (1.22) and show that its L^∞ normalization is essentially supported on $[0, 1] \times [0, 1]$, in the sense that its absolute value decays rapidly (at scale 1) away from this unit square in both directions.

The advantage of the representations (1.21), (1.22) is that the *inner functions*

$$(x, \xi) \mapsto a_\varphi(x, \xi)$$

and

$$(x, \xi) \mapsto b_\varphi(x2^{-k} - n, 2^k\xi - m)$$

are essentially supported on their corresponding Heisenberg boxes $[0, 1] \times [0, 1]$ and $I \times \omega$. These Heisenberg boxes are also sometimes called the phase-space portraits of the corresponding original functions φ. More generally, if $T : \mathbb{R}^2 \mapsto \mathbb{R}^2$ is a smooth bijective map that does not distort the area too much, one can sometimes even consider functions given by

$$\int_{\mathbb{R}} a(T^{-1}(x, \xi))e^{2\pi i x\xi}\, d\xi, \tag{1.23}$$

where $a(x, \xi)$ is as before essentially supported on $[0, 1] \times [0, 1]$, and say that their phase-space portrait is the Heisenberg region $T([0, 1] \times [0, 1])$. This will have area comparable to 1 (because of the properties of T), but its shape may be far from the shape of a dyadic rectangle. Notice that in the previous case,

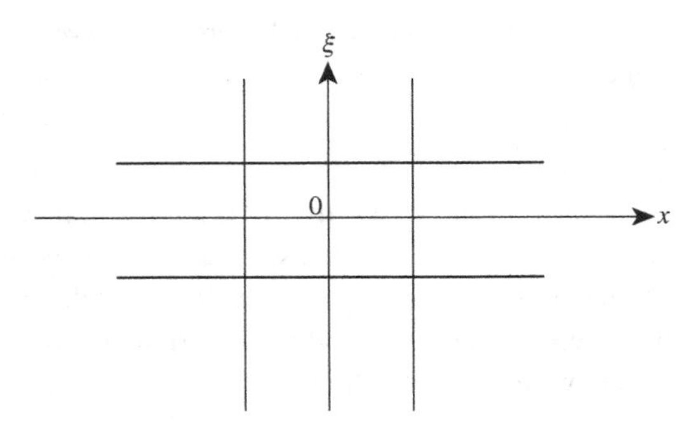

Figure 1.2. Phase-space portraits of δ_m and $e^{2\pi i m x}$ for $m \in \mathbb{Z}$.

(1.22), we had $T^{-1}(x, \xi) = (x2^{-k} - n, 2^k\xi - m)$ and $T([0, 1] \times [0, 1]) = I \times \omega$. More interesting curved examples will be discussed in detail in the next section. As before, the general intuition is that two functions whose phase-space portraits are sufficiently far from each other are almost orthogonal.

In some situations the phase-space portraits can be degenerate; there are quite a few fundamental examples of this kind. Consider φ smooth, identically equal to 1 on the interval $[-1/4, 1/4]$, and supported on $[-1/2, 1/2]$. Clearly the phase-space portrait of $\varphi(x/2^k)$ is the rectangle $[-2^{k-1}, 2^{k-1}] \times [-1/2^{k-1}, 1/2^{k-1}]$. Since the limit as $k \to \infty$ of $\varphi(x/2^k)$ is the constant function 1, it is natural to say that its phase-space portrait is the limit of the above rectangles, which is the line $\xi = 0$ in the phase plane $x\xi$. Notice that this is consistent with the fact that 1 is supported on the whole real line and does not oscillate at all. A similar argument proves that the phase-space portrait of δ_0 (the Dirac distribution centered at the origin) is given by the *dual* line $x = 0$. Notice also that this is consistent with the fact that $\widehat{\delta_0} = 1$. Both these lines should be regarded as having area 1 (see Figure 1.2).

Given an arbitrary function $f(x)$, it is also natural to define its *phase-space projection* onto the Heisenberg box $I \times \omega$ by the formula

$$\int_{\mathbb{R}} b(x, \xi)\widehat{f}(\xi)e^{2\pi i x\xi}\, d\xi, \tag{1.24}$$

where $b(x, \xi)$ is a smooth function supported (or essentially supported) on $I \times \omega$.

To see that this is consistent with our intuition, let us assume for simplicity that we are in the previous, $[0, 1] \times [0, 1]$, case, for which (1.24) becomes $(f * \varphi)\widetilde{\varphi}$. It is clear that this function and its Fourier transform are essentially supported on the interval $[0, 1]$.

One can similarly define phase-space projections onto phase-space regions that are more complicated than rectangles but whose areas are at least as large as 1 (to avoid a contradiction of the Heisenberg principle). Since there are many ways to partition the phase space smoothly, there are also many phase-space portraits of a given function. However, for concrete functions, there are some that are more natural than others.

1.5. The phase-space portraits of $e^{2\pi i x^2}$ and $e^{2\pi i x^3}$

We can now describe the phase-space portraits of the functions $e^{2\pi i x^2}$ and $e^{2\pi i x^3}$ (see Figures 1.3–1.5); these will be used later on. Consider a function φ, whose Heisenberg box is $[0, 1] \times [0, 1]$, that is supported on the interval $[0, 1]$, for simplicity. We would first like to understand the phase-space portrait of $\varphi(2^{-k}x)e^{2\pi i x^2}$. If we now denote by $\widetilde{\varphi}$ the same function as that used before, we can write

$$\begin{aligned}
\widetilde{\varphi}(2^{-k}x)\varphi(2^{-k}x)e^{2\pi i x^2} &= 2^k \int_{\mathbb{R}} \widetilde{\varphi}(2^{-k}x)\widehat{\varphi}(2^k\xi)e^{2\pi i x^2}e^{2\pi i x\xi}\, d\xi \\
&= 2^k \int_{\mathbb{R}} \widetilde{\varphi}(2^{-k}x)\widehat{\varphi}(2^k\xi)e^{2\pi i x(\xi+x)}\, d\xi \\
&= 2^k \int_{\mathbb{R}} \widetilde{\varphi}(2^{-k}x)\widehat{\varphi}(2^k(\xi - x))e^{2\pi i x\xi}\, d\xi \\
&= 2^k \int_{\mathbb{R}} \widetilde{\varphi} \otimes \widehat{\varphi}(T^{-1}(x,\xi))e^{2\pi i x\xi}\, d\xi,
\end{aligned}$$

where $T^{-1}(x, \xi) = (2^{-k}x, 2^k(\xi - x))$. Using the intuition discussed before, it is natural to say that the phase-space portrait of the function $\varphi(2^{-k}x)e^{2\pi i x^2}$ is the Heisenberg region $T([0, 1] \times [0, 1])$ in phase space, which is precisely the parallelogram of area 1 whose base is the interval $[0, 2^k]$, whose height is 2^{-k}, and which lies along the line $\xi = x$ as in Figure 1.3. Now if one starts

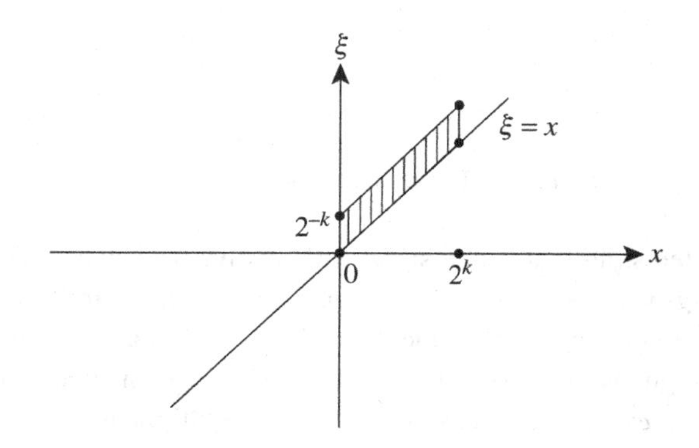

Figure 1.3. Phase-space portrait of $\varphi(2^{-k}x)e^{2\pi i x^2}$, where φ is adapted to $[0, 1]$.

instead with a function φ supported on $[-1/2, 1/2]$ and then sends k to infinity, one deduces as before that the phase-space portrait of $e^{2\pi i x^2}$ is the degenerate line $\xi = x$ (see Figure 1.4). Notice that this is consistent with the fact that the

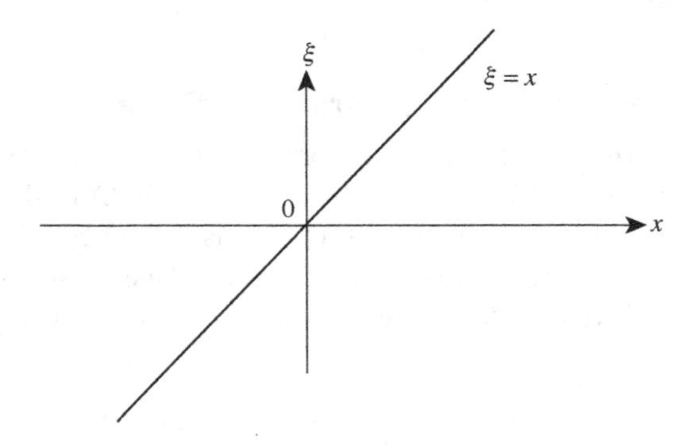

Figure 1.4. Phase-space portrait of $e^{2\pi i x^2}$.

Fourier transform of the function φ is itself. A similar argument proves that for every $n \geq 2$ the phase-space portrait of $e^{2\pi i x^n}$ is the degenerate curve $\xi = x^{n-1}$ (see Figure 1.5 for the case $n = 3$). As before, all these lines in the phase plane should be thought of as having area 1.

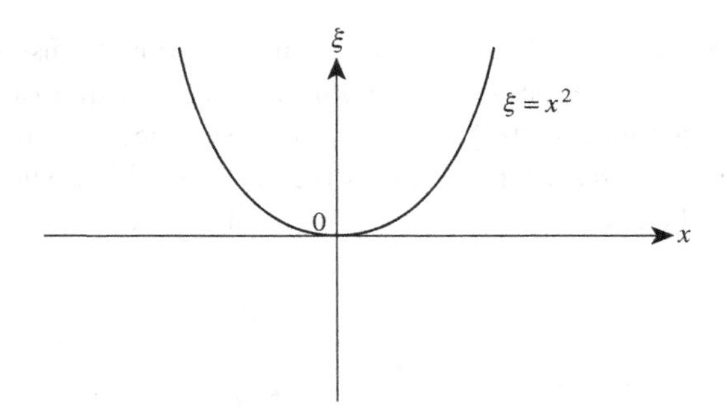

Figure 1.5. Phase-space portrait of $e^{2\pi i x^3}$.

It is also interesting that if one starts with a wave packet ϕ associated with the Heisenberg box $I \times [0, 1/|I|]$ then the phase-space portrait of $\phi(x)e^{2\pi i x^n}$ is the curved region along the line $\xi = x^{n-1}$ whose base is the interval I and whose height is $|I|^{-1}$; see Figure 1.6 for $n = 3$. In other words, multiplication with $e^{2\pi i x^n}$ has the same effect as multiplication with a linear phase function of the type $e^{2\pi i \mu x}$. Because of this fact, multiplication with

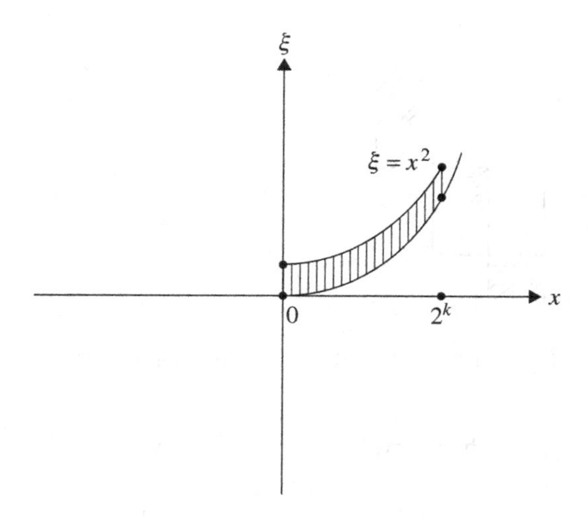

Figure 1.6. Phase-space portrait of $\varphi(2^{-k}x)e^{2\pi i x^3}$, where φ is adapted to [0, 1].

$e^{2\pi i x^2}$ is sometimes called quadratic modulation, multiplication with $e^{2\pi i x^3}$ cubic modulation, and so on. Finally, we will describe two alternative phase-space portraits of these two functions, which contain wave packets localized at scale 1.

Proposition 1.10 (Scale-1 phase-space portrait of $e^{2\pi i x^2}$) *There exists a sequence of smooth functions $(\varphi_n)_{n\in\mathbb{Z}}$ such that:*

(i) $\operatorname{supp}\varphi_n \subseteq [n-1, n+1]$, $n \in \mathbb{Z}$;
(ii) $|\varphi_n^{(\ell)}(x)| \leq C_\ell$ *uniformly in* $n \in \mathbb{Z}$

and such that

$$e^{2\pi i x^2} = \sum_{n\in\mathbb{Z}} e^{4\pi i n x}\varphi_n(x).$$

Proof First, choose a sequence of functions $(\widetilde{\varphi}_n)_{n\in\mathbb{Z}}$ satisfying the support and the derivative conditions above and also having the property that

$$\sum_{n\in\mathbb{Z}} \widetilde{\varphi}_n = 1.$$

As a consequence,

$$e^{2\pi i x^2} = \sum_{n\in\mathbb{Z}} e^{2\pi i x^2}\widetilde{\varphi}_n(x).$$

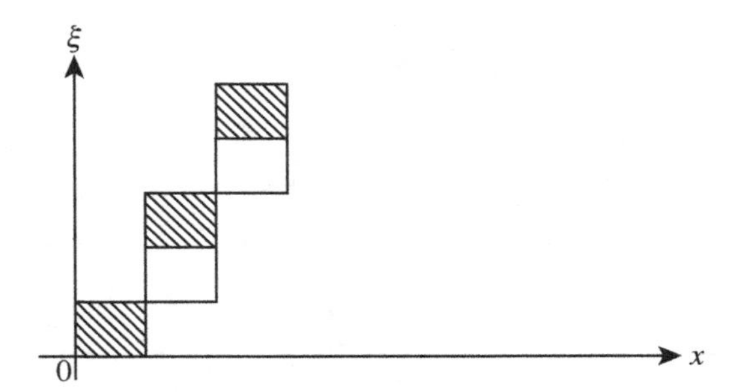

Figure 1.7. Scale-1 phase-space portrait of $e^{2\pi i x^2}$.

Fix n and, for x inside the interval $[n-1, n+1]$, rewrite $e^{2\pi i x^2}$ as follows:

$$e^{2\pi i x^2} = e^{2\pi i f(n) x} e^{2\pi i (x^2 - f(n)x)},$$

with $f(n)$ chosen so that $(x^2 - f(n)x)'$ is of the form $\mu(x-n)$ for a certain μ that is independent of n. It is easy to see that the only possibility is $\mu = 2$ and $f(n) = 2n$.

In particular, we have

$$e^{2\pi i x^2} = \sum_{n\in\mathbb{Z}} e^{4\pi i n x} e^{2\pi i (x^2 - 2nx)} \widetilde{\varphi}_n(x) =: \sum_{n\in\mathbb{Z}} e^{4\pi i n x} \varphi_n(x),$$

as desired. □

Since, for each n, φ_n is a wave packet associated with $[n, n+1] \times [0, 1]$, it follows that the function $e^{2\pi i x^2}$ becomes an infinite sum of wave packets whose Heisenberg boxes are aligned along the line $\xi = 2x$, as in Figure 1.7. Notice also that the function $x \mapsto 2x$ is precisely the derivative of $x \mapsto x^2$. One also observes that the new line $\xi = 2x$ is different from the previous one, $\xi = x$. This is a consequence of the fact that the supports of two consecutive functions of the φ_n type intersect each other.

Proposition 1.11 (Scale-1 phase-space portrait of $e^{2\pi i x^3}$) *There exists a sequence of smooth functions $(\varphi_n)_{n\in\mathbb{Z}}$ such that:*

(i) supp $\varphi_n \subseteq [n-1, n+1]$, $n \in \mathbb{Z}$;
(ii) $|\varphi_n^{(\ell)}(x)| \leq C_\ell$ uniformly in $n \in \mathbb{Z}$

and such that

$$e^{2\pi i x^3} = \sum_{n\in\mathbb{Z}} e^{6\pi i n x^2} e^{-6\pi i n^2 x} \varphi_n(x). \tag{1.25}$$

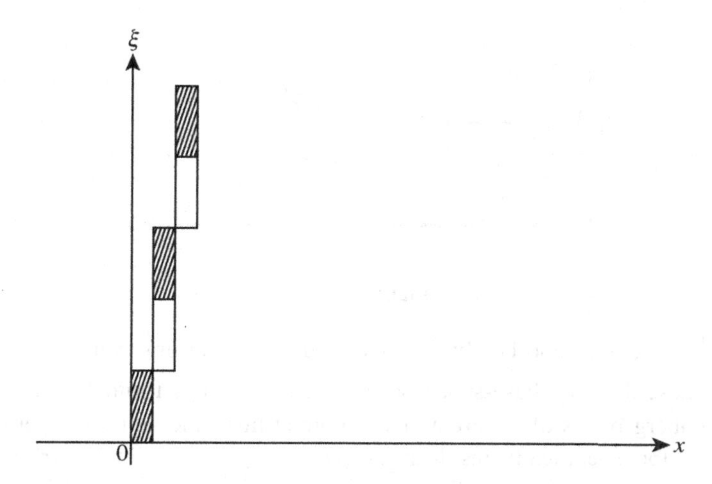

Figure 1.8. The phase-space portrait of $e^{6\pi i n x^2}$.

Proof The proof follows the same method as before. First, using the previous decomposition, split $e^{2\pi i x^3}$ as follows:

$$e^{2\pi i x^3} = \sum_{n\in\mathbb{Z}} e^{2\pi i x^3} \widetilde{\varphi}_n(x).$$

Then fix n and, on the interval $[n-1, n+1]$, rewrite $e^{2\pi i x^3}$ as

$$e^{2\pi i x^3} = e^{2\pi i f(n)x^2} e^{2\pi i g(n)x} e^{2\pi i (x^3 - f(n)x^2 - g(n)x)},$$

with $f(n)$ and $g(n)$ chosen so that the derivative $(x^3 - f(n)x^2 - g(n)x)'$ is of the form $\mu(x-n)^2$. An easy computation shows that here the only possibility is $f(n) = 3n$, $g(n) = -3n^2$, and $\mu = 3$. Using these, one obtains the desired decomposition. □

As before, we would like to have a geometric description of the above formula. Clearly, this amounts to understanding the contribution of the term $e^{6\pi i n x^2}$ for x near n. Since $e^{6\pi i n x^2} = e^{2\pi i(\sqrt{3n}x)^2}$ (assuming first that $n > 0$), its phase-space portrait can be easily obtained by dilating the corresponding portrait of $e^{2\pi i x^2}$ by a factor $\sqrt{3n}$ (see Figure 1.8). The Heisenberg boxes are now of size $\frac{1}{\sqrt{3n}} \times \sqrt{3n}$ and they are aligned along the oblique line $\xi = 6nx$. This line crosses the vertical line $x = n$ at a height $\xi = 6n^2$. Since we also have a factor of the form $e^{-6\pi i n^2 x}$ in (1.25), this implies that near $x = n$ the phase-space portrait of $e^{2\pi i x^3}$ is represented by a Heisenberg box of size $\frac{1}{\sqrt{3n}} \times \sqrt{3n}$ whose frequency interval contains the point $3n^2$, as in Figure 1.9. A similar argument proves that, for $n < 0$, the corresponding Heisenberg box has exactly the same

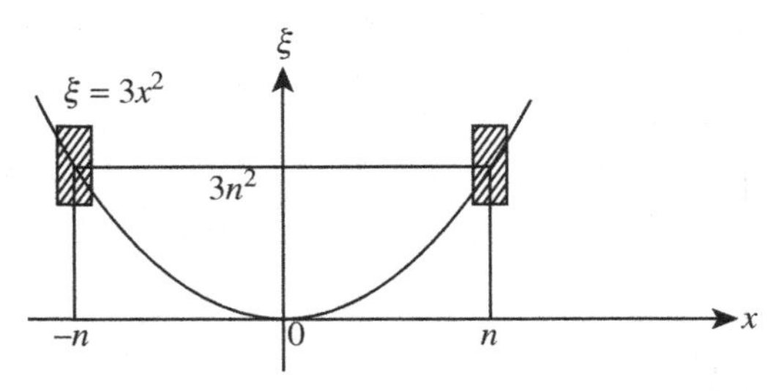

Figure 1.9

size, $\frac{1}{\sqrt{3|n|}} \times \sqrt{3|n|}$, and height $3n^2$ but is adapted to an interval near $x = n$. All this describe the phase-space portrait of $e^{2\pi i x^3}$; it is formed by a union of Heisenberg boxes of different sizes, aligned this time along the parabola $\xi = 3x^2$. More specifically, inside the strip $n - 1 \leq x \leq n + 1$ the corresponding Heisenberg boxes have all the same size, $\frac{1}{\sqrt{3|n|}} \times \sqrt{3|n|}$, and lie along $\xi = 6nx$. Notice also that $(x^3)' = 3x^2$.

1.6. Asymptotics for the Airy function

It therefore remains for us to prove Lemma 1.1. The phase-space portraits described in the previous section will help considerably.

Fix $\mu < 0$ and consider the expression

$$Ai(\mu) = \int_{\mathbb{R}} e^{2\pi i x^3} e^{2\pi i \mu x}\, dx.$$

By looking at the phase-space portrait of $e^{2\pi i x^3}$, one can see that the main contribution to the integral comes from those Heisenberg boxes whose frequency intervals contain $-\mu$ (because only then do the oscillations cancel). Equivalently, the main contribution comes from the Heisenberg boxes near $x = \pm\sqrt{|\mu|/3}$ (see Figure 1.10). As we have seen, they all have spatial sizes

$$O\left(\frac{1}{\sqrt{3}\sqrt[4]{|\mu|/3}}\right) = O\left(\frac{1}{|\mu|^{1/4}}\right),$$

which proves that indeed, as desired,

$$|Ai(\mu)| \lesssim \frac{1}{|\mu|^{1/4}} \quad \text{as } \mu \to -\infty.$$

Then, we observe that for $\mu > 0$ the line $\xi = -\mu$ does not intersect the parabola $\xi = 3x^2$ and, because of this, we have in fact the better estimate

$$|Ai(\mu)| \lesssim \frac{1}{|\mu|^m},$$

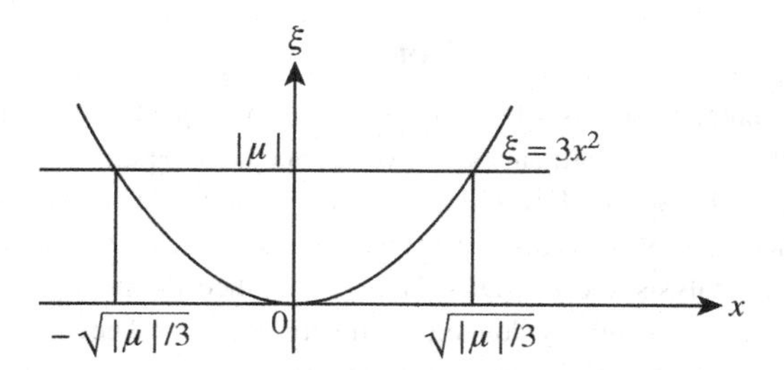

Figure 1.10

for any positive integer m, by the standard integration-by-parts argument. To prove the second part of the lemma, we write

$$
\begin{aligned}
D^\alpha Ai(\mu) &= (2\pi)^\alpha \int_{\mathbb{R}} |x|^\alpha e^{2\pi i x^3} e^{2\pi i \mu x}\, dx \\
&= (2\pi)^\alpha \int_{\mathbb{R}} \Phi_0(x)|x|^\alpha e^{2\pi i x^3} e^{2\pi i \mu x}\, dx \\
&\quad + (2\pi)^\alpha \int_{\mathbb{R}} (1 - \Phi_0(x))|x|^\alpha e^{2\pi i x^3} e^{2\pi i \mu x}\, dx,
\end{aligned}
$$

where $\Phi_0(x)$ is a smooth function supported on the interval $[-3, 3]$ and which equals 1 on $[-2, 2]$. Observe that the first integral is uniformly bounded regardless of α, while the second has the property that the inner function $x \mapsto |x|^\alpha$ is now smooth on the support of $1 - \Phi_0(x)$, which does not contain the origin. In particular, for a given negative and large enough μ, the main contribution comes again from the Heisenberg boxes near $x = \pm\sqrt{|\mu|/3}$, and a similar argument proves that this contribution is now

$$
O\left(\left(\sqrt{|\mu|}\right)^\alpha \frac{1}{|\mu|^{1/4}}\right) = O\left(|\mu|^{\alpha/2-1/4}\right).
$$

If, however, μ is negative and small then the main contribution comes from the Heisenberg boxes near the origin, which are finitely many. Finally, if μ is positive then, as pointed out before, we have even better estimates. These show that in order for $|D^\alpha Ai(\mu)|$ to be bounded, one does indeed need $\alpha \in \left[0, \frac{1}{2}\right]$, as desired.

Exercise 1.6 Fill in the details in the above argument. More precisely, show that the main contribution to the estimate for the Airy function $Ai(\mu)$ comes from those wave packets in the phase-space decomposition of $e^{2\pi i x^3}$ whose Heisenberg boxes contain the frequency $-\mu$, as claimed.

Notes

The material presented in Sections 1.2 and 1.3 follows the paper by Christ and Weinstein [20], where such dispersive estimates are proved. See also the earlier works by Ponce and Vega [102] and by Strauss [107]. For more about the Korteweg–de Vries equation see Kenig, Ponce, and Vega [66]. Wave packets and phase-space portraits play an important role in analysis and in PDEs. Here we defined them for functions of one variable but they can be similarly defined for functions in $\mathbb{R}^n$. They are at the heart of what is sometimes called *microlocal analysis* and will in fact appear in every chapter of this second volume, but more prominently in Chapters 5–7, where the bilinear Hilbert transform and the Carleson operator are discussed.

Curved phase-space portraits were first considered by Egorov [38], and they play an important role in the analysis of PDEs. See also an article concerning the uncertainty principle by Fefferman [41], where many other aspects are discussed in some detail. More recently, polynomially modulated wave packets appeared also in Stein and Wainger [106] and in Lie [76].

In the context of Banach spaces, Leibnitz rules first appeared in Kato and Ponce [62] and [20]. For other topics related to them, see also [66], Bony [9], Taylor [111], Bahouri, Chemin and Danchin [4], and Linares and Ponce [78]. The general quasi-Banach form of the Leibnitz rules described here, seems to be new. See also Muscalu, Pipher, Tao *et al.* [93] for a related discussion.

Problems

Problem 1.1 Rederive the conserved quantities discussed in Section 1.1 using integration by parts.

Problem 1.2 Prove the inequality (1.16), using arguments similar to those used to prove (1.15).

Problem 1.3 For any $k \geq 2$ consider the function $\exp(2\pi i x^k)$. Generalize the results of the Sections 1.5 and 1.6 (which correspond to $k = 2, 3$) by constructing a scale-1 phase-space portrait for this function. Then use the intuition that comes from this, also as before, to obtain the asymptotics for its inverse Fourier transform.

2

Classical paraproducts

The reader should be convinced by now about the usefulness of the Leibnitz rule (1.1). In this chapter we concentrate on obtaining a through understanding of this rule. As we will see, (1.1) can be reduced to an important theorem in harmonic analysis, the so-called Coifman–Meyer theorem on *paraproducts*. Let us restate (1.1):

$$\|D^\alpha(fg)\|_r \lesssim \|D^\alpha f\|_{p_1} \|g\|_{q_1} + \|f\|_{p_2} \|D^\alpha g\|_{q_2} \tag{2.1}$$

for every $\alpha > 0$, for $1 < p_i, q_i \leq \infty$ satisfying $1/r = 1/p_i + 1/q_i$ for $i = 1, 2$, and for $1/(1+\alpha) < r < \infty$.

A few comments are in order. First, since the bi-linear paraproducts that will appear later on are in fact bilinear Calderón–Zygmund operators (also known as bilinear Marcinkiewicz–Mikhlin–Hörmander multipliers or as bilinear Mikhlin multipliers), the natural range of estimates that one expects them to satisfy is given by Hölder inequalities. Since both the constituent functions of such an operator belong to some L^p space for $1 < p < \infty$, the corresponding index for the *target* space L^r satisfies the condition $1/2 < r < \infty$. That is why some of the inequalities (2.1) are valid even when the target space corresponds to an index r that is less than 1. It is timely to remember that these L^r spaces are no longer Banach spaces for $r < 1$. They are what are sometimes called *quasi-Banach* spaces, since the corresponding *quasi-norm* $\|h\|_r := (\int_{\mathbb{R}} |h(x)|^r \, dx)^{1/r}$ satisfies only a weaker form of the triangle inequality, of the type $\|h_1 + h_2\|_r \leq C(\|h_1\|_r + \|h_2\|_r)$, as one can easily check.

Let us also comment on the sharpness of the constraint condition $1/(1+\alpha) < r < \infty$. Notice that, on the one hand, as α gets close to zero (2.1) becomes close to a Hölder inequality that is true in the entire range $1/2 < r < \infty$. On the other hand, in this limit our condition gives $1 \leq r < \infty$, a fact that could be considered as an indication that (2.1) might not be sharp under such a constraint. Even though it is very tempting to argue in this way, the situation is more delicate than that. This sharp difference between the cases $\alpha = 0$ and $\alpha > 0$ is fundamentally related to the fact that the function $|\xi|^\alpha$ that appears in the definition of the α derivatives is smooth for $\alpha = 0$, but it is merely continuous if $\alpha > 0$ in a neighborhood of the origin. Consider the following particular case. Let φ be a Schwartz function having the property that $\operatorname{supp} \widehat{\varphi} \subseteq [-1, 1]$ and define $f(x) := \varphi(x)e^{2\pi i 5x}$ and $g(x) := \varphi(x)e^{-2\pi i 5x}$. Observe that $\operatorname{supp} \widehat{f} \subseteq [4, 6]$ and $\operatorname{supp} \widehat{g} \subseteq [-6, -4]$. As a consequence, the corresponding right-hand side of (2.1) is bounded by a universal constant. However, $fg = \varphi^2$ is a function whose Fourier support lies inside $[-2, 2]$, an interval that contains the origin. In particular, one has the sharp decay estimate

$$|D^\alpha(\varphi^2)(x)| \lesssim \frac{1}{(1 + |x|)^{1+\alpha}} \tag{2.2}$$

for every $\alpha > 0$ and every $x \in \mathbb{R}$. Now, in order for the function $1/(1 + |x|)^{1+\alpha}$ to be r-integrable, one has to have $r(1 + \alpha) > 1$; this explains the necessity of our constraint condition.

Exercise 2.1 Prove the inequality (2.2) and show that it is sharp, in the sense that the function on the left-hand side could not decay at a faster rate.

Coming back to the original inequalities (2.1), observe that if $1 \le r < \infty$ then they are true for any $\alpha > 0$, while if we want them to hold for any $1/2 < r < \infty$ we need α to be greater than 1.

Before beginning a detailed study of (2.1) we mention that the proof of the Coifman–Meyer theorem that will be presented is not the simplest or the shortest possible but it is clear and robust. The backbone of this proof will be used later, in Chapter 8, where we study flag paraproducts. We also hope that this approach will help the reader to understand the similarities and differences between paraproducts and other analytical objects to be studied in the following chapters, such as the bilinear Hilbert transform and the Carleson maximal operator.

2.1. Paraproducts

To prove (2.1) the first thing that needs to be done is to *mollify* the nonlinearity $D^\alpha(fg)$ in a natural bilinear way. The classical Littlewood–Paley decompositions will be helpful. Let us recall them carefully from Lemma I.8.1.

Let $\varphi \in S(\mathbb{R})$ be a Schwartz function such that $\operatorname{supp} \widehat{\varphi} \subseteq [-2, 2]$ and $\widehat{\varphi}(\xi) = 1$ on $[-1, 1]$. Then define $\psi \in S(\mathbb{R})$ to be that Schwartz function whose Fourier transform satisfies

$$\widehat{\psi}(\xi) := \widehat{\varphi}(\xi) - \widehat{\varphi}(2\xi).$$

Observe that supp $\widehat{\psi} \subseteq [-2, \frac{1}{2}] \cup [\frac{1}{2}, 2]$. Then, for every $k \in \mathbb{Z}$, define $\widehat{\psi_k} \in S(\mathbb{R})$ by

$$\widehat{\psi_k}(\xi) := \widehat{\psi}\left(\frac{\xi}{2^k}\right)$$

and observe similarly that

$$\operatorname{supp} \widehat{\psi_k} \subseteq [-2^{k+1}, -2^{k-1}] \cup [2^{k-1}, 2^{k+1}].$$

Since it is easy to see that

$$1 = \sum_{k \in \mathbb{Z}} \widehat{\psi_k}(\xi)$$

for almost every $\xi \in \mathbb{R}$, one obtains as a consequence the following Littlewood–Paley decomposition of a function f:

$$f = \sum_{k_1 \in \mathbb{Z}} f * \psi_{k_1}. \tag{2.3}$$

One also has

$$g = \sum_{k_2 \in \mathbb{Z}} g * \psi_{k_2}.$$

In particular,[1]

$$\begin{aligned} f \cdot g &= \sum_{k_1, k_2} (f * \psi_{k_1})(g * \psi_{k_2}) \\ &= \sum_{k_1 \ll k_2} (f * \psi_{k_1})(g * \psi_{k_2}) + \sum_{k_2 \ll k_1} (f * \psi_{k_1})(g * \psi_{k_2}) \\ &\quad + \sum_{k_1 \simeq k_2} (f * \psi_{k_1})(g * \psi_{k_2}). \end{aligned} \tag{2.4}$$

The first two terms of (2.4) are very similar. One can rewrite the first, for instance, as follows:

$$\begin{aligned} \sum_{k_1 \ll k_2} (f * \psi_{k_1})(g * \psi_{k_2}) &= \sum_{k_2} \left(\sum_{k_1 \ll k_2} f * \psi_{k_1} \right) (g * \psi_{k_2}) \\ &=: \sum_{k_2} (f * \varphi_{k_2})(g * \psi_{k_2}) = \sum_{k} (f * \varphi_k)(g * \psi_k), \end{aligned} \tag{2.5}$$

[1] Note that, here and elsewhere, $k_a \ll k_b$ means $k_a < k_b - 100$ and $k_a \simeq k_b$ means $k_a - 100 \leq k_b \leq k_a + 100$.

where φ_k is also a Schwartz function, with the property that

$$\operatorname{supp}\widehat{\varphi_k} \subseteq \left[-2^{k-10}, 2^{k-10}\right].$$

Therefore (2.5) can be written in the equivalent form (i.e., *completed* as)

$$\sum_k ((f * \varphi_k)(g * \psi_k)) * \widetilde{\psi}_k, \tag{2.6}$$

where $\widetilde{\psi}_k \in S(\mathbb{R})$ satisfies $\operatorname{supp}\widehat{\widetilde{\psi}_k} \subseteq \left[-2^{k+2}, -2^{k-2}\right] \cup \left[2^{k-2}, 2^{k+2}\right]$.

Expressions of the type (2.6) are called *paraproducts* and, as we shall see, they play an important role in the analysis of (2.1). Since one can clearly treat every term of (2.4) in a similar manner, (2.4) represents the standard way of decomposing a product of two functions as a sum of paraproducts. There are analogous decompositions for products of an arbitrary number of functions. We invite the reader to compare (2.6) with the formula that appeared at the beginning of subsection 9.5.3 in Vol. I.

If one denotes by $\Pi(f, g)$ the expression in (2.6), one can further write

$$\begin{aligned}
D^\alpha\left(\Pi(f, g)\right) &= \sum_k ((f * \varphi_k)(g * \psi_k)) * D^\alpha \widetilde{\psi}_k \\
&=: \sum_k ((f * \varphi_k)(g * \psi_k)) * 2^{k\alpha} \widetilde{\widetilde{\psi}}_k \\
&= \sum_k \left((f * \varphi_k)\left(g * 2^{k\alpha}\psi_k\right)\right) * \widetilde{\widetilde{\psi}}_k \\
&=: \sum_k \left((f * \varphi_k)\left(g * D^\alpha \widetilde{\widetilde{\widetilde{\psi}}}_k\right)\right) * \widetilde{\widetilde{\psi}}_k \\
&= \sum_k \left((f * \varphi_k)\left(D^\alpha g * \widetilde{\widetilde{\widetilde{\psi}}}_k\right)\right) * \widetilde{\widetilde{\psi}}_k \\
&=: \widetilde{\Pi}\left(f, D^\alpha g\right),
\end{aligned} \tag{2.7}$$

where the Schwartz functions $\widetilde{\widetilde{\psi}}_k$ and $\widetilde{\widetilde{\widetilde{\psi}}}_k$ are defined in a natural way by

$$\widehat{\widetilde{\widetilde{\psi}}}_k(\xi) := \widehat{\widetilde{\psi}}_k(\xi)\left|\frac{\xi}{2^k}\right|^\alpha$$

and

$$\widehat{\widetilde{\widetilde{\widetilde{\psi}}}}_k(\xi) := \widehat{\psi_k}(\xi)\left(\frac{2^k}{|\xi|}\right)^\alpha$$

respectively. Observe that in order to obtain (2.7) we took advantage of the fact that the support of $\widehat{\psi_k}$ is away from zero; if it had been otherwise we could not have defined $\widehat{\widetilde{\widetilde{\psi}}}_k$ in the way we did. It is also important to notice that, since the supports of $\widehat{\widetilde{\psi}}_k$ are away from zero, the functions $\widetilde{\widetilde{\psi}}_k$ are smooth. Since Π and $\widetilde{\Pi}$ are very similar objects, equality (2.7) shows that paraproducts have the capacity of "absorbing" derivatives.

The third term of (2.4) is more difficult, since now (assuming that $k_1 = k_2 = k$ for simplicity) the sum of the supports is given by

$$\operatorname{supp} \widehat{\psi_k} + \operatorname{supp} \widehat{\psi_k}$$

and might contain the origin. As a consequence, if one performs a similar calculation on that term (first "completing" it and then taking the α derivative), the analogue $\widetilde{\widetilde{\varphi}}_k$ of the function $\widetilde{\widetilde{\psi}}_k$ is no longer smooth. More precisely, one can think of $\widetilde{\widetilde{\varphi}}_k$ as being the L^1-normalized dilation of a given function $\widetilde{\widetilde{\varphi}}$ (corresponding to $k = 0$), which satisfies only the weaker decay estimate

$$|\widetilde{\widetilde{\varphi}}(x)| \lesssim \frac{1}{(1 + |x|)^{1+\alpha}}. \tag{2.8}$$

Recall that this is precisely the estimate from (2.2).

Let us denote by $\Pi^\alpha(f, g)$ an expression of the type

$$\sum_k ((f * \psi_k)(g * \psi_k)) * \varphi_k^\alpha,$$

where the φ_k^α are obtained as before, by rescaling from a given function, and satisfy the weaker decay estimate above.

All these calculations show that proving (2.1) can be reduced to the problem of obtaining Hölder-type estimates for bilinear expressions of the type $(F, G) \mapsto \Pi(F, G)$ and of the type $(F, G) \mapsto \Pi^\alpha(F, G)$.

Let us first assume that $1 < p, q, r < \infty$ are such that $1/p + 1/q = 1/r$ and r' is defined by $1/r + 1/r' = 1$. Then one can write

$$\begin{aligned}
\|\Pi(f, g)\|_r &= \left| \int_{\mathbb{R}} \Pi(f, g)(x) h(x)\, dx \right| \\
&= \left| \int_{\mathbb{R}} \sum_{k \in \mathbb{Z}} (f * \varphi_k)(x)\, (g * \psi_k)(x)\, \big(h * \widetilde{\psi}_k\big)(x)\, dx \right| \\
&\le \int_{\mathbb{R}} \sum_k |f * \varphi_k(x)|\, |g * \psi_k(x)|\, \big|h * \widetilde{\psi}_k(x)\big|\, dx
\end{aligned}$$

$$\le \int_{\mathbb{R}} \left(\sup_k |f * \varphi_k(x)|\right)\left(\sum_k |g * \psi_k(x)|^2\right)^{1/2}$$
$$\times \left(\sum_k \left|h * \widetilde{\psi}_k(x)\right|^2\right)^{1/2} dx$$
$$\lesssim \int_{\mathbb{R}} Mf(x)Sg(x)Sh(x)\,dx,$$

where clearly h is an appropriately chosen function with $\|h\|_{r'} = 1$, M is the Hardy–Littlewood maximal operator, and S is the square-function of Littlewood and Paley. Since it is well known that both M and S are bounded operators on L^s spaces for $1 < s < \infty$ (see the first volume of the book, more precisely Proposition I.2.9 and Theorem I.8.3), this shows that one has

$$\|\Pi(f, g)\|_r \lesssim \|f\|_p \, \|g\|_q. \tag{2.9}$$

A similar estimate is available for Π^α, since the corresponding Hardy–Littlewood maximal operator associated with the family of functions φ_k^α still satisfies the usual L^s estimates. As we pointed out earlier, this proves the Leibnitz rule (2.1) in the particular case when all the indices involved are strictly between 1 and ∞.

2.2. Discretized paraproducts

The analogous inequality (2.9) in the quasi-Banach case where $1/(1+\alpha) < r \le 1$ is more subtle. In order to understand it, we need to discretize both $\Pi(f, g)(x)$ and $\Pi^\alpha(f, g)(x)$ in the x variable.

To do this, we first consider the associated trilinear form $\Lambda(f, g, h)$ of $\Pi(f, g)$ and write

$$\Lambda(f, g, h) := \int_{\mathbb{R}} \Pi(f, g)(x)h(x)\,dx.$$
$$= \sum_{k\in\mathbb{Z}} \int_{\mathbb{R}} (f * \varphi_k)(x)(g * \psi_k)(x)\big(h * \widetilde{\psi}_k\big)(x)\,dx. \tag{2.10}$$

Fix $k \in \mathbb{Z}$ and consider the corresponding term in (2.10). It can be written as

$$2^{-k} \int_{\mathbb{R}} (f * \varphi_k)(2^{-k}y)(g * \psi_k)(2^{-k}y)(h * \tilde{\psi}_k)(2^{-k}y)\,dy$$
$$= 2^{-k} \sum_{n\in\mathbb{Z}} \int_0^1 (f * \varphi_k)(2^{-k}(n + \beta))$$
$$\times (g * \psi_k)(2^{-k}(n + \beta))(h * \tilde{\psi}_k)(2^{-k}(n + \beta))\,d\beta. \tag{2.11}$$

Then, the term $(f * \varphi_k)(2^{-k}(n+\beta))$ can be written as follows:

$$(f * \varphi_k)(2^{-k}(n+\beta)) = \int_{\mathbb{R}} f(t)\varphi_k(2^{-k}(n+\beta)-t)\,dt$$
$$= 2^{k/2}\int_{\mathbb{R}} f(t)2^{-k/2}\varphi_k(2^{-k}(n+\beta)-t)\,dt$$
$$= 2^{k/2}\langle f, \varphi_I^{1,\beta}\rangle,$$

where $\varphi_I^{1,\beta}$ is defined by

$$\varphi_I^{1,\beta}(t) := 2^{-k/2}\overline{\varphi_k(2^{-k}(n+\beta)-t)}$$

and I is the dyadic interval $[2^{-k}n, 2^{-k}(n+1)]$.

Similarly, we define $\varphi_I^{2,\beta}$ and $\varphi_I^{3,\beta}$ in such a way that the second and third factors in (2.11) become $2^{k/2}\langle g, \varphi_I^{2,\beta}\rangle$ and $2^{k/2}\langle h, \varphi_I^{3,\beta}\rangle$ respectively.

As a consequence, (2.11) can be written as

$$\int_0^1 \sum_{|I|=2^{-k}} \frac{1}{|I|^{1/2}}\langle f, \varphi_I^{1,\beta}\rangle\langle g, \varphi_I^{2,\beta}\rangle\langle h, \varphi_I^{3,\beta}\rangle\,d\beta, \tag{2.12}$$

where the sum runs over all dyadic intervals of size $|I| = 2^{-k}$. If one does this for every scale $k \in \mathbb{Z}$, one obtains a similar expression,

$$\int_0^1 \sum_{I} \frac{1}{|I|^{1/2}}\langle f, \varphi_I^{1,\beta}\rangle\langle g, \varphi_I^{2,\beta}\rangle\langle h, \varphi_I^{3,\beta}\rangle\,d\beta, \tag{2.13}$$

for (2.10), where this time I runs over all dyadic intervals.

We may also remark that the functions $\varphi_I^{j,\beta}$ for $j = 1, 2, 3$ are all L^2-normalized bump functions adapted to I and whose Fourier supports are included either in $[-2^{k-10}, 2^{k-10}]$ in the $j = 1$ case or in $[-2^{k+2}, -2^{k-2}] \cup [2^{k-2}, 2^{k+2}]$ in the $j = 2, 3$ cases, when $|I| = 2^{-k}$. In particular, the Cartesian products of these intervals form the Heisenberg boxes naturally associated with these families of functions, as in Figures 2.1 and 2.2.

Definition 2.1 A family of L^2-normalized adapted bump functions $(\varphi_I)_I$ is said to be *nonlacunary*[2] if and only if for every I one has

$$\operatorname{supp}\widehat{\varphi_I} \subseteq \left[-4|I|^{-1}, 4|I|^{-1}\right].$$

A family of L^2-normalized adapted bump functions $(\varphi_I)_I$ is said to be *lacunary* if and only if for any I one has

$$\operatorname{supp}\widehat{\varphi_I} \subseteq \left[-4|I|^{-1}, -\tfrac{1}{4}|I|^{-1}\right] \cup \left[\tfrac{1}{4}|I|^{-1}, 4|I|^{-1}\right].$$

[2] A lacuna is a gap.

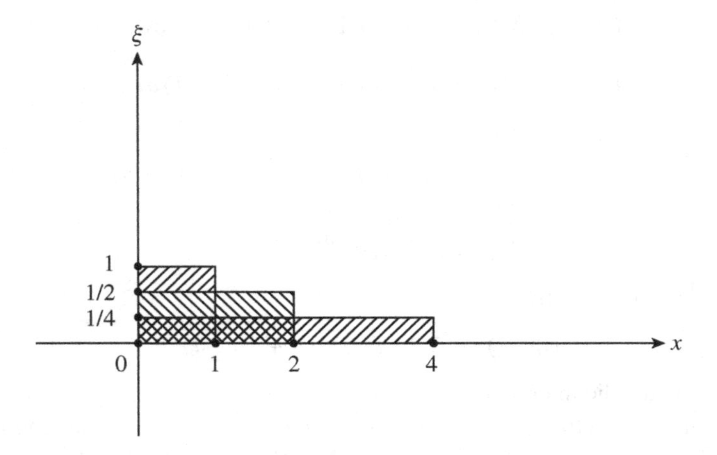

Figure 2.1. Heisenberg boxes of a nonlacunary family.

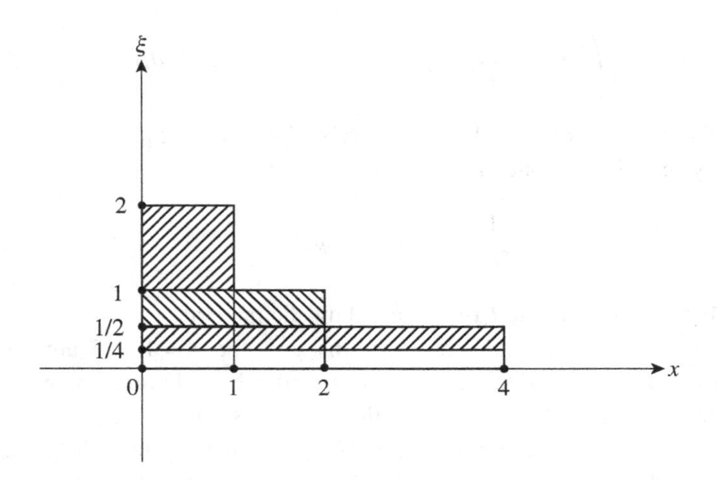

Figure 2.2. Heisenberg boxes of a lacunary family.

It is important to realize that these lacunary L^2-normalized bump functions $(\varphi_I)_I$ are the smooth analogue of the Haar functions $(h_I)_I$ studied in Section I.8.4. Similarly, a nonlacunary L^2-normalized sequence $(\varphi_I)_I$ can be thought of as a smooth analogue of $(\chi_I/|I|^{1/2})_I$.

Definition 2.2 Let $\mathcal{J}$ be a finite set of dyadic intervals. A bilinear expression of the type

$$\Pi_{\mathcal{J}}(f, g) = \sum_{I \in \mathcal{J}} c_I \frac{1}{|I|^{1/2}} \langle f, \varphi_I^1 \rangle \langle g, \varphi_I^2 \rangle \varphi_I^3 \tag{2.14a}$$

is called a *bilinear discretized paraproduct* if and only if $(c_I)_I$ is a bounded sequence of complex numbers and at least two of the families of L^2-normalized bump functions $(\varphi_I^j)_I$ for $j = 1, 2, 3$ are lacunary in the sense of Definition 2.1; cf. Definition I.9.12.

A similar discretization procedure applied to $\Pi^\alpha(f, g)$ gives rise to expressions of the form

$$\Pi^\alpha_{\mathcal{J}}(f, g) = \sum_{I \in \mathcal{J}} c_I \frac{1}{|I|^{1/2}} \langle f, \varphi_I^1 \rangle \langle g, \varphi_I^2 \rangle \varphi_I^{3,\alpha}, \tag{2.14b}$$

where the families $(\varphi_I^j)_I$ for $j = 1, 2$ are lacunary while $(\varphi_I^{3,\alpha})_I$ is nonlacunary and satisfies only the weaker decay condition

$$\left| |I|^{1/2} \varphi_I^{3,\alpha}(x) \right| \lesssim \frac{1}{(1 + \mathrm{dist}(x, I)/|I|)^{1+\alpha}},$$

for every $I \in J$.

The main theorem of the chapter is the following.

Theorem 2.3 *Any bilinear discretized paraproduct $\Pi_{\mathcal{J}}$ has a bounded mapping $L^p \times L^q$ to L^r as long as $1 < p, q \leq \infty$, $1/p + 1/q = 1/r$, and $0 < r < \infty$. Moreover, the implicit constants in the bounds depend only on p, q, r and are independent of the cardinality of $\mathcal{J}$, provided that the sequence $(c_I)_I$ in* (2.14a) *is bounded by a universal constant.*

The analogous result for $\Pi^\alpha_{\mathcal{J}}$ is contained in the following.

Theorem 2.4 *For any $\alpha > 0$, $\Pi^\alpha_{\mathcal{J}}$ maps $L^p \times L^q$ into L^r boundedly, as long as $1 < p, q \leq \infty$, $1/p + 1/q = 1/r$ and $1/(1 + \alpha) < r < \infty$. Moreover, as before the implicit boundedness constants depend only on p, q, r and are independent of the cardinality of $\mathcal{J}$, provided that the sequence $(c_I)_I$ in* (2.14b) *is bounded by a universal constant.*

The constraint $1/(1 + \alpha) < r < \infty$ appears only in the second statement, Theorem 2.3 being completely general. The result in Theorem 2.3 is a discretized version of the Coifman–Meyer theorem. As we shall see, the method of proof of these two theorems together with (2.13) will guarantee that (2.9) and an analogous statement for Π^α hold also for $1/(1 + \alpha) < r \leq 1$ and this, as mentioned earlier, is enough to prove our desired (2.1) in the general case.

First let us observe that the proof of the $1 < p, q, r < \infty$ case is as easy as the proof of (2.9) in the same range of exponents.

Indeed, one can write

$$\begin{aligned}
\|\Pi_{\mathcal{J}}(f,g)\|_r &= \left|\int_{\mathbb{R}} \Pi_{\mathcal{J}}(f,g)(x)h(x)\,dx\right| \\
&\lesssim \sum_{I\in\mathcal{J}} \frac{1}{|I|^{1/2}} |\langle f,\varphi_I^1\rangle||\langle g,\varphi_I^2\rangle||\langle h,\varphi_I^3\rangle| \\
&= \sum_{I\in\mathcal{J}} \frac{|\langle f,\varphi_I^1\rangle|}{|I|^{1/2}} \frac{|\langle g,\varphi_I^2\rangle|}{|I|^{1/2}} \frac{|\langle h,\varphi_I^3\rangle|}{|I|^{1/2}} |I| \\
&= \int_{\mathbb{R}} \sum_{I\in\mathcal{J}} \frac{|\langle f,\varphi_I^1\rangle|}{|I|^{1/2}} \frac{|\langle g,\varphi_I^2\rangle|}{|I|^{1/2}} \frac{|\langle h,\varphi_I^3\rangle|}{|I|^{1/2}} \chi_I(x)\,dx \\
&\leq \int_{\mathbb{R}} \left(\sup_{I\in\mathcal{J}} \frac{|\langle f,\varphi_I^1\rangle|}{|I|^{1/2}} \chi_I(x)\right) \left(\sum_{I\in\mathcal{J}} \frac{|\langle g,\varphi_I^2\rangle|^2}{|I|} \chi_I(x)\right)^{1/2} \\
&\qquad \times \left(\sum_{I\in\mathcal{J}} \frac{|\langle h,\varphi_I^3\rangle|^2}{|I|} \chi_I(x)\right)^{1/2} dx \\
&\leq \int_{\mathbb{R}} Mf(x)Sg(x)Sh(x)\,dx, \qquad (2.15)
\end{aligned}$$

where, as before, M is the Hardy–Littlewood maximal operator, S is the discrete Littlewood–Paley square-function operator, while h is again a well chosen function in $L^{r'}$ so that $\|h\|_{r'} = 1$. We have also assumed that the families $(\varphi_I)_I^j$ are lacunary for $j = 2, 3$. Then

$$\|\Pi_{\mathcal{J}}(f,g)\|_r \lesssim \|f\|_p \|g\|_q$$

holds because of the known boundedness properties of M and S. Also as before, one observes that a similar bound can be obtained in the case of $\Pi_{\mathcal{J}}^{\alpha}(f,g)$.

2.3. Discretized Littlewood–Paley square-function operator

Since such discrete Littlewood–Paley square-function operators will appear quite often in later chapters, we will prove their boundedness properties in the following theorem. *Continuous* square functions were discussed thoroughly in Chapter I.8 but it is instructive to include a direct proof of the *discrete* result here.

Theorem 2.5 *Let $\mathcal{J}$ be a finite family of dyadic intervals and $(\varphi_I)_I$ a lacunary family of L^2-normalized bump functions.*

Then the following discretized Littlewood–Paley square-function operator S, defined by

$$Sf(x) = \left(\sum_{I \in \mathcal{J}} \frac{|\langle f, \varphi_I \rangle|^2}{|I|} \chi_I(x) \right)^{1/2},$$

maps L^s boundedly into itself for any $1 < s < \infty$ and also L^1 into $L^{1,\infty}$. Moreover, the implicit constants in the bounds will depend on s but not on the cardinality of $\mathcal{J}$.

Proof We first observe that S maps L^2 into L^2. Indeed, one has

$$\|Sf\|_2 = \left(\sum_{I \in \mathcal{J}} |\langle f, \varphi_I \rangle|^2 \right)^{1/2}$$

and since $(\varphi_I)_I$ is a lacunary family, it is enough to prove that

$$\sum_{\substack{I \in \mathcal{J} \\ |I| = \text{const}}} |\langle f, \varphi_I \rangle|^2 \lesssim \|f\|_2^2. \tag{2.16}$$

The left-hand side of (2.16) is equal to

$$\begin{aligned}
\left| \sum_{|I|=\text{const}} \langle f, \varphi_I \rangle \overline{\langle f, \varphi_I \rangle} \right| &= \left| \sum_{|I|=\text{const}} \langle f, \varphi_I \rangle \langle \overline{f}, \overline{\varphi}_I \rangle \right| \\
&= \left| \left\langle \overline{f}, \sum_{|I|=\text{const}} \overline{\langle f, \varphi_I \rangle \varphi_I} \right\rangle \right| \\
&\leq \|f\|_2 \left\| \sum_{|I|=\text{const}} \langle f, \varphi_I \rangle \varphi_I \right\|_2
\end{aligned}$$

and so, to prove (2.16), it is enough to show that

$$\left\| \sum_{|I|=\text{const}} \langle f, \varphi_I \rangle \varphi_I \right\|_2 \lesssim \left(\sum_{|I|=\text{const}} |\langle f, \varphi_I \rangle|^2 \right)^{1/2}. \tag{2.17}$$

Now, the square of the left-hand side of (2.17) is equal to

$$\begin{aligned}
&\left| \left\langle \sum_{|I|=\text{const}} \langle f, \varphi_I \rangle \varphi_I, \sum_{|J|=\text{const}} \langle f, \varphi_J \rangle \varphi_J \right\rangle \right| \\
&\leq \sum_{|I|,|J|=\text{const}} |\langle f, \varphi_I \rangle| \, |\langle f, \varphi_J \rangle| \, |\langle \varphi_I, \varphi_J \rangle|
\end{aligned}$$

$$= \sum_{n=0}^{\infty} \sum_{\substack{|I|=|J|=\text{const} \\ \frac{\text{dist}(I,J)}{|I|}=n}} |\langle f, \varphi_I \rangle| \, |\langle f, \varphi_J \rangle| \, |\langle \varphi_I, \varphi_J \rangle|$$

$$\lesssim \sum_{n=0}^{\infty} \sum_{\substack{|I|=|J|=\text{const} \\ \frac{\text{dist}(I,J)}{|I|}=n}} |\langle f, \varphi_I \rangle| \, |\langle f, \varphi_J \rangle| \frac{1}{(1+n)^{10}}$$

$$\lesssim \sum_{n=0}^{\infty} \left(\sum_{|I|=\text{const}} |\langle f, \varphi_I \rangle|^2 \right)^{1/2} \left(\sum_{|J|=\text{const}} |\langle f, \varphi_J \rangle|^2 \right)^{1/2} \frac{1}{(1+n)^{10}}$$

$$\lesssim \sum_{|I|=\text{const}} |\langle f, \varphi_I \rangle|^2,$$

as desired.

We now show that $S : L^1 \to L^{1,\infty}$, in other words that

$$|\{x \in \mathbb{R} \mid Sf(x) > \lambda\}| \lesssim \frac{1}{\lambda} \|f\|_1,$$

for every $\lambda > 0$. Fix such a $\lambda > 0$ and perform a Calderón–Zygmund decomposition of the function f *at the level* λ, as in Lemma I.2.17.

Pick a dyadic interval J such that

$$\frac{1}{|J|} \int_J |f(x)| dx > \lambda$$

and such that J is maximal with respect to inclusion. Denote by Ω the union of all such maximal dyadic intervals. Clearly, because of their maximality they are all disjoint. Also, one has

$$|\Omega| = \sum_J |J| < \frac{1}{\lambda} \sum_J \int_J |f(x)| \, dx \leq \frac{1}{\lambda} \|f\|_1.$$

We also note that, by construction, $|f(z)| \leq \lambda$ for every $z \in \Omega^c$. Now we split our function f, setting

$$f = g + b,$$

where

$$g = f \chi_{\Omega^c} + \sum_J \left(\frac{1}{|J|} \int_J f(x) \, dx \right) \chi_J$$

and

$$b = f - g = \sum_J b_J$$

with

$$b_J := \left(f - \frac{1}{|J|} \int_J f(x)\, dx \right) \chi_J.$$

Clearly, the support of every b_J lies inside the interval J.

Notice that

$$\|g\|_{L^\infty} \lesssim \lambda \tag{2.18}$$

since

$$\left| \frac{1}{|J|} \int_J f(x)\, dx \right| \le \frac{1}{|J|} \int_J |f(x)|\, dx \le 2 \frac{1}{|\widetilde{J}|} \int_{\widetilde{J}} |f(x)|\, dx \le 2\lambda,$$

where $\widetilde{J}$ is the unique dyadic interval with the property that $J \subseteq \widetilde{J}$ and $|\widetilde{J}| = 2|J|$.

It is also important to observe that $\int_{\mathbb{R}} b_J(z)\, dz = 0$, by definition, and that

$$\begin{aligned} \int_{\mathbb{R}} |b_J(z)|\, dz &= \int_J |b_J(z)|\, dz \\ &\le \int_J |f(z)|\, dz + \left(\frac{1}{|J|} \int_J |f(x)|\, dx \right) |J| \lesssim \int_J |f(z)|\, dz \lesssim \lambda |J|. \end{aligned} \tag{2.19}$$

Using all this information, one can write

$$\begin{aligned} &|\{x \in \mathbb{R} \mid Sf(x) > \lambda\}| \\ &\quad \lesssim \left| \left\{ x \in \mathbb{R} \,\middle|\, Sg(x) > \frac{\lambda}{2} \right\} \right| + \left| \left\{ x \in \mathbb{R} \,\middle|\, Sb(x) > \frac{\lambda}{2} \right\} \right| \\ &\quad := I + II. \end{aligned} \tag{2.20}$$

To estimate I, we use the fact that S is bounded on L^2 and so

$$\begin{aligned} &\left| \left\{ x \in \mathbb{R} \,\middle|\, Sg(x) > \frac{\lambda}{2} \right\} \right| \lesssim \frac{1}{\lambda^2} \|Sg\|_2^2 \\ &\lesssim \frac{1}{\lambda^2} \|g\|_2^2 = \frac{1}{\lambda^2} \int_{\mathbb{R}} |g(x)|^2\, dx \lesssim \frac{1}{\lambda^2} \lambda \int_{\mathbb{R}} |g(x)|\, dx \\ &= \frac{1}{\lambda} \|g\|_1 \le \frac{1}{\lambda} \left(\int_{\Omega^c} |f(x)|\, dx + \sum_J \int_J |f(x)|\, dx \right) = \frac{1}{\lambda} \|f\|_1, \end{aligned}$$

as desired. To estimate II one has to proceed more carefully. We write

$$\left|\left\{x \,\middle|\, Sb(x) > \frac{\lambda}{2}\right\}\right| = \left|\left\{x \in \bigcup_J 5J \,\middle|\, Sb(x) > \frac{\lambda}{2}\right\}\right|$$
$$+ \left|\left\{x \in \left(\bigcup_J 5J\right)^c \,\middle|\, Sb(x) > \frac{\lambda}{2}\right\}\right| := II_1 + II_2.$$

In the above, $5J$ stands for the interval having the same center as J but five times as long as J.

The term II_1 can be estimated using

$$\left|\bigcup_J 5J\right| \le \sum_J 5|J| = 5|\Omega| \lesssim \frac{1}{\lambda}\|f\|_1;$$

the term II_2 can be estimated by

$$\left|\left\{x \in \left(\bigcup_J 5J\right)^c \,\middle|\, Sb(x) > \frac{\lambda}{2}\right\}\right| \lesssim \frac{1}{\lambda}\int_{(\cup_J 5J)^c} Sb(x)\,dx$$
$$\lesssim \frac{1}{\lambda}\sum_J \int_{(\cup_J 5J)^c} Sb_J(x)\,dx \le \frac{1}{\lambda}\sum_J \int_{(5J)^c} Sb_J(x)\,dx. \tag{2.21}$$

Fix J. We claim that the following inequality holds:

$$\int_{(5J)^c} Sb_J(x)\,dx \lesssim \lambda|J|. \tag{2.22}$$

If (2.22) were true then (2.21) would be smaller than

$$\frac{1}{\lambda}\lambda\sum_J |J| = |\Omega| \lesssim \frac{1}{\lambda}\|f\|_1,$$

which would complete the proof. Thus, it remains to show (2.22). Its left-hand side can be rewritten as follows:

$$\int_{(5J)^c} Sb_J(x)\,dx = \int_{(5J)^c} \left(\sum_I \frac{|\langle b_J, \varphi_I\rangle|^2}{|I|}\chi_I(x)\right)^{1/2} dx$$
$$\le \int_{(5J)^c} \sum_I \frac{|\langle b_J, \varphi_I\rangle|}{|I|^{1/2}}\chi_I(x)\,dx. \tag{2.23}$$

Clearly, the only intervals I that appear in the summation in (2.23) are those for which $I \cap (5J)^c \neq \phi$. Thus, one we split (2.23) as follows:

$$\sum_{|I|\leq|J|} \int_{(5J)^c} \frac{|\langle b_J, \varphi_I\rangle|}{|I|^{1/2}} \chi_I(x)\, dx + \sum_{|I|>|J|} \int_{(5J)^c} \frac{|\langle b_J, \varphi_I\rangle|}{|I|^{1/2}} \chi_I(x)\, dx := A + B. \tag{2.24}$$

To estimate A, fix I in such a way that $I \cap (5J)^c \neq \phi$ and $|I| \leq |J|$ and observe that

$$\begin{aligned} |\langle b_J, \varphi_I\rangle| &\leq \int_{\mathbb{R}} |b_J(z)||\varphi_I(z)|\, dz \\ &\lesssim \frac{1}{|I|^{1/2}} \left(1 + \frac{\text{dist}\,(I, J)}{|I|}\right)^{-5} \int_{\mathbb{R}} |b_J(z)|\, dz. \end{aligned}$$

Using this, we deduce that A is smaller than

$$\sum_{|I|\leq|J|} \left(1 + \frac{\text{dist}(I, J)}{|I|}\right)^{-5} \|b_J\|_1 \lesssim \lambda|J| \sum_{|I|\leq|J|} \left(1 + \frac{\text{dist}(I, J)}{|I|}\right)^{-5}, \tag{2.25}$$

using (2.19). It is not difficult to see that if one sums (2.25) over dyadic intervals I with $I \cap (5J)^c \neq \phi$ and $|I| \leq |J|$ then the result is smaller than $\lambda|J|$, as desired.

Finally, to estimate B one has to take advantage of the fact that $\int_{\mathbb{R}} b_J(z)\, dz = 0$, which is an important property that has not been used so far.

First, we notice that B can be estimated by

$$\sum_{|I|>|J|} |\langle b_J, \widetilde{\varphi}_I\rangle|, \tag{2.26}$$

where $\widetilde{\varphi}_I := |I|^{1/2} \varphi_I$ is L^∞-normalized. Then we observe that

$$\begin{aligned} |\langle b_J, \widetilde{\varphi}_I\rangle| &= \left| \int_J b_J(z) \widetilde{\varphi}_I(z)\, dz \right| \\ &= \left| \int_J b_J(z) (\widetilde{\varphi}_I(z) - \widetilde{\varphi}_I(c_J))\, dz \right|, \end{aligned} \tag{2.27}$$

where c_J is the center of J.

However, by the mean-value theorem one has that

$$|\widetilde{\varphi}_I(z) - \widetilde{\varphi}_I(c_J)| \lesssim |J||I|^{-1} \left(1 + \frac{\text{dist}(I, J)}{|I|}\right)^{-5} \tag{2.28}$$

for any $z \in J$. Using (2.28), one can bound (2.26) by

$$\sum_{|I|>|J|} \frac{|J|}{|I|}\left(1+\frac{\operatorname{dist}(I,J)}{|I|}\right)^{-5} \|b_J\|_1. \tag{2.29}$$

One should recall that the intervals I that participate in the above summation also have the property that $I \cap (5J)^c \neq \phi$.

Since $\|b_J\|_1 \lesssim \lambda|J|$ it is enough to show that

$$\sum_{\substack{|I|>|J| \\ I\cap(5J)^c\neq\phi}} \frac{|J|}{|I|}\left(1+\frac{\operatorname{dist}(I,J)}{|I|}\right)^{-5} \lesssim 1,$$

and this can be checked very easily. This ends the proof of the $L^1 \to L^{1,\infty}$ boundedness of S. By interpolation with L^2 bounds we can also obtain its $L^p \to L^p$ boundedness for any $1 < p < 2$. To obtain the L^p estimates in the case $p > 2$, one has to use the Khinchine inequality and write

$$\begin{aligned}
\|Sf\|_p &= \left\| \left(\sum_I \frac{|\langle f, \varphi_I\rangle|^2}{|I|} \chi_I \right)^{1/2} \right\|_p \\
&\lesssim \left\| \left(\sum_I |\langle f, \varphi_I\rangle h_I|^2 \right)^{1/2} \right\|_p \\
&\lesssim \left(\int_{\mathbb{R}} \int_0^1 \left| \sum_I r_I(t)\langle f, \varphi_I\rangle h_I(x) \right|^p dxdt \right)^{1/p},
\end{aligned} \tag{2.30}$$

where $(r_I)_I$ is the Rademacher system and $(h_I)_I$ is the Haar system described in Chapter I.8.

Fix $t \in [0, 1]$ and consider the linear operator

$$f \mapsto \sum_I r_I(t)\langle f, \varphi_I\rangle h_I. \tag{2.31}$$

Using an argument, similar to that before, based on the Calderón–Zygmund decomposition, one can prove that this linear operator is bounded in L^p for $1 < p \leq 2$ and, by duality, also bounded for any $1 < p < \infty$, with a bound independent of $t \in [0, 1]$. Using this fact in (2.30) completes the proof of the boundedness of our square-function operator in the general case. □

Exercise 2.2 Prove that the linear operator (2.31) is indeed bounded on every L^p space, for $1 < p < \infty$.

2.4. Dualization of quasi-norms

The reader may recall that part of the reason why the estimates on paraproducts turned out to be so easy in the Banach case was related to the possibility of applying the duality between L^r and $L^{r'}$ for $1/r + 1/r' = 1$ and $1 < r < \infty$. Using it we reduced the estimates on the bilinear paraproducts to an analysis of their corresponding trilinear forms. Clearly, in the quasi-Banach case, since for $0 < r < 1$ one has that $(L^r)^* = \{0\}$, such a line of argument is very unlikely to succeed. However, our strategy will not be to prove these L^r estimates directly but instead to demonstrate some weaker Lorentz $L^{r,\infty}$ variants of them and after that to use some kind of *multilinear Marcinkiewicz interpolation* result that will imply the original strong inequalities.

Recall that for every $0 < r < \infty$, the so-called weak-L^r Lorentz space, $L^{r,\infty}$, is defined to be the collection of all measurable functions f with the property that

$$\|f\|_{r,\infty} := \sup_{\lambda>0} \lambda \, |\{x \,|\, |f(x)| > \lambda\}|^{1/r} < \infty.$$

The pleasant surprise that one has with these spaces is that even though they are still quasi-Banach for $0 < r \leq 1$ their quasi-norms can be dualized. The precise way in which this can be done is explained in detail in the following *duality lemma*, which will be helpful later on.

Lemma 2.6 *Let $0 < r \leq 1$ and $A > 0$. Then the following statements are equivalent:*

(i) $\|f\|_{r,\infty} \leq A$;
(ii) *for every set E with $0 < |E| < \infty$, there exists a subset $E' \subseteq E$ with $|E'| \simeq |E|$ and $|\langle f, \chi_{E'}\rangle| \lesssim A\,|E|^{1/r'}$, where $1/r + 1/r' = 1$.*

(Note that, for $r \neq 1$, r' is a negative number.)

Proof As previously stated,

$$\|f\|_{r,\infty} = \sup_{\lambda>0} \lambda \, |\{x \,|\, |f(x)| > \lambda\}|^{1/r}.$$

Clearly, one can assume without loss of generality that f is real-valued. To prove that the first statement implies the second, let E be fixed and define

$$\Omega = \left\{x \,\middle|\, |f(x)| \geq CA|E|^{-1/r}\right\}.$$

Since $f \in L^{r,\infty}$ one has

$$|\Omega| \leq \frac{1}{\left(CA|E|^{-1/r}\right)^r} \|f\|_{r,\infty}^r$$
$$\leq \frac{|E|}{C^r A^r} A^r = \frac{|E|}{C^r} < \frac{|E|}{100},$$

if C is a sufficiently large constant. Then, simply define $E' := E \setminus \Omega$ and observe that $|E'| \simeq |E|$. Moreover, one has

$$|\langle f, \chi_{E'} \rangle| \leq \int_{E'} |f(x)| dx \leq CA|E|^{-1/r}|E'|$$
$$\lesssim A|E|^{-1/r}\,|E| = A|E|^{1/r'},$$

as desired.

For the converse we need to prove that $\|f\|_{r,\infty} \lesssim A$, in other words that

$$\lambda^r\,|\{x \mid |f(x)| > \lambda\}| \lesssim A^r \tag{2.32}$$

for every $\lambda > 0$. First, set $E := \{x \mid f(x) > \lambda\}$. We know that there exists $E' \subseteq E$ with $|E'| \simeq |E|$ and such that

$$|\langle f, \chi_{E'} \rangle| \lesssim A|E|^{1/r'}.$$

Since the left-hand side of this inequality is larger than $\lambda|E|$, this implies that

$$\lambda|E| \lesssim A|E|^{1/r'},$$

which is equivalent to

$$\lambda^r|E| \lesssim A^r. \tag{2.33}$$

Then one can similarly estimate the set $F := \{x \mid -f(x) > \lambda\}$ and obtain

$$\lambda^r|F| \lesssim A^r. \tag{2.34}$$

Using (2.33) and (2.34) one obtains (2.32). □

2.5. Two particular cases of Theorem 2.3

We are still not ready to prove the general case of Theorem 2.3; first we will consider two simpler particular cases in order to motivate the approach that we are going to use.

Recall that

$$\Pi_{\mathcal{J}}(f, g) = \sum_{I \in \mathcal{J}} c_I \frac{1}{|I|^{1/2}} \langle f, \varphi_I^1 \rangle \langle g, \varphi_I^2 \rangle \varphi_I^3.$$

Our goal is to prove that this operator is bounded from $L^1 \times L^1$ into $L^{1/2,\infty}$.

Case 1. Assume that the numbers $(c_I)_I$ are zero except for that corresponding to a fixed interval I_0. In this case our bilinear operator becomes

$$(f, g) \mapsto c_{I_0} \frac{1}{|I_0|^{1/2}} \langle f, \varphi_{I_0}^1 \rangle \langle g, \varphi_{I_0}^2 \rangle \varphi_{I_0}^3. \tag{2.35}$$

We will prove that this operator is bounded even from $L^1 \times L^1$ into $L^{1/2}$.

One observes first that

$$\left| \langle f, \varphi_{I_0}^1 \rangle \right| \lesssim \|f\|_1 \, |I_0|^{-1/2}$$

and, similarly,

$$\left| \langle g, \varphi_{I_0}^2 \rangle \right| \lesssim \|g\|_1 \, |I_0|^{-1/2}.$$

In particular, the expression (2.35) is pointwise bounded by $|I_0|^{-3/2} \|f\|_1 \times \|g\|_1 \, |\varphi_{I_0}^3|$. As a consequence, its $L^{1/2}$ quasi-norm must be smaller than

$$\|f\|_1 \, \|g\|_1 \, |I_0|^{-3/2} |I_0|^{-1/2} \, |I_0|^2 = \|f\|_1 \, \|g\|_1,$$

as desired.

Case 2. Assume now that all the numbers $(c_I)_I$ are zero except for those corresponding to a fixed scale. Then we have, for our bilinear operator,

$$(f, g) \mapsto \sum_{|I| = \text{const}} c_I \frac{1}{|I|^{1/2}} \langle f, \varphi_I^1 \rangle \langle g, \varphi_I^2 \rangle \varphi_I^3. \tag{2.36}$$

The argument that follows is completely positive; the lacunarity property will not be needed in this case.

Let f, g be such that $\|f\|_1 = \|g\|_1 = 1$. Using Lemma 2.6, to prove that this operator maps $L^1 \times L^1$ into $L^{1/2,\infty}$, it is enough to show that, given $E \subseteq \mathbb{R}$ with $0 < |E| < \infty$, there exists $E' \subseteq E$ such that $|E'| \simeq |E|$ and

$$\left| \sum_{|I| = \text{const}} c_I \frac{1}{|I|^{1/2}} \langle f, \varphi_I^1 \rangle \langle g, \varphi_I^2 \rangle \langle h, \varphi_I^3 \rangle \right| \lesssim |E|^{-1}, \tag{2.37}$$

where $h := \chi_{E'}$. Using the scaling invariance of expression (2.37), we can assume that $|E| = 1$. Clearly, this means that we have to change the scale of the intervals, but this does not affect our argument. If $|E| = 1$, (2.37) simplifies to

$$\left| \sum_{|I|=\text{const}} c_I \frac{1}{|I|^{1/2}} \langle f, \varphi_I^1 \rangle \langle g, \varphi_I^2 \rangle \langle h, \varphi_I^3 \rangle \right| \lesssim 1. \tag{2.38}$$

Define an "exceptional" set Ω (i.e., a set that for convenience one wants to remove when estimating the measure of a certain set that contains it) by

$$\Omega := \{x | Mf(x) > C\} \cup \{x | Mg(x) > C\},$$

where M is the Hardy–Littlewood maximal operator. Since both f and g are L^1-normalized, if C is a sufficiently large constant then one has $|\Omega| < 1/2$. Next, we simply set

$$E' := E \backslash \Omega$$

and claim that this E' satisfies (2.38). To establish this claim we first split our collection of intervals $\mathcal{J}$ as follows:

$$\mathcal{J} = \bigcup_{d \geq 0} \mathcal{J}_d,$$

where $\mathcal{J}_d$ contains the dyadic intervals in $\mathcal{J}$ having the property that

$$1 + \frac{\text{dist}\,(I, \Omega^c)}{|I|} \simeq 2^d.$$

We will now show that for any $d \geq 0$ one has

$$\left| \sum_{I \in \mathcal{J}_d} c_I \frac{1}{|I|^{1/2}} \langle f, \varphi_I^1 \rangle \langle g, \varphi_I^2 \rangle \langle h, \varphi_I^3 \rangle \right| \lesssim 2^{-100d} \tag{2.39}$$

and this will be enough. To obtain (2.39), we further decompose $\mathcal{J}_d$, writing

$$\mathcal{J}_d = \bigcup_{n_1} \mathcal{J}_{d,1}^{n_1} \tag{2.40}$$

where $\mathcal{J}_{d,1}^{n_1}$ contains the intervals in $\mathcal{J}_d$ with the property that, taking the average

$$\frac{|\langle f, \varphi_I^1 \rangle|}{|I|^{\frac{1}{2}}} \simeq 2^{-n_1}. \tag{2.41}$$

Similarly, using the functions g and h one can also obtain the decompositions

$$\mathcal{J}_d = \bigcup_{n_2} \mathcal{J}_{d,2}^{n_2} \tag{2.42}$$

and

$$\mathcal{J}_d = \bigcup_{n_3} \mathcal{J}_{d,3}^{n_3}. \tag{2.43}$$

As a consequence, the left-hand side of (2.39) can be decomposed as

$$\sum_{n_1,n_2,n_3} \sum_{I \in \mathcal{J}_d^{n_1,n_2,n_3}} c_I \frac{1}{|I|^{1/2}} \langle f, \varphi_I^1 \rangle \langle g, \varphi_I^2 \rangle \langle h, \varphi_I^3 \rangle, \tag{2.44}$$

where $\mathcal{J}_d^{n_1,n_2,n_3}$ is defined to be the intersection $\mathcal{J}_{d,1}^{n_1} \cap \mathcal{J}_{d,2}^{n_2} \cap \mathcal{J}_{d,3}^{n_3}$. Then, the absolute value of (2.44) can be estimated by

$$\sum_{n_1,n_2,n_3} \sum_{I \in \mathcal{J}_d^{n_1,n_2,n_3}} \frac{|\langle f, \varphi_I{}^1 \rangle|}{|I|^{1/2}} \frac{|\langle g, \varphi_I{}^2 \rangle|}{|I|^{1/2}} \frac{|\langle h, \varphi_I{}^3 \rangle|}{|I|^{1/2}} |I|$$

$$\lesssim \sum_{n_1,n_2,n_3} 2^{-n_1} 2^{-n_2} 2^{-n_3} \sum_{I \in \mathcal{J}_d^{n_1,n_2,n_3}} |I|. \tag{2.45}$$

Clearly, from the definition of $\mathcal{J}_d^{n_1,n_2,n_3}$ and using also the fact that the intervals I are disjoint, since they all have the same length, we have that

$$\sum_{I \in \mathcal{J}_d^{n_1,n_2,n_3}} |I| \le \sum_{I \in \mathcal{J}_d^{n_1}} |I|$$

$$\lesssim 2^{n_1} \sum_{I \in \mathcal{J}_d^{n_1}} \int_{\mathbb{R}} |f(x)| \widetilde{\chi}_I(x)\, dx \lesssim 2^{n_1} \|f\|_1 = 2^{n_1},$$

where $\widetilde{\chi}_I$ is a smooth rapidly decreasing function that is L^∞-normalized and adapted to the interval I. Similarly, one also has

$$\sum_{I \in \mathcal{J}_d^{n_1,n_2,n_3}} |I| \lesssim 2^{n_2}$$

and

$$\sum_{I \in \mathcal{J}_d^{n_1,n_2,n_3}} |I| \lesssim 2^{n_3}.$$

However, since every such I has the property that it is in $\mathcal{J}_d$, one has

$$2^{-n_1} \lesssim 2^d, \quad 2^{-n_2} \lesssim 2^d, \quad 2^{-n_3} \lesssim 2^{-200d}. \tag{2.46}$$

This is a consequence of (2.41) and of the definition of the exceptional set Ω. The decay in the last inequality comes from the fact that the function h is supported inside Ω^c.

Exercise 2.3 Check carefully the details of (2.46).

Putting all this information together one finally estimates (2.45) as

$$\sum_{n_1,n_2,n_3} 2^{-n_1}2^{-n_2}2^{-n_3}2^{n_1/3}2^{n_2/3}2^{n_3/3} = \sum_{n_1,n_2,n_3} 2^{-2n_1/3}2^{-2n_2/3}2^{-2n_3/3}$$
$$\lesssim 2^{2d/3}2^{2d/3}2^{-2\times 200d/3} \lesssim 2^{-100d}$$

which clearly proves (2.39), as desired. This ends our discussion of case 2.

In the general case one has an arbitrary number of scales and it is clear that one cannot hope to sum the contribution of each separately, since now these intervals may overlap significantly. The key to handling the general case will be the fact that at least two positions in the discrete paraproduct are lacunary and, as a consequence, different scales correspond to disjoint frequency intervals.

Fix a dyadic interval I_0. A natural way to estimate

$$\left|\sum_{I\subseteq I_0} c_I \frac{1}{|I|^{1/2}}\langle f,\varphi_I^1\rangle\langle g,\varphi_I^2\rangle\langle h,\varphi_I^3\rangle\right|$$

is by using Cauchy–Schwarz; one then obtains

$$\left(\sup_{I\subseteq I_0}\frac{|\langle f,\varphi_I^1\rangle|}{|I|^{1/2}}\right)\left(\sum_{I\subseteq I_0}|\langle g,\varphi_I^2\rangle|^2\right)^{1/2}\left(\sum_{I\subseteq I_0}|\langle h,\varphi_I^3\rangle|^2\right)^{1/2}$$
$$= \left(\sup_{I\subseteq I_0}\frac{|\langle f,\varphi_I^1\rangle|}{|I|^{1/2}}\right)\frac{1}{|I_0|^{1/2}}\left(\sum_{I\subseteq I_0}|\langle g,\varphi_I^2\rangle|^2\right)^{1/2}$$
$$\times \frac{1}{|I_0|^{1/2}}\left(\sum_{I\subseteq I_0}|\langle h,\varphi_I^3\rangle|^2\right)^{1/2}|I_0|. \tag{2.47}$$

Clearly, each of the three factors in (2.47) is some kind of an average over the interval I_0, similar to those considered in case 2. The factor involving f is an L^1-type average, while the other two are L^2-type averages. The plan now is to devise a stopping-time argument, similar to that used earlier but based on these new more complex averages.

Since our goal is to obtain estimates in L^p when P is close to 1, it may seem that working with L^2 averages will not be helpful. However, the so-called John–Nirenberg inequality will enable us to control these L^2 averages by the corresponding L^1 averages.

This inequality is the last technical ingredient needed to prove our general estimates on paraproducts; see Chapter I.7.

2.6. The John–Nirenberg inequality

We begin this section by relating its main result, Theorem 2.7, to other theorems involving bounded mean oscillation space (BMO) and the John–Nirenberg inequality discussed in Vol. I. The so-called dyadic BMO ([0, 1]) was defined in Section I.8.4 to be the space of all measurable functions f supported in [0, 1] satisfying $\int_0^1 f(x)dx = 0$ and having also the property that

$$\sup_{I_0}\left(\frac{1}{|I_0|}\int_{I_0}|f(y)-f_{I_0}|^2\,dy\right)^{1/2} < \infty, \tag{2.48}$$

where f_{I_0} denotes the average

$$\frac{1}{|I_0|}\int_{I_0} f(y)\,dy$$

and the supremum is taken over all dyadic intervals $I_0 \subseteq [0, 1]$. It was pointed out in Exercise I.7.9 that as a consequence of the *classical John–Nirenberg inequality*, Theorem I.7.17, one has that

$$\sup_{I_0}\left(\frac{1}{|I_0|}\int_{I_0}|f(y)-f_{I_0}|^p dy\right)^{1/p} \simeq \sup_{I_0}\left(\frac{1}{|I_0|}\int_{I_0}|f(y)-f_{I_0}|^q dy\right)^{1/q} \tag{2.49}$$

for any two indices $1 < p, q < \infty$. If one takes also into account Theorem I.8.20 it can be seen that (2.49) is equivalent to

$$\sup_{I_0}\frac{1}{|I_0|^{1/p}}\left\|\left(\sum_{I\subseteq I_0}\frac{|\langle f,h_I\rangle|^2}{|I|}\chi_I(x)\right)^{1/2}\right\|_p$$

$$\simeq \sup_{I_0}\frac{1}{|I_0|^{1/q}}\left\|\left(\sum_{I\subseteq I_0}\frac{|\langle f,h_I\rangle|^2}{|I|}\chi_I(x)\right)^{1/2}\right\|_q, \tag{2.50}$$

where $(h_I)_I$ is the Haar system studied in Section I.8.4.

The next theorem can be seen as an abstract extension of (2.50). It will be very useful in estimating many averages that appear later in this second volume.

Theorem 2.7 *Let $\mathcal{J}$ be a finite family of dyadic intervals. For any positive real number r and any sequence of complex numbers $(a_I)_{I\in\mathcal{J}}$ one defines*

$\|(a_I)_I\|_{\mathrm{BMO}(r)}$ *as follows:*

$$\|(a_I)_I\|_{\mathrm{BMO}(r)} := \sup_{I_0 \in \mathcal{J}} \frac{1}{|I_0|^{1/r}} \left\| \left(\sum_{I \subseteq I_0} \frac{|a_I|^2}{|I|} \chi_I(x) \right)^{1/2} \right\|_r .$$

Then, for any $0 < p < q < \infty$, *one has*

$$\|(a_I)_I\|_{\mathrm{BMO}(p)} \simeq \|(a_I)_I\|_{\mathrm{BMO}(q)}. \tag{2.51}$$

Proof Fix $0 < p < q < \infty$. We will show that

$$\|(a_I)_I\|_{\mathrm{BMO}(q)} \lesssim \|(a_I)_I\|_{\mathrm{BMO}(p,\infty)}, \tag{2.52}$$

where $\|(a_I)_I\|_{\mathrm{BMO}(p,\infty)}$ is defined in the same way but by using the space $L^{p,\infty}$ instead of the previous L^p. Clearly, this is enough given that the weak-L^p norms are smaller than the L^p norms. Denote the left-hand side of (2.52) by B and the right-hand side by A. The goal is then to show that

$$B \lesssim A. \tag{2.53}$$

Clearly, from the definition of B there exists an interval $I_0 \in \mathcal{J}$ such that

$$\frac{1}{|I_0|^{1/q}} \left\| \left(\sum_{\substack{I \in \mathcal{J} \\ I \subseteq I_0}} \frac{|a_I|^2}{|I|} \chi_I \right)^{1/2} \right\|_q = B$$

or, equivalently,

$$B\,|I_0|^{1/q} = \left\| \left(\sum_{I \subseteq I_0} \frac{|a_I|^2}{|I|} \chi_I \right)^{1/2} \right\|_q . \tag{2.54}$$

From the definition of A, we know that

$$\left\| \left(\sum_{I \subseteq I_0} \frac{|a_I|^2}{|I|} \chi_I \right)^{1/2} \right\|_{p,\infty} \leq A\,|I_0|^{1/p} .$$

In particular, this implies that

$$\left| \left\{ x \,\middle|\, \left(\sum_{I \subseteq I_0} \frac{|a_I|^2}{|I|} \chi_I(x) \right)^{1/2} > CA \right\} \right| \leq \left(\frac{A\,|I_0|^{1/p}}{CA} \right)^p = \frac{|I_0|}{C^p} < \frac{1}{M}\,|I_0| , \tag{2.55}$$

where M is a large constant, holds if C itself is large enough. We denote by E the set

$$\left\{ x \;\middle|\; \left(\sum_{I \subseteq I_0} \frac{|a_I|^2}{|I|} \chi_I(x) \right)^{1/2} > CA \right\}.$$

Then one can write, using (2.54),

$$B^q |I_0| = \left\| \left(\sum_{I \subseteq I_0} \frac{|a_I|^2}{|I|} \chi_I(x) \right)^{1/2} \right\|_q^q = \int_{\mathbb{R}} \left(\sum_{I \subseteq I_0} \frac{|a_I|^2}{|I|} \chi_I(x) \right)^{q/2} dx. \quad (2.56)$$

Now we decompose our set E as follows:

$$E = \bigcup_{I_{\max} \in \mathcal{J}_{\max}} I_{\max},$$

where $I_{\max}$ runs over the dyadic intervals in $\mathcal{J}$ having the property that

$$\left(\sum_{\substack{I \subseteq I_0 \\ I_{\max} \subseteq I}} \frac{|a_I|^2}{|I|} \right)^{1/2} > CA \quad (2.57)$$

and is maximal with this property, and $\mathcal{J}_{\max}$ is the collection of these intervals. Clearly, all the intervals $I_{\max}$ are disjoint. As a consequence, (2.56) can be split as follows:

$$\int_E \left(\sum_{I \subseteq I_0} \frac{|a_I|^2}{|I|} \chi_I(x) \right)^{q/2} dx + \int_{E^c} \left(\sum_{I \subseteq I_0} \frac{|a_I|^2}{|I|} \chi_I(x) \right)^{q/2} dx =: I + II.$$

To estimate II is easy since, from the definition of E, the integrand is pointwise smaller than CA on E^c and this gives a contribution $(CA)^q |I_0|$, which is helpful in (2.56) since we are aiming to prove (2.53).

To estimate I, one first writes it as

$$I = \sum_{I_{\max}} \int_{I_{\max}} \left(\sum_{I \subseteq I_0} \frac{|a_I|^2}{|I|} \chi_I(x) \right)^{q/2} dx. \quad (2.58)$$

Fix $I_{\max} \in \mathcal{J}_{\max}$. Now the corresponding term in (2.58) can be estimated by a constant depending on q times the sum

$$\int_{I_{\max}} \left(\sum_{\substack{I \subseteq I_0 \\ I_{\max} \subseteq I \\ I \neq I_{\max}}} \frac{|a_I|^2}{|I|} \chi_I(x) \right)^{q/2} dx + \int_{I_{\max}} \left(\sum_{I \subseteq I_{\max}} \frac{|a_I|^2}{|I|} \chi_I(x) \right)^{q/2} \cdot dx. \quad (2.59)$$

To estimate the first expression in (2.59) is easy, since by the maximality of $I_{\max}$ one has the opposite of (2.57), and this gives a contribution $(CA)^q|I_{\max}|$. Since all the intervals $I_{\max}$ are disjoint, summing over them gives an upper bound of the type $(CA)^q|I_0|$, which again is helpful in (2.56).

Finally, the second term in (2.59) can be written as

$$|I_{\max}|\,\frac{1}{|I_{\max}|}\int_{I_{\max}}\left(\sum_{I\subseteq I_{\max}}\frac{|a_I|^2}{|I|}\chi_I(x)\right)^{q/2}dx \le |I_{\max}|\,B^q,$$

this time using the definition of B. If one sums over the intervals $I_{\max}$, one obtains an upper bound of the type

$$\sum_{I_{\max}}|I_{\max}|B^q = B^q|E| \le B^q\frac{1}{M}|I_0|,$$

by using (2.55). Putting all these together we obtain

$$B^q|I_0| \le \widetilde{C}A^q|I_0| + \frac{1}{M}B^q|I_0|$$

for large enough M; clearly this proves the inequality $B \lesssim A$, as desired. □

As a consequence of the proof of Theorem 2.7 we obtain the following result.

Corollary 2.8 *One has*

$$\|(a_I)_{I\in\mathcal{J}}\|_{\mathrm{BMO}(q)} \simeq \|(a_I)_{I\in\mathcal{J}}\|_{\mathrm{BMO}(1,\infty)}$$

for any $1 < q < \infty$.

2.7. $L^{1,\infty}$ sizes and $L^{1,\infty}$ energies

We can now start presenting some definitions and lemmas which will finally lead to a complete proof of Theorem 2.3. Clearly, because of Lemma 2.6 we need to understand how to estimate trilinear forms of the type

$$\sum_{I\in\mathcal{J}} c_I\frac{1}{|I|^{1/2}}\langle f,\varphi_I^1\rangle\langle g,\varphi_I^2\rangle\langle h,\varphi_I^3\rangle.$$

Since part of the argument does not depend on the functions f, g, h, we choose to present it in an abstract setting, which will be helpful later in the book. More precisely, we will present a way of estimating expressions of the type

$$\sum_{I\in\mathcal{J}}\frac{1}{|I|^{1/2}}\,a_I^1\,a_I^2\,a_I^3. \tag{2.60}$$

Since this expression is a discretized paraproduct, we know that at least two of the families $(\varphi_I^j)_I$ are lacunary; to be specific, we will assume them to be those corresponding to the indices $j = 2, 3$.

The next definition introduces two important and very useful concepts, those of *size* and *energy*. Before defining them we mention that the reason why we described the previous particular cases of Theorem 2.3 was to make this section and the next seem as natural and intuitive as possible. The reader may remember that when we considered the second case of Theorem 2.3 (where all the intervals have the same length and as a consequence are all disjoint) we selected dyadic intervals I with the property that

$$\frac{|\langle f, \varphi_I \rangle|}{|I|^{1/2}} \simeq 2^{-\mu}$$

for some fixed integer μ. One should think of these averages as being the "baby versions" of the upcoming *sizes*. For an even better perspective, we recall (2.47), together with Theorem 2.7. If we denote by $\mathcal{C}_\mu$ the collection all the intervals I, we may also recall that we had to estimate sums of the type $\sum_{I \in \mathcal{C}_\mu} |I|$. Equivalently, sums of the type $2^{-\mu} \sum_{I \in \mathcal{C}_\mu} |I|$ appeared naturally. One should think of the supremum of these expressions as being the "baby variants" of the *energies*. In our particular case, it is not difficult to see that they are all smaller than $\|f\|_1$.

Definition 2.9 Let $\widetilde{\mathcal{J}} \subseteq \mathcal{J}$ be a fixed family of dyadic intervals.

For $j = 1$ we define

$$\text{size}^{(1)}_{\widetilde{\mathcal{J}}}\left(\left(a_I^1\right)_{I \in \widetilde{\mathcal{J}}}\right) := \sup_{I \in \widetilde{\mathcal{J}}} \frac{|a_I^1|}{|I|^{1/2}}$$

while for $j = 2, 3$ we define

$$\text{size}^{(j)}_{\widetilde{\mathcal{J}}}\left(\left(a_I^j\right)_{I \in \widetilde{\mathcal{J}}}\right) := \sup_{I_0 \in \widetilde{\mathcal{J}}} \frac{1}{|I_0|} \left\| \left(\sum_{\substack{I \subseteq I_0 \\ I \in \widetilde{\mathcal{J}}}} \frac{|a_I^j|^2}{|I|} \chi_I \right)^{1/2} \right\|_{1,\infty}.$$

Then, for $j = 1$ we define

$$\text{energy}^{(1)}_{\widetilde{\mathcal{J}}}\left(\left(a_I^1\right)_{I \in \widetilde{\mathcal{J}}}\right) := \sup_{n \in \mathbb{Z}} 2^n \sup_{\mathbb{D}} \left(\sum_{I \in \mathbb{D}} |I| \right),$$

where $\mathbb{D}$ ranges over all collections of disjoint dyadic intervals $I \in \widetilde{\mathcal{J}}$ having the property that

$$\frac{|a_I^1|}{|I|^{1/2}} \geqslant 2^n. \tag{2.61}$$

For $j = 2, 3$ we define

$$\text{energy}^{(j)}_{\widetilde{\mathcal{J}}}\left(\left(a_I^j\right)_{I\in\widetilde{\mathcal{J}}}\right) := \sup_{n\in\mathbb{Z}} 2^n \sup_{\mathbb{D}} \left(\sum_{I\in\mathbb{D}} |I|\right),$$

where this time $\mathbb{D}$ ranges over all collections of disjoint dyadic intervals $I_0 \in \widetilde{\mathcal{J}}$ having the property that

$$\frac{1}{|I_0|}\left\|\left(\sum_{\substack{I\in\widetilde{\mathcal{J}}\\ I\subseteq I_0}} \frac{|a_I^j|^2}{|I|}\chi_I\right)^{1/2}\right\|_{1,\infty} \geqslant 2^n. \tag{2.62}$$

Now taking into account where these $(a_I^j)_I$ sequences come from, it is easy to see that the sizes are averages of the corresponding functions f, g, h. More specifically, since the expressions within the $L^{1,\infty}$ quasi-norms are localized square functions, by using their $L^1 \to L^{1,\infty}$ boundedness one can deduce that they will be smaller than the L^1 averages of f, g, or h respectively, as desired.

Because of the abstract nature of the John–Nirenberg inequality in Theorem 2.7, we could have used any L^p averages in the definition of the size. However, it is important to keep the weak-L^1 averages in (2.62) when we define the energies, since we want them to be L^1-type quantities as well.

To explain in more detail what the energy operator means, we will fix n and $\mathbb{D}$ and observe that in the case $j = 1$ (for instance) we have, as a consequence of the inequality

$$\frac{|a_I^1|}{|I|^{1/2}} \geqslant 2^n,$$

that $\left|\langle f, \varphi_I^1\rangle\right|/|I|^{1/2} \geqslant 2^n$ and this implies that

$$2^n|I| \leq \left|\langle f, |I|^{1/2}\varphi_I^1\rangle\right|. \tag{2.63}$$

Now, since φ_I^1 is L^2-normalized it follows that $|I|^{1/2}\varphi_I^1$ is L^∞-normalized and, since all the intervals $I \in \mathbb{D}$ are disjoint, we can intuitively think of these functions as having disjoint support, which means that, heuristically at least, the energies are smaller than the L^1 norms of the corresponding functions.

We will see later that one can naturally define L^p-adapted energies for any $1 < p < \infty$, and that they also are useful.

2.8. Stopping-time decompositions

Suppose now that the function f has the property that $\|f\|_1 = 1$ and that $\mathcal{C}_\mu$ again contains intervals (of the same length and therefore disjoint) with the property that $|\langle f, \varphi_I\rangle|/|I|^{1/2} \simeq 2^{-\mu}$. Recall from the inequality after (2.45) that as a consequence

$$\sum_{I\in\mathcal{C}_\mu} |I| \lesssim 2^\mu \sum_I \int_{\mathbb{R}} |f(y)|\widetilde{\chi}_I(y)\,dy \lesssim 2^\mu,$$

since the function f is L^1-normalized. If one then varies the parameter μ, one obtains a decomposition of the set of all intervals (of the same particular length) as a disjoint union of collections of the type $\mathcal{C}_\mu$.

This simple argument is the germ of the following lemma, which will play an important role in our argument.

Lemma 2.10 *Let $\mathcal{J}$ be a finite family of dyadic intervals, let $j = 1, 2, 3$, and let $\mathcal{J}' \subseteq \mathcal{J}$ such that*

$$\text{size}^{(j)}_{\mathcal{J}'}\left(\left(a^j_I\right)_I\right) \leq 2^{-n_0}\,\text{energy}^{(j)}_{\mathcal{J}}\left(\left(a^j_I\right)_I\right) \tag{2.64}$$

for a certain fixed integer n_0. Then there exists a decomposition $\mathcal{J}' = \mathcal{J}'' \cup \mathcal{J}'''$ such that

$$\text{size}^{(j)}_{\mathcal{J}''}((a^j_I)_I) \leq 2^{-n_0-1}\,\text{energy}^{(j)}_{\mathcal{J}}\left(\left(a^j_I\right)_I\right) \tag{2.65}$$

and such that $\mathcal{J}'''$ can be written as a disjoint union of subsets $T \in \mathbb{T}$ such that for every $T \in \mathbb{T}$ there exists a dyadic interval I_T in T having the properties that every $I \in T$ satisfies $I \subseteq I_T$ and also

$$\sum_{T\in\mathbb{T}} |I_T| \lesssim 2^{n_0}. \tag{2.66}$$

Proof Let us point out that the factor

$$\text{energy}^{(j)}_{\mathcal{J}}\left(\left(a^j_I\right)_I\right)$$

should be considered as a *normalization factor*. If our sequence is divided by it then it will have *total energy* 1. We should think of this as analogous to the fact that the function f in the previous section had L^1 norm 1.

Assume first that $j = 1$. We proceed as follows. Choose an interval $I \in \mathcal{J}'$ such that $|I|$ is as large as possible and such that

$$\frac{|a^1_I|}{|I|^{1/2}} > 2^{-n_0-1}\,\text{energy}^{(1)}_{\mathcal{J}}\left((a^1_I)_I\right). \tag{2.67}$$

Now collect all the intervals $I' \in \mathcal{J}'$ with $I' \subseteq I$ in a set T. Then define $I_T := I$, look at the remaining intervals in $\mathcal{J}' \setminus T$, and repeat the procedure. Clearly, since the cardinality of $\mathcal{J}$ is finite this algorithm ends after finitely many steps, producing the subsets $T \in \mathbb{T}$. Next define $\mathcal{J}''' := \bigcup_{T \in \mathbb{T}} T$ and $\mathcal{J}'' := \mathcal{J}' \setminus \mathcal{J}'''$. By construction (2.65) is automatically satisfied, and it remains to check (2.66). Here we need to observe that all the intervals $(I_T)_{T \in \mathbb{T}}$ are disjoint by construction. From this and (2.67) we deduce that

$$2^{-n_0-1} \operatorname{energy}^{(1)}_{\mathcal{J}} ((a^1_I)_I) \left(\sum_{T \in \mathbb{T}} |I_T| \right) \leq \operatorname{energy}^{(1)}_{\mathcal{J}'} ((a^1_I)_I) \leq \operatorname{energy}^{(1)}_{\mathcal{J}} ((a^1_I)_I),$$

from which (2.66) follows immediately. The other case, when $j = 2, 3$, is very similar. The only difference is that this time we pick intervals $I \in \mathcal{J}'$ such that $|I|$ is as large as possible and having the property that

$$\frac{1}{|I|} \left\| \left(\sum_{\substack{I' \subseteq I \\ I' \in \mathcal{J}'}} \frac{|a^j_{I'}|}{|I'|} \chi_{I'} \right)^{1/2} \right\|_{1,\infty} > 2^{-n_0-1} \operatorname{energy}^{(j)}_{\mathcal{J}} \left((a^j_I)_I \right).$$

After this, the argument is identical to that in the previous case. □

Now, if one iterates the above Lemma 2.10, one obtains the following.

Corollary 2.11 *Let $j = 1, 2, 3$ and let $\mathcal{J}$ be a finite family of dyadic intervals. Then there exists a partition*

$$\mathcal{J} = \bigcup_{n \in \mathbb{Z}} \mathcal{J}^{n,j}$$

such that for any $n \in \mathbb{Z}$ one has

$$2^{-n-1} E_j \leq \operatorname{size}^{(j)}_{\mathcal{J}^{n,j}} \left((a^j_I)_I \right) \leq \min \left(2^{-n} E_j, S_j \right),$$

where for simplicity we write

$$S_j := \operatorname{size}^{(j)}_{\mathcal{J}} \left((a^j_I)_I \right)$$

and

$$E_j := \operatorname{energy}^{(j)}_{\mathcal{J}} \left(\left(a^j_I \right)_I \right).$$

Also, as before one can write each $\mathcal{J}^{n,j}$ as a disjoint union of subsets $T \in \mathbb{T}^{n,j}$ having the property that

$$\sum_{T \in \mathbb{T}^{n,j}} |I_T| \lesssim 2^n.$$

2.9. Generic estimate of the trilinear paraproduct form

The following proposition uses all this. It describes how one can estimate generic expressions such as that in (2.60).

Proposition 2.12 *Let $\mathcal{J}$ be as before. Then,*

$$\left| \sum_{I \in \mathcal{J}} \frac{1}{|I|^{1/2}} a_I^1 a_I^2 a_I^3 \right| \lesssim \prod_{j=1}^{3} \left(\text{size}_{\mathcal{J}}^{(j)} \left((a_I^j)_I \right) \right)^{1-\theta_j} \left(\text{energy}_{\mathcal{J}}^{(j)} \left((a_I^j)_I \right) \right)^{\theta_j} \tag{2.68}$$

for any $\theta \leq \theta_1, \theta_2, \theta_3 < 1$ such that $\theta_1 + \theta_2 + \theta_3 = 1$, where the implicit constants depend on $\theta_1, \theta_2, \theta_3$ only.

Proof The idea is to run a stopping-time argument based on Corollary 2.11. By applying this corollary to our situation, we can decompose and estimate the left-hand side of (2.68) by

$$\sum_{n_1, n_2, n_3} \sum_{T \in \mathbb{T}^{n_1, n_2, n_3}} \sum_{I \in T} \frac{1}{|I|^{\frac{1}{2}}} |a_I^1| \, |a_I^2| \, |a_I^3|, \tag{2.69}$$

where $\mathbb{T}^{n_1, n_2, n_3}$ contains sets T of the type $T_1 \cap T_2 \cap T_3$ with $T_j \in \mathbb{T}^{n_j, j}$.

Fix such a set T and look at the corresponding term in (2.69). It can be estimated as

$$\sup_{I \in T} \frac{|a_I^1|}{|I|^{1/2}} \prod_{j \neq 1} \left(\sum_{I \in T} |a_I^j|^2 \right)^{1/2}$$

$$= \left(\sup_{I \in T} \frac{|a_I^1|}{|I|^{1/2}} \right) \frac{1}{|I_T|^{1/2}} \left(\sum_{I \in T} |a_I^2|^2 \right)^{1/2} \frac{1}{|I_T|^{1/2}} \left(\sum_{I \in T} |a_I^3|^2 \right)^{1/2} |I_T|$$

$$\lesssim \prod_{j=1}^{3} \text{size}_T^{(j)} \left(\left(a_I^j \right)_I \right) |I_T|$$

by using the John–Nirenberg inequality from Corollary 2.8. As a consequence, we can estimate (2.69) further as

$$E_1 E_2 E_3 \sum_{n_1, n_2, n_3} 2^{-n_1} 2^{-n_2} 2^{-n_3} \sum_{T \in \mathbb{T}^{n_1, n_2, n_3}} |I_T|, \tag{2.70}$$

where, according to Corollary 2.11, the summations run over those indices n_1, n_2, n_3 for which

$$2^{-n_j} \lesssim \frac{S_j}{E_j}, \tag{2.71}$$

$j = 1, 2, 3$. However, Corollary 2.11 allows us to estimate

$$\sum_{T \in \mathbb{T}^{n_1,n_2,n_3}} |I_T|$$

in three different ways, by using

$$\sum_{T \in \mathbb{T}^{n_1,n_2,n_3}} |I_T| \lesssim 2^{n_1}, 2^{n_2}, 2^{n_3}.$$

In particular, one has

$$\sum_{T \in \mathbb{T}^{n_1,n_2,n_3}} |I_T| \lesssim 2^{n_1\theta_1} 2^{n_2\theta_2} 2^{n_3\theta_3} \tag{2.72}$$

whenever $0 \leq \theta_1, \theta_2, \theta_3 < 1$ and $\theta_1 + \theta_2 + \theta_3 = 1$. Using all this information we can estimate (2.70) further by

$$\begin{aligned} &E_1 E_2 E_3 \sum_{n_1,n_2,n_3} 2^{-n_1(1-\theta_1)} 2^{-n_2(1-\theta_2)} 2^{-n_3(1-\theta_3)} \\ &\lesssim E_1 E_2 E_3 \left(\frac{S_1}{E_1}\right)^{1-\theta_1} \left(\frac{S_2}{E_2}\right)^{1-\theta_2} \left(\frac{S_3}{E_3}\right)^{1-\theta_3} \\ &= \prod_{j=1}^{3} S_j^{1-\theta_j} E_j^{\theta_j}, \end{aligned}$$

as desired. □

In our particular case we have $a_I^1 = \langle f, \varphi_I^1 \rangle$, $a_I^2 = \langle g, \varphi_I^1 \rangle$, and $a_I^3 = \langle h, \varphi_I^3 \rangle$ (see the start of Section 2.7) and so, in order to be able to improve our results we need to learn how to estimate the sizes and energies for this case.

2.10. Estimates for sizes and energies

If I is a dyadic interval, we will write the approximate cutoff function as

$$\widetilde{\chi}_I(x) := \left(1 + \frac{\operatorname{dist}(x, I)}{|I|}\right)^{-100}. \tag{2.73}$$

Lemma 2.13 *If F is an L^1 function and $j = 1, 2, 3$ then*

$$\operatorname{size}_{\mathcal{J}}^{(j)}((\langle F, \varphi_I^j \rangle_I) \lesssim \sup_{I \in \mathcal{J}} \frac{1}{|I|} \int_{\mathbb{R}} |F| \widetilde{\chi}_I^M \, dx$$

for every $M > 0$, where the implicit constants depend on M.

Proof The case $j = 1$ follows directly from the definition, so we need to discuss only the cases $j = 2, 3$.

Fix $I_0 \in \mathcal{J}$. We will prove that

$$\left\| \left(\sum_{I \subseteq I_0} \frac{|\langle F, \varphi_I^j \rangle|^2}{|I|} \chi_I \right)^{1/2} \right\|_{1,\infty} \lesssim \int_{\mathbb{R}} |F| \widetilde{\chi}_{I_0}^M \, dx, \tag{2.74}$$

and this will be enough.

Split the real line as a disjoint union of intervals $(I_n)_{n\in\mathbb{Z}}$ having the same length as I_0 (I_n for $n > 0$ lies to the right of I_0, while I_n for $n < 0$ lies to the left of I_0).

By using the boundedness of the discretized square functions from L^1 into $L^{1,\infty}$, as stated in Theorem 2.5, one can see that the left-hand side of (2.74) becomes smaller than

$$\sum_{n=-2}^{2} \left\| \left(\sum_{I \subseteq I_0} \frac{|\langle F\chi_{I_n}, \varphi_I^j \rangle|^2}{|I|} \chi_I \right)^{1/2} \right\|_{1,\infty}$$

$$+ \sum_{n \neq 0, \pm 1, \pm 2} \left\| \left(\sum_{I \subseteq I_0} \frac{|\langle F\chi_{I_n}, \varphi_I^j \rangle|^2}{|I|} \chi_I \right)^{1/2} \right\|_{1}$$

$$\lesssim \sum_{n=-2}^{2} \| F\chi_{I_n} \|_1 + \sum_{n \neq 0, \pm 1, \pm 2} \left\| \sum_{I \subseteq I_0} \frac{|\langle F\chi_{I_n}, \varphi_I^j \rangle|}{|I|^{1/2}} \chi_I \right\|_1$$

$$= \sum_{n=-2}^{2} \| F\chi_{I_n} \|_1 + \sum_{n \neq 0, \pm 1, \pm 2} \sum_{I \subseteq I_0} \langle |F|\chi_{I_n}, |I|^{1/2} |\varphi_I^j| \rangle. \tag{2.75}$$

Since the first first term in (2.75) is clearly smaller than the right-hand side of (2.74), we need to consider only the second. Fix $n \neq 0, \pm 1, \pm 2$. The corresponding sum in (2.75) is

$$\sum_{I \subseteq I_0} \langle |F|\chi_{I_n}, |I|^{1/2} |\varphi_I^j| \rangle. \tag{2.76}$$

Since the new functions $|I|^{1/2}|\varphi_I^j|$ are now L^∞-normalized, it is not difficult to see that (2.76) can be estimated by

$$\frac{1}{n^M} \| F\chi_{I_n} \|_1,$$

since the smaller the lengths of the intervals I, the smaller their contribution to (2.76) becomes. In the end, if one sums all these expressions then one clearly gets an upper bound of the form $\| F \widetilde{\chi}_{I_0}^M \|$, as desired. □

Lemma 2.14 *If F is an L^1 function and $j = 1, 2, 3$ then*

$$\text{energy}_{\mathcal{J}}^{(j)}\left(\left(\langle F, \varphi_I^j\rangle\right)_I\right) \lesssim \|F\|_1.$$

Proof We will consider the cases $j = 2, 3$ since the case $j = 1$ is simpler and follows the same ideas. Let $n \in \mathbb{Z}$ and $\mathbb{D}$ be such that the supremum in Definition 2.9 is attained. Then, one can write

$$\begin{aligned}
\text{energy}_{\mathcal{J}}^{(j)}\left(\left(\langle F, \varphi_I^j\rangle\right)_I\right) &= 2^n \left(\sum_{I\in\mathbb{D}} |I|\right) \\
&= 2^n \left\|\sum_{I\in\mathbb{D}} \chi_I\right\|_1 = 2^n \left\|\sum_{I\in\mathbb{D}} \chi_I\right\|_{1,\infty} \\
&\le \left\|\sum_{I\in\mathbb{D}} \frac{1}{|I|} \left\|\left(\sum_{I'\subseteq I} \frac{|\langle F, \varphi_{I'}^j\rangle|^2}{|I'|}\chi_{I'}\right)^{1/2}\right\|_{1,\infty} \chi_I\right\|_{1,\infty} \\
&\lesssim \left\|\sum_{I\in\mathbb{D}} \frac{\langle |F|, \widetilde{\chi}_I^{100}\rangle}{|I|}\chi_I\right\|_{1,\infty} \lesssim \|MF\|_{1,\infty} \lesssim \|F\|_1,
\end{aligned}$$

where M is the usual Hardy–Littlewood maximal operator. □

2.11. L^p bounds for the first discrete model

We can finally complete the proof of the Theorem 2.3. We will first show that the operator

$$\Pi_{\mathcal{J}}(f, g) = \sum_{I\in\mathcal{J}} c_I \frac{1}{|I|^{\frac{1}{2}}} \langle f, \varphi_I^1\rangle\langle g, \varphi_I^2\rangle\varphi_I^3$$

maps $L^1 \times L^1$ to $L^{1/2,\infty}$.

Fix $f, g \in L^1$ such that $\|f\|_1 = \|g\|_1 = 1$. Let $E \subseteq \mathbb{R}$ such that $|E| = 1$ (as discussed earlier, by scaling invariance one can always assume that this is the case). The goal is to find $E' \subseteq E$ such that $|E'| \simeq 1$ and such that

$$\sum_{I\in\mathcal{J}} \frac{1}{|I|^{1/2}} |\langle f, \varphi_I^1\rangle||\langle g, \varphi_I^2\rangle||\langle h, \varphi_I^3\rangle| \lesssim 1,$$

where $h := \chi_{E'}$. As before, define the "exceptional" set

$$\Omega = \{x | M(f)(x) > C\} \cup \{x | M(g)(x) > C\}$$

and choose $C > 0$ large enough that $|\Omega| < \frac{1}{10}$. Then set $E' := E \setminus \Omega$. After that, split the collection of dyadic intervals $\mathcal{J}$ as follows:

$$\mathcal{J} = \bigcup_{d \geq 0} \mathcal{J}_d,$$

where $\mathcal{J}_d$ contains all the intervals in $\mathcal{J}$ such that $1 + \text{dist}(I, \Omega^c)/|I| \simeq 2^d$. It is clearly enough to show that the left-hand side of the above inequality, when summed over $I \in \mathcal{J}_d$, becomes smaller than 2^{-d}, say. At this point we simply have to apply Proposition 2.12. One observes that, because of Lemmas 2.13 and 2.14, one has

$$\text{size}^{(1)}_{\mathcal{J}_d}((\langle f, \varphi^1_I \rangle)_I) \lesssim 2^d,$$
$$\text{size}^{(2)}_{\mathcal{J}_d}((\langle g, \varphi^2_I \rangle)_I) \lesssim 2^d,$$

and

$$\text{size}^{(3)}_{\mathcal{J}_d}((\langle h, \varphi^3_I \rangle)_I) \lesssim 2^{-10d}.$$

Also, $\text{energy}^1_{\mathcal{J}_d}$, $\text{energy}^2_{\mathcal{J}_d}$, and $\text{energy}^3_{\mathcal{J}_d}$ are all bounded by 1, since f, g, and h are L^1-normalized.

Then, Proposition 2.12 in the particular case $\theta_1 = \theta_2 = \theta_3 = \frac{1}{3}$ gives an upper bound

$$2^{2d/3} 2^{2d/3} 2^{-20d/3} \lesssim 2^{-d}$$

and this completes the proof of the fact that discrete paraproducts do indeed map $L^1 \times L^1$ into $L^{1/2,\infty}$.

A similar argument allows one to prove that $\Pi_{\mathcal{J}} : L^{\widetilde{p}} \times L^{\widetilde{q}} \to L^{\widetilde{r},\infty}$ for $\widetilde{p}$ and $\widetilde{q}$ larger than but arbitrarily close to 1. To be a little more precise, now one has to work with $L^{\widetilde{p}}$-adapted and $L^{\widetilde{q}}$-adapted energies, as will be described in the proof of Theorem 2.4 in the next section.

The general case in Theorem 2.3 then follows by symmetry using standard interpolation arguments. More precisely, let us first recall that, if $\Lambda_{\mathcal{J}}(f, g, h)$ denotes the trilinear form associated with $\Pi_{\mathcal{J}}$, the two adjoint operators $\Pi^{*1}_{\mathcal{J}}$, $\Pi^{*2}_{\mathcal{J}}$ attached to $\Pi_{\mathcal{J}}$ are defined by the equalities

$$\int_{\mathbb{R}} \Pi^{*1}_{\mathcal{J}}(g, h)(x) f(x)\, dx = \Lambda_{\mathcal{J}}(f, g, h)$$

and

$$\int_{\mathbb{R}} \Pi^{*2}_{\mathcal{J}}(f, h)(x) g(x)\, dx = \Lambda_{\mathcal{J}}(f, g, h)$$

respectively. By interpolating between the previous Banach estimates and the $L^{\widetilde{p}} \times L^{\widetilde{q}} \to L^{\widetilde{r},\infty}$ estimates proved earlier, one obtains that $\Pi_{\mathcal{J}}$ maps $L^p \times L^q$ into L^r for every $1 < p, q < \infty$ and $1/p + 1/q = 1/r$. By symmetry, since $\Pi^{*1}_{\mathcal{J}}$ and $\Pi^{*2}_{\mathcal{J}}$ are also discretized paraproducts, they satisfy similar estimates and these together imply the general statement in Theorem 2.3 (for instance, to prove that $\Pi_{\mathcal{J}}$ maps $L^p \times L^\infty$ into L^p is equivalent to proving that $\Pi^{*2}_{\mathcal{J}}$ maps $L^p \times L^{p'}$ into L^1).

Let us also remark that Theorem 2.3 holds not only for the discrete operators $\Pi_{\mathcal{J}}$ but also for the original paraproduct Π. This is an easy consequence of the facts that, on the one hand, the trilinear form of Π can be written as an average of the trilinear forms of operators of the type $\Pi_{\mathcal{J}}$ (as we proved earlier in (2.13)) and, on the other hand, the exceptional sets considered before were defined independently of the averaging parameter. Of course, these facts should be coupled with standard limiting arguments based on the fact that the bounds in Theorem 2.3 are independent of the cardinality of $\mathcal{J}$.

2.12. L^p bounds for the second discrete model

The proof of Theorem 2.4 is all that remains to complete the general case of the Leibnitz rule (2.1). Given that the functions $(\varphi_I^{3,\alpha})_I$ have only limited decay, one has to proceed with a little more care than in the proof of Theorem 2.3. Recall also the definition of $\Pi^{\alpha}_{\mathcal{J}}$ from (2.14b).

Fix $1 < p, q < \infty$ such that $r > 1/(1+\alpha)$. We would like to prove that $\Pi^{\alpha}_{\mathcal{J}}$ maps $L^p \times L^q$ to $L^{r,\infty}$. If we can prove this then, as before, symmetry arguments (i.e. the similar estimates for the two adjoints of the operator) together with standard interpolation theory will finish the proof of the theorem.

If μ is an arbitrary real number, we denote by $\mu+$ any real number strictly larger than μ and arbitrarily close to it and by $\mu-$ any real number strictly smaller than μ and arbitrarily close to it.

For every $I \in \mathcal{J}$ one can decompose the real line $\mathbb{R}$ in a natural way as a disjoint union of dyadic shells: $\mathbb{R} = I \cup \bigcup_{k=0}^{\infty} \left(2^{k+1}I \setminus 2^k I\right)$. Using this, one can smoothly localize the function $\varphi_I^{3,\alpha}$ to these regions and split it as follows:

$$\varphi_I^{3,\alpha} = \sum_{k=0}^{\infty} \frac{1}{2^{k(\alpha-)}} \varphi_I^{3,\alpha,k},$$

where $\varphi_I^{3,\alpha,k}$ is L^2-normalized, supported inside $2^{k+2}I$, and weakly adapted to I in the sense that it satisfies the decay estimate

$$\left| |I|^{1/2} \varphi_I^{3,\alpha,k}(x) \right| \lesssim \frac{1}{(1 + \mathrm{dist}(x, I)/|I|)^{1+}}.$$

As a consequence, $\Pi^\alpha_{\mathcal{J}}$ can be split as

$$\Pi^\alpha_{\mathcal{J}}(f, g) = \sum_{k=0}^{\infty} \frac{1}{2^{k(\alpha-)}} \sum_{I \in \mathcal{J}} \frac{1}{|I|^{1/2}} \langle f, \varphi_I^1 \rangle \langle g, \varphi_I^2 \rangle \varphi_I^{3,\alpha,k}.$$

Now let $f \in L^p$ with $\|f\|_p = 1$ and $g \in L^q$ with $\|g\|_q = 1$. Also, let $E \subseteq \mathbb{R}$ such that $|E| = 1$ (by the scaling invariance of $\Pi^\alpha_{\mathcal{J}}(f, g)$ one can always assume this). The goal then is to find a subset $E' \subseteq E$ with $|E'| \simeq 1$ and such that

$$\left| \int_{\mathbb{R}} \Pi^\alpha_{\mathcal{J}}(f, g)(x) h(x)\, dx \right| \lesssim 1,$$

where $h = \chi_{E'}$. To prove this, it would clearly be enough to show that, for each $k \geq 0$,

$$\left| \sum_{I \in \mathcal{J}} \frac{1}{|I|^{1/2}} \langle f, \varphi_I^1 \rangle \langle g, \varphi_I^2 \rangle \langle h, \varphi_I^{3,\alpha,k} \rangle \right| \lesssim 2^{\beta k},$$

for some $\beta < \alpha$.

We define now an "exceptional" set as follows. First, for any $k \geq 0$ we define Ω_k by

$$\Omega_k = \left\{ x \mid Mf(x) > C_1 2^{\mu_1 k} \right\} \cup \left\{ x \mid Mg(x) > C_2 2^{\mu_2 k} \right\},$$

with μ_1, μ_2 such that $\mu_1 p > 1$ and $\mu_2 q > 1$. Then, define $\widetilde{\Omega}_k$ by

$$\widetilde{\Omega}_k = \left\{ x \,\middle|\, M(\chi_{\Omega_k})(x) > \frac{1}{2^{k+10}} \right\}$$

and finally define the set Ω by $\Omega = \bigcup_{k=0}^{\infty} \widetilde{\Omega}_k$. Since M is bounded from L^p into $L^{p,\infty}$ and from L^q into $L^{q,\infty}$, we have that

$$|\Omega_k| \lesssim \frac{1}{C_1^p 2^{\mu_1 p k}} + \frac{1}{C_2^p 2^{\mu_2 q k}}$$

and, since $|\widetilde{\Omega}_k| \lesssim 2^{k+10} |\Omega_k|$, it follows that $|\Omega| \leq 1/100$ if C_1 and C_2 are chosen large enough.

Then, as usual, we set $E' := E \setminus \Omega$. Fix $k \geq 0$. Since the support of $\varphi_I^{3,\alpha,k}$ lies inside $2^{k+2} I$, one has that $2^{k+2} I \cap \Omega^c \neq \emptyset$ and, as a consequence,

$$I \cap \Omega_k^c \neq \emptyset.$$

In other words, all the intervals I that participate in the summation

$$\sum_{I \in \mathcal{J}} \frac{1}{|I|^{1/2}} \langle f, \varphi_I^1 \rangle \langle g, \varphi_I^2 \rangle \langle h, \varphi_I^{3,\alpha,k} \rangle$$

must have this property, and so they intersect a set where everything is controlled.

To be able finally to estimate the above expression for our fixed $k \geq 0$, we need to use the generic statement in Proposition 2.12. However, since the functions involved are L^p- and L^q-normalized, it is more natural to use L^p- and L^q-adapted energies instead of the $L^{1,\infty}$ energies used before.

The first energy, $E_1(p)$, is defined by

$$E_1(p) := \sup_{n \in \mathbb{Z}} 2^n \sup_{\mathbb{D}} \left(\sum_{I \in \mathbb{D}} |I| \right)^{1/p},$$

where $\mathbb{D}$ ranges over all collections of disjoint dyadic intervals I_0 with the property $I_0 \cap \Omega_k^c \neq \emptyset$ and for which one has

$$\frac{1}{|I_0|^{1/p}} \left\| \left(\sum_{I \subseteq I_0} \frac{\langle f, \varphi_I^1 \rangle^2}{|I|} \right)^{1/2} \right\|_p \geq 2^n.$$

The second energy, $E_2(q)$, is defined simlarly but in terms of the function g.

Then, the analogue of Proposition 2.12 (which can be proved in exactly the same way) allows us to estimate the trilinear form associated to the fixed k as

$$S_1^{1-\theta_1 p} S_2^{1-\theta_2 q} S_3^{1-\theta_3} E_1(p)^{\theta_1 p} E_2(q)^{\theta_2 q} E_3^{\theta_3}$$

where S_1, S_2, S_3, E_3 are the previous $L^{1,\infty}$ sizes and energies, for every $0 \leq \theta_1, \theta_2, \theta_3 < 1$ with $\theta_1 + \theta_2 + \theta_3 = 1$. It is not difficult to observe (as before) that S_3, $E_1(p)$, $E_2(q)$, and E_3 are all $O(1)$ while $|S_1| \lesssim 2^{\mu_1 k}$ and $|S_2| \lesssim 2^{\mu_2 k}$.

Using the above allows us to estimate our expression by $O(2^{\beta k})$ for a certain $\beta < \alpha$, as desired, if one picks μ_1, μ_2 such that both $p\mu_1$ and $q\mu_2$ are very close to 1.

Let us end this section with the remark that, exactly as in the case of Theorem 2.3, Theorem 2.4 holds not only for the discrete operators $\Pi_{\mathcal{J}}^\alpha$ but also for the original operators Π^α. This completes the proof of (2.1) in the most general case.

Exercise 2.4 Check the estimates for the L^p-adapted energies used above and also the analogue of Proposition 2.12 that is available for them.

2.13. The general Coifman–Meyer theorem

Let us now come back to the original definition of a paraproduct, and recall formula (2.6):

$$(f, g) \mapsto \sum_k [(f * \varphi_k)(g * \psi_k)] * \widetilde{\psi}_k. \tag{2.77}$$

A simple calculation shows that this expression can also be rewritten as

$$\int_{\mathbb{R}^2} \left(\sum_k \widehat{\varphi_k}(\xi_1)\widehat{\psi_k}(\xi_2)\widehat{\widetilde{\psi}_k}(\xi_1 + \xi_2) \right) \widehat{f}(\xi_1)\widehat{g}(\xi_2) e^{2\pi i x(\xi_1+\xi_2)}\, d\xi_1 d\xi_2$$

$$=: \int_{\mathbb{R}^2} m(\xi_1, \xi_2)\widehat{f}(\xi_1)\widehat{g}(\xi_2) e^{2\pi i x(\xi_1+\xi_2)}\, d\xi_1 d\xi_2. \tag{2.78}$$

Exercise 2.5 Prove equality (2.78).

It is also not difficult to see that the multiplier

$$m(\xi_1, \xi_2) := \sum_k \widehat{\varphi_k}(\xi_1)\widehat{\psi_k}(\xi_2)\widehat{\widetilde{\psi}}_k(\xi_1 + \xi_2)$$

satisfies the classical Marcinkiewicz–Mikhlin–Hörmander condition

$$|\partial^\alpha m(\xi)| \lesssim \frac{1}{|\xi|^{|\alpha|}} \tag{2.79}$$

for sufficiently many multi-indices α, where ξ denotes the vector $\xi := (\xi_1, \xi_2)$.

It is then natural to ask whether bilinear multipliers such as those defined by (2.78) satisfy the same L^p estimates as the above paraproducts, for any symbol m satisfying (2.79). The answer to this question is affirmative, and this is the original statement of the Coifman–Meyer theorem. If m is a symbol satisfying (2.79), we denote by T_m the bilinear operator defined by the formula (2.78). Notice that for $m = 1$ the expression $T_m(f, g)$ becomes the product of the functions f and g.

From now on by a *paraproduct* of the functions f and g we will mean *any* expression of the type $T_m(f, g)$.

Theorem 2.15 *The bilinear multiplier operator T_m defined above maps $L^p \times L^q$ into L^r provided that $1 < p, q \leq \infty$, $1/p + 1/q = 1/r$ and $0 < r < \infty$.*

Proof In a few words, we will show that this theorem can be reduced to the Theorem 2.3. Before going any further, let us note two important facts about classical multipliers satisfying (2.79).

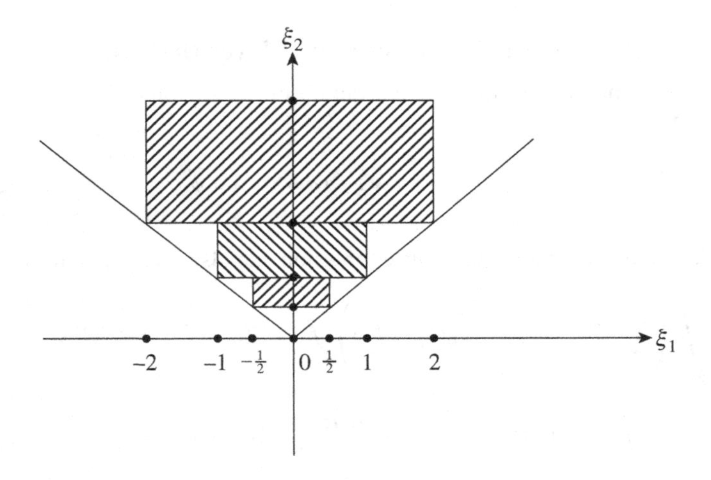

Figure 2.3. Conical support for the symbol of a paraproduct.

First, if for any positive real number $\lambda > 0$ one denotes by m_λ the *dilated symbol* defined by $m_\lambda(\xi) := m(\xi/\lambda)$ then m_λ satisfies (2.79) uniformly in λ, as one can easily check. As a consequence of this fact, one can see that essentially the operator T_m commutes with dilations. More precisely, it is easy to see that if one dilates both functions f and g with the same parameter λ then (modulo replacing m by m_λ) this is equivalent to dilating $T_m(f, g)$ with λ. It is not difficult to see that T_m commutes with translations as well.

Second, if one smoothly restricts m to a Whitney square with respect to the origin (i.e. a square whose sides are parallel to the coordinate axes and whose side length is comparable with its distance to the origin) then m is essentially constant there, in the sense that its Fourier coefficients decay very fast and also uniformly with respect to the side length of the square (we will see the details of this later).

Let us recall now the basic decomposition (2.4), which allowed us to split the product fg as a sum of *paraproducts*:[3]

$$fg = \sum_{k_1 \ll k_2} (f * \psi_{k_1})(g * \psi_{k_2}) + \sum_{k_2 \ll k_1} (f * \psi_{k_1})(g * \psi_{k_2})$$
$$+ \sum_{k_1 \simeq k_2} (f * \psi_{k_1})(g * \psi_{k_2}).$$

[3] Recall that, as mentioned in connection with (2.4), $k_a \ll k_b$ means $k_a < k_b - 100$ and $k_a \simeq k_b$ means $k_a - 100 \leq k_b \leq k_a + 100$.

This can be rewritten as

$$\begin{aligned}
&\int_{\mathbb{R}^2} \widehat{f}(\xi_1)\widehat{g}(\xi_2)e^{2\pi i x(\xi_1+\xi_2)}\,d\xi_1 d\xi_2 \\
&\quad = \int_{\mathbb{R}^2} \left(\sum_{k_1 \ll k_2} \widehat{\psi}_{k_1}(\xi_1)\widehat{\psi}_{k_2}(\xi_2)\right) \widehat{f}(\xi_1)\widehat{g}(\xi_2)e^{2\pi i x(\xi_1+\xi_2)}\,d\xi_1 d\xi_2 \\
&\quad\quad + \int_{\mathbb{R}^2} \left(\sum_{k_2 \ll k_1} \widehat{\psi}_{k_1}(\xi_1)\widehat{\psi}_{k_2}(\xi_1)\right) \widehat{f}(\xi_1)\widehat{g}(\xi_2)e^{2\pi i x(\xi_1+\xi_2)}\,d\xi_1 d\xi_2. \\
&\quad\quad + \int_{\mathbb{R}^2} \left(\sum_{k_1 \simeq k_2} \widehat{\psi}_{k_1}(\xi_1)\widehat{\psi}_{k_2}(\xi_2)\right) \widehat{f}(\xi_1)\widehat{g}(\xi_2)e^{2\pi i x(\xi_1+\xi_2)}\,d\xi_1 d\xi_2.
\end{aligned}$$

In other words, the function 1 viewed as a function of the variables ξ_1 and ξ_2 could be split as follows:

$$1 = \sum_{k_1 \ll k_2} \widehat{\psi}_{k_1}(\xi_1)\widehat{\psi}_{k_2}(\xi_2) + \sum_{k_2 \ll k_1} \widehat{\psi}_{k_1}(\xi_1)\widehat{\psi}_{k_2}(\xi_2) + \sum_{k_1 \simeq k_2} \widehat{\psi}_{k_1}(\xi_1)\widehat{\psi}_{k_2}(\xi_2). \tag{2.80}$$

In particular, any symbol m satisfying (2.79) can be decomposed as

$$\begin{aligned}
m(\xi_1, \xi_2) &= \sum_{k_1 \ll k_2} \widehat{\psi}_{k_1}(\xi_1)\widehat{\psi}_{k_2}(\xi_2)m(\xi_1, \xi_2) \\
&\quad + \sum_{k_2 \ll k_1} \widehat{\psi}_{k_1}(\xi_1)\widehat{\psi}_{k_2}(\xi_2)m(\xi_1, \xi_2) \\
&\quad + \sum_{k_1 \simeq k_2} \widehat{\psi}_{k_1}(\xi_1)\widehat{\psi}_{k_2}(\xi_2)m(\xi_1, \xi_2).
\end{aligned}$$

Thus, our operator T_m also splits as a sum of three corresponding operators. We will analyze only the first term since the other two can be treated similarly. It is given by the formula

$$\begin{aligned}
&\int_{\mathbb{R}^2} \left(\sum_{k_1 \ll k_2} \widehat{\psi}_{k_1}(\xi_1)\widehat{\psi}_{k_2} m(\xi_1, \xi_2)\right) \widehat{f}(\xi_1)\widehat{g}(\xi_2)e^{2\pi i x(\xi_1+\xi_2)}\,d\xi_1 d\xi_2 \\
&\quad =: \int_{\mathbb{R}^2} \left(\sum_{k} \widehat{\varphi}_{k}(\xi_1)\widehat{\psi}_{k}(\xi_2)m(\xi_1, \xi_2)\right) \widehat{f}(\xi_1)\widehat{g}(\xi_2)e^{2\pi i x(\xi_1+\xi_2)}\,d\xi_1 d\xi_2 \\
&\quad = \sum_{k} \int_{\mathbb{R}^2} \widehat{\varphi}_{k}(\xi_1)\widehat{\psi}_{k}(\xi_2)m(\xi_1, \xi_2)\widehat{f}(\xi_1)\widehat{g}(\xi_2)e^{2\pi i x(\xi_1+\xi_2)}\,d\xi_1 d\xi_2. \quad (2.81)
\end{aligned}$$

Note that Figure 2.3 gives a geometric description of the above decomposition. Fix $k \in \mathbb{Z}$. Since the support of $\widehat{\psi_k}$ is a union of two intervals of size 2^k lying to the right and to the left of the origin, one can assume without loss of generality that supp $\widehat{\psi_k}$ lies within an interval of size 2^k and whose distance to the origin is also of size 2^k (splitting $\widehat{\psi_k}$ into $\widehat{\psi_k^+} + \widehat{\psi_k^-}$, it can be seen that each of these two terms has this property).

In particular supp $\widehat{\varphi_k} \otimes \widehat{\psi_k}$ lies inside a cube of side length 2^k whose distance to the origin is also of size 2^k. The smooth restriction of the symbol $m(\xi_1, \xi_2)$ to that cube, i.e. $m(\xi_1, \xi_2)$ times an appropriate smooth bump function supported on a slightly larger cube, is denoted by $m_k(\xi_1, \xi_2)$ and can be decomposed as a double Fourier series:

$$m_k(\xi_1, \xi_2) = \sum_{n_1, n_2 \in \mathbb{Z}} C^k_{n_1, n_2} e^{2\pi i n_1 \xi_1 / 2^k} e^{2\pi i n_2 \xi_2 / 2^k}, \tag{2.82}$$

where the Fourier coefficients $C^k_{n_1, n_2}$ are given by

$$C^k_{n_1, n_2} = \frac{1}{2^{2k}} \int_{\mathbb{R}^2} m_k(\xi, \eta) e^{-2\pi i n_1 \xi / 2^k} e^{-2\pi i n_2 \eta / 2^k} \, d\xi d\eta.$$

By taking advantage of (2.79) one can see that

$$\begin{aligned} \left|C^k_{n_1, n_2}\right| &= \left| \int_{\mathbb{R}^2} m_k(2^k \xi, 2^k \eta) \, e^{-2\pi i n_1 \xi} e^{-2\pi i n_2 \eta} \, d\xi d\eta \right| \\ &\lesssim \frac{1}{(1 + |n_1| + |n_2|)^M}, \end{aligned} \tag{2.83}$$

by the usual integration-by-parts argument performed sufficiently many times.

Exercise 2.6 Check (2.83) carefully.

If one does this for any particular k, one can rewrite (2.81) as

$$\sum_{n_1, n_2 \in \mathbb{Z}} \sum_{k \in \mathbb{Z}} C^k_{n_1, n_2} \int_{\mathbb{R}^2} \widehat{\varphi}_{k, n_1}(\xi_1) \widehat{\psi}_{k, n_2}(\xi_2) \widehat{f}(\xi_1) \widehat{g}(\xi_2) e^{2\pi i x(\xi_1 + \xi_2)} \, d\xi_1 d\xi_2, \tag{2.84}$$

where $\widehat{\varphi}_{k, n_1}(\xi_1) := \widehat{\varphi_k}(\xi_1) e^{2\pi i n_1 \xi_1 / 2^k}$ and $\widehat{\psi}_{k, n_2}(\xi_2) := \widehat{\psi_k}(\xi_2) e^{2\pi i n_2 \xi_2 / 2^k}$.

Then, (2.84) can be rewritten as

$$\sum_{n_1, n_2 \in \mathbb{Z}} \sum_{k \in \mathbb{Z}} C^k_{n_1, n_2} (f * \varphi_{k, n_1})(g * \psi_{k, n_2})$$

and, exactly as before, one can complete (rewrite) it as

$$\sum_{n_1,n_2\in\mathbb{Z}} \sum_{k\in\mathbb{Z}} C^k_{n_1,n_2}\big((f * \varphi_{k,n_1})(g * \psi_{k,n_2})\big) * \widetilde{\psi}_k. \tag{2.85}$$

If we fix $(n_1, n_2) \in \mathbb{Z}^2$ and consider only the inner sum, we see that the corresponding expression is a classical paraproduct and, as we know, it satisfies the desired L^p estimates.

Moreover, a careful inspection of the paraproducts proof shows that, because of the presence of the indices n_1, n_2 in (2.85), one could in principle have to contend with an upper bound of the form $(1 + |n_1| + |n_2|)^{100}$, but clearly even this would be acceptable because of the rapid decay in (2.83). □

Finally, since the generic multipliers m satisfying (2.79) have the property that $K = \widehat{m}$ is a Calderón–Zygmund kernel, one can easily see that T_m can also be written as

$$p.v. \int_{\mathbb{R}^2} f(x - t_1)\, g(x - t_2)\, K(t_1, t_2)\, dt_1 dt_2 \tag{2.86}$$

and, because of this, these operators are sometimes referred to as *multilinear* singular integrals. The reader should recall the *classical linear* singular integrals treated in the first volume of the book.

Exercise 2.7 Prove formula (2.86).

2.14. Bilinear pseudodifferential operators

We end this chapter by studying some pseudodifferential variants of T_m, defined as follows.

Let $a(x, \xi)$ be a symbol satisfying

$$|\partial_x^\beta \partial_\xi^\alpha a(x, \xi)| \lesssim \frac{1}{(1 + |\xi|)^{|\alpha|}} \tag{2.87}$$

for sufficiently many multi-indices α and β. Denote by T_a the bilinear operator defined by

$$T_a(f, g)(x) = \int_{\mathbb{R}^2} a(x, \xi_1, \xi_2)\widehat{f}(\xi_1)\widehat{g}(\xi_2)e^{2\pi i x(\xi_1+\xi_2)}\, d\xi_1 d\xi_2.$$

One should compare these operators with the analogous *linear pseudodifferential* operators treated in Chapter I.12.

One has

Theorem 2.16 *The operator T_a maps $L^p \times L^q$ into L^r boundedly for any $1 < p, q \leq \infty$ with $1/p + 1/q = 1/r$ and $0 < r < \infty$.*

Proof We will show that this theorem can be essentially reduced to the Coifman–Meyer theorem or, more precisely, to a localized variant of it. The fact that in (2.87) the derivatives with respect to the x variable do not contribute to the right-hand side suggests that if one smoothly restricts the symbol $a(x, \xi)$ to any interval of length 1 then the new, restricted, symbol should be essentially constant in x. The way in which one expresses this fact analytically is by decomposing the restricted symbol as a Fourier series and observing that the Fourier coefficients decay rapidly away from the origin.

We thus proceed as follows. First, we pick a sequence of smooth functions $(\varphi_n)_{n\in\mathbb{Z}}$ such that $\operatorname{supp} \varphi_n \subseteq [n-1, n+1]$ and

$$\sum_{n\in\mathbb{Z}} \varphi_n = 1.$$

As a consequence, T_a can be split as follows:

$$T_a = \sum_{n\in\mathbb{Z}} T_a{}^n,$$

where

$$T_a^n(f, g)(x) := T_a(f, g)(x)\, \varphi_n(x).$$

We claim that for every $n \in \mathbb{Z}$ one has

$$\left\| T_a^n(f, g) \right\|_r \lesssim \left\| f \widetilde{\chi}_{I_n} \right\|_p \left\| g \widetilde{\chi}_{I_n} \right\|_q$$

when I_n is the interval $[n, n+1]$; $\widetilde{\chi}_{I_n}$ was defined in (2.73). If one can prove this claim, the theorem would be proved since then one could write

$$\begin{aligned}\|T_a(f, g)\|_r &\lesssim \left(\sum_{n\in\mathbb{Z}} \|T_a^n(f, g)\|_r^r \right)^{1/r} \lesssim \left(\sum_{n\in\mathbb{Z}} \left\| f \widetilde{\chi}_{I_n} \right\|_p^r \left\| g \widetilde{\chi}_{I_n} \right\|_q^r \right)^{1/r} \\ &\lesssim \left(\sum_{n\in\mathbb{Z}} \|f \widetilde{\chi}_{I_n}\|_p^p \right)^{1/p} \left(\sum_{n\in\mathbb{Z}} \|g \widetilde{\chi}_{I_n}\|_q^q \right)^{1/q} \lesssim \|f\|_p \|g\|_q,\end{aligned}$$

as desired. We are therefore left with proving our claim.

Fix $n_0 \in \mathbb{Z}$. The symbol of the operator $T_a^{n_0}$, which is the function $a(x, \xi)\varphi_{n_0}(x)$, can also be written as $a(x, \xi)\widetilde{\varphi}_{n_0}(x)\varphi_{n_0}(x)$, where $\widetilde{\varphi}_{n_0}$ is a smooth function supported on the interval $[n-2, n+2]$ and which equals 1 on the support of φ_{n_0}. In particular, by taking advantage of (2.87) one can rewrite the symbol as

$$\left(\sum_{\ell \in \mathbb{Z}} m_\ell(\xi) e^{2\pi i x \ell}\right) \varphi_{n_0}(x)$$

simply by splitting part of it, namely $a(x, \xi)\widetilde{\varphi}_{n_0}(x)$, as a Fourier series with respect to the x variable. The condition (2.87) guarantees that

$$\|\partial^\alpha m_\ell(\xi)\| \lesssim \frac{1}{(1+|\ell|)^M} \frac{1}{(1+|\xi|)^{|\alpha|}} \tag{2.88}$$

for a large number M and sufficiently many multi-indices α. Because of this large decay in ℓ, it will be enough to consider the operator corresponding to $\ell = 0$, which is given by

$$(f, g) \mapsto \left(\int_{\mathbb{R}^2} m_0(\xi) \widehat{f}(\xi_1) \widehat{g}(\xi_2) e^{2\pi i x(\xi_1+\xi_2)} \, d\xi_1 d\xi_2\right) \varphi_{n_0}(x), \tag{2.89}$$

where m_0 satisfies

$$|\partial^\alpha m_0(\xi)| \lesssim \frac{1}{(1+|\xi|)^{|\alpha|}}. \tag{2.90}$$

This is what we meant when we said, at the beginning of the proof, that this theorem can be reduced to the Coifman–Meyer theorem. As one can see, the operator in (2.89) is simply a localization of a Coifman–Meyer bilinear operator. The discretization procedure described earlier allows one to reduce operators such as

$$(f, g) \mapsto \int_{\mathbb{R}^2} m_0(\xi) \widehat{f}(\xi_1) \widehat{g}(\xi_2) e^{2\pi i x(\xi_1+\xi_2)} \, d\xi_1 d\xi_2$$

to averages of discrete operators of the form

$$\int_0^1 \sum_{I \in \mathcal{J}} c_I \frac{1}{|I|^{1/2}} \langle f, \varphi_I^{1,\beta} \rangle \langle g, \varphi_I^{2,\beta} \rangle \varphi_I^{3,\beta}(x) \, d\beta. \tag{2.91}$$

In the present case, since m_0 satisfies (2.90), which is stronger than (2.79), we can assume in (2.91) that the summation runs over dyadic intervals having the property that $|I| \leq 1$.

Exercise 2.8 Convince yourself that, as a consequence of the more restrictive inequality (2.90), one can assume that the summation in (2.91) runs over only those dyadic intervals having the property that $|I| \leq 1$.

It is therefore enough to analyze operators of the type defined by

$$(f,g) \mapsto \left(\int_0^1 \sum_{\substack{I \in \mathcal{J} \\ |I| \leq 1}} C_I \frac{1}{|I|^{1/2}} \langle f, \varphi_I^{1,\beta} \rangle \langle g, \varphi_I^{2,\beta} \rangle \varphi_I^{3,\beta}(x)\, d\beta \right) \varphi_{n_0}(x) \tag{2.92}$$

and to prove our previous claim,

$$\|T_a^{n_0}(f,g)\|_r \lesssim \|f \widetilde{\chi}_{I_{n_0}}\|_p \|g \widetilde{\chi}_{I_{n_0}}\|_q \tag{2.93}$$

for such operators.

By translation invariance one can assume that $n_0 = 0$ in (2.93). We will use the previous notation $T_a^0(f,g)$ for the operator given by (2.92).

We then split T_a^0 as follows:

$$T_a^0 = T_{a,I}^0 + T_{a,II}^0,$$

where

$$T_{a,I}^0(f,g)(x) := \left(\int_0^1 \sum_{\substack{I \in \mathcal{J} \\ |I| \leq 1 \\ I \subseteq 5I_0}} C_I \frac{1}{|I|^{1/2}} \langle f, \varphi_I^{1,\beta} \rangle \langle g, \varphi_I^{2,\beta} \rangle \varphi_I^{3,\beta}(x)\, d\beta \right) \varphi_0(x) \tag{2.94}$$

and

$$T_{a,II}^0(f,g)(x) := \left(\int_0^1 \sum_{\substack{I \in \mathcal{J} \\ |I| \leq 1 \\ I \subseteq (5I_0)^c}} C_I \frac{1}{|I|^{1/2}} \langle f, \varphi_I^{1,\beta} \rangle \langle g, \varphi_I^{2,\beta} \rangle \varphi_I^{3,\beta}(x)\, d\beta \right) \varphi_0(x). \tag{2.95}$$

The term $T_{a,II}^0$. The operator $T_{a,II}^0$ can be considered as an error term and its analysis will be simple. For every $n \in \mathbb{Z}$ consider the contribution in (2.95) coming from those intervals having the property that $I \subseteq I_n$. We claim that the L^r quasi-norm of the corresponding expression can be estimated by

$$\frac{1}{(1+|n|)^M} \left\| f \widetilde{\chi}_{I_n}^M \right\|_p \left\| g \widetilde{\chi}_{I_n}^M \right\|_q, \tag{2.96}$$

for a large constant M. Assuming (2.96), one can use the triangle inequality if $r \geq 1$ or the subadditivity of $\|\cdot\|_r^r$ if $0 < r < 1$ to sum the contributions and obtain (2.93).

To prove (2.96) is straightforward, since for each fixed $I \subseteq I_n$ the corresponding *one-term operator* has an L^r quasi-norm smaller than

$$\frac{1}{(1 + \mathrm{dist}(I, I_0)/|I|)^M} \left\| f \widetilde{\chi}_{I_n}^M \right\|_p \left\| g \widetilde{\chi}_{I_n}^M \right\|_q . \tag{2.97}$$

Summing the contributions in (2.97) gives (2.96).

Exercise 2.9 Check carefully both (2.97) and (2.96).

The term $T^0_{a,I}$. This is the main term. This time we decompose the functions f, g as follows,

$$f = \sum_{n_1} f \chi_{I_{n_1}}$$

and

$$g = \sum_{n_2} g \chi_{I_{n_2}},$$

and insert the two sums into the formula for $T^0_{a,I}$. If both n_1 and n_2 are not far from zero, the Coifman–Meyer theorem guarantees that (2.93) holds for each term. If, however, either n_1 or n_2 is large then a simpler argument, similar to that used before, provides a decay factor of the type

$$\frac{1}{(1 + |n_1|)^M} \frac{1}{(1 + |n_2|)^M}$$

in front of the right-hand side of (2.93) and this is enough to conclude the proof of (2.93). □

Exercise 2.10 Complete the details of the proof for the term $T^0_{a,I}$, considered above.

Finally we mention that even though here we have shown only the connection of paraproducts to the Leibnitz rules, it is fair to say that these objects are everywhere in the field of nonlinear dispersive PDEs, in particular since they provide a very natural way of mollifying various nonlinearities, as we have seen.

Notes

The main theorems of the chapter are essentially due to Coifman and Meyer [25, 26] with various generalizations by Kenig and Stein [65] and by Grafakos and Torres [55]. Lemma 2.6 was taken from Muscalu, Tao, and Thiele [90]. The proof of Theorem 2.16 is from the unpublished manuscript Muscalu [82] (see also Bernicot [6] for an independent, somewhat similar, idea). The treatment of the Coifman–Meyer theorem described in this chapter follows Muscalu [84] closely. See also Auscher, Hoffman, Muscalu *et al.* [3] and Muscalu, Pipher, Tao *et al.* [93]. Other proofs can also be found in the lecture notes of Christ [15] and Thiele [116]. The size and energy terminology goes back to Muscalu, Tao, and Thiele [94]. Related topics in discretized analysis can be found in Frazier and Jawert [44]. Other applications of paraproducts to nonlinear PDEs can be found in Bony [9]. For extensions of the Coifman–Meyer theorem to Hardy spaces, see Coifman and Grafakos [23], Grafakos [48], Grafakos and Torres [54], and Grafakos and Kalton [50]. More recent works on paraproducts include Kovać [67], Muscalu [85], and Do, Muscalu, and Thiele [37].

Problems

Problem 2.1 Extend the Leibnitz rule and the Coifman–Meyer theorem in a natural way to Euclidean spaces of arbitrary dimension $\mathbb{R}^d$.

Problem 2.2 Extend the Leibnitz rule and the Coifman–Meyer theorem in a natural way from the bilinear setting to a general-n linear setting, for every $n \geq 2$.

Problem 2.3 Show that every $L^2(\mathbb{R})$ function admits a decomposition of the type

$$f = \int_0^1 \sum_I \langle f, \varphi_I^{1,\alpha} \rangle \varphi_I^{2,\alpha} \, d\alpha,$$

where both families $\varphi_I^{j,\alpha}$ for $j = 1, 2$ are L^2-normalized and lacunary in the sense of Definition 2.2.

Problem 2.4 (Bony [9]) Let u be a function in the Hölder class C^α, for $0 < \alpha < 1$, and F a C^∞ function. Show that there exists a paraproduct Π such that

$$F(u) = \Pi(u, F'(u)) + E(u)$$

with $E(u) \in C^{2\alpha}$.

Problem 2.5 Let J be a finite collection of dyadic intervals and Π_J a discretized paraproduct given by

$$\Pi_J(f, g) = \sum_{I \in J} c_I \frac{1}{|I|^{1/2}} \langle f, \varphi_I^1 \rangle \langle g, \varphi_I^2 \rangle \varphi_I^3.$$

Assume in addition that the family $(\varphi_I^3)_I$ is *lacunary*. Prove that in this case Π_J maps $L^p \times L^{p'}$ into the Hardy space H^1 for any $1 < p < \infty$ where, as usual, $1/p + 1/p' = 1$.

Hint: Using the duality between H^1 and BMO, estimate the trilinear form $\Lambda_J(f, g, h)$ where h is a BMO function. Perform a *double* (instead of *triple*) stopping-time procedure for the functions f, g and note that the corresponding size of h is bounded by its BMO norm; or, more directly, take advantage of the L^p-adapted energies introduced in Section 2.11.

Problem 2.6 (Coifman and Meyer [26]) Let m be a classical symbol in the plane $\mathbb{R}^2$ and as usual denote by T_m the bilinear multiplier with symbol m given by (2.78). Assume that $m(\xi, -\xi) = 0$ for every $\xi \neq 0$. Prove that in this case T_m maps $L^p \times L^{p'}$ into the Hardy space H^1 for any $1 < p < \infty$.

Hint: Reduce the problem carefully to the previous one.

Problem 2.7 (Kenig, Ponce, and Vega [66]) Let $0 < \alpha < 1$. Prove the following inequality:

$$\|D^\alpha(fg) - fD^\alpha g\|_p \lesssim \|g\|_\infty \|D^\alpha f\|_p$$

for any $1 < p < \infty$.

Does the inequality remain true for arbitrary $\alpha \geq 1$?

Problem 2.8 ([66]) Let $0 < \alpha, \alpha_1, \alpha_2 < 1$ be such that $\alpha = \alpha_1 + \alpha_2$. Prove the inequality

$$\|D^\alpha(fg) - fD^\alpha g - gD^\alpha f\|_r \lesssim \|D^{\alpha_1} g\|_p \, \|D^{\alpha_2} f\|_q$$

for any $1 < p, q \leq \infty$ and $1 \leq r < \infty$ with $1/p + 1/q = 1/r$. Does the inequality remain true for arbitrary $\alpha \geq 1$?

Problem 2.9 (Coifman, Lions, Meyer *et al.* [31]) Let $F : \mathbb{R}^2 \to \mathbb{R}^2$ be a vector field with the property that all the entries of the differential matrix DF are $L^2(\mathbb{R}^2)$ functions. Prove that $\det(DF)$ belongs to the Hardy space $H^1(\mathbb{R}^2)$. Then generalize this to $\mathbb{R}^n$.

Problem 2.10 (Coifman, Rochberg, and Weiss [28]) Let H be the Hilbert transform and b an arbitrary BMO function. Regard b also as an operator giving multiplication by the function b. Prove that the commutator $[H, b]$ maps L^p into L^p for every $1 < p < \infty$. Then generalize this to $\mathbb{R}^n$ by showing that one can replace H by any Riesz transform R_j for $1 \leq j \leq n$.

Problem 2.11 (div curl lemma, [31]) Let $F, G : \mathbb{R}^n \to \mathbb{R}^n$ be two vector fields whose entries belong to $L^2(\mathbb{R}^n)$. Clearly, their scalar product $F \cdot G$ is an $L^1(\mathbb{R}^n)$ function. Show that if in addition one assumes that $\operatorname{div} F(x) = \operatorname{curl} G(x) = 0$ then the scalar product belongs to the Hardy space $H^1(\mathbb{R}^n)$.

Hint: Consider the $n = 2$ case first. Show that since G is curl free one can find another function g such that $G_1 = R_1 g$ and $G_2 = R_2 g$, where R_1, R_2 are the corresponding Riesz transforms. After that, try to reduce the problem to the previous one by taking advantage of the fact that F is div free.

Problem 2.12 (The original proof from [26]) Let m be a classical symbol in the plane. As we have seen, it is not difficult to prove that the associated bilinear multiplier T_m maps $L^p \times L^q$ into L^r as long as $1/p + 1/q = 1/r$ and all the indices p, q, r are strictly between 1 and ∞. Reprove the fact that T_m maps $L^1 \times L^1$ into $L^{1/2,\infty}$, by using two distinct Calderón–Zygmund decompositions of the functions f and g and also the kernel representation formula (2.86).

Hint: In the more classical *linear case* of Calderón–Zygmund operators, the L^2 estimate is easy and, by using it and a Calderón–Zygmund decomposition, one can prove the more intricate $L^1 \to L^{1,\infty}$ bound. Here, in the bilinear case, the idea is to use the easy Banach estimates mentioned before, together with two Calderón–Zygmund decompositions, to obtain the harder estimate $L^1 \times L^1 \to L^{1/2,\infty}$.)

Problem 2.13 (Thiele [112])

(a) Let I_0 be a fixed dyadic interval and M a fixed positive integer. Denote by $\mathcal{I}_{\mathcal{M}}$ the set of all dyadic intervals $I \subseteq I_0$ with the property that $|I| \geq |I_0|/2^M$. Show that

$$\frac{1}{|I_0|^{1/2}} \left\| \sum_{I \in \mathcal{I}_{\mathcal{M}}} \langle f, h_I \rangle h_I \right\|_2 \lesssim \max\left(\max_{|I|=|I_0|/2^M} \frac{\langle |f|, \chi_I \rangle}{|I|}, \frac{\langle |f|, \chi_{I_0} \rangle}{|I_0|} \right), \tag{2.98}$$

where $(h_I)_I$ is the L^2-normalized Haar system. This inequality should be thought of as a quantified variant of the trivial fact that the left-hand side of (2.98) is less than the L^∞ norm of the function f.

(b) A finite collection of dyadic intervals $\mathcal{I}$ is said to be *convex* if and only if, for every dyadic interval $I_1 \subseteq I_2 \subseteq I_3$, the fact that $I_1, I_3 \in \mathcal{I}$ implies that $I_2 \in \mathcal{I}$ as well. Show that there is a natural generalization of (2.98) for arbitrary convex collections $\mathcal{I}$.

(c) Use these facts to show that there is a natural *triple* stopping-time argument, similar to that used before and involving averages $\langle |f|, \chi_I \rangle / |I|$, to prove the usual estimates, this time for *dyadic paraproducts* of the type

$$\sum_I \frac{1}{|I|^{1/2}} \langle f, h_I \rangle \left\langle g, \frac{\chi_I}{|I|^{1/2}} \right\rangle h_I.$$

3

Paraproducts on polydisks

Suppose that we have two Schwartz functions f and g, defined in the plane $\mathbb{R}^2$. If $\alpha, \beta > 0$ then the Leibnitz rule (2.1) (coupled with the Fubini and Hölder inequalities) allows us to write

$$\|D_1^\alpha(fg)\|_r \lesssim \|D_1^\alpha f\|_{p_1}\|g\|_{q_1} + \|f\|_{p_2}\|D_1^\alpha g\|_{q_2} \tag{3.1}$$

and

$$\|D_2^\beta(fg)\|_r \lesssim \|D_2^\beta f\|_{p_1}\|g\|_{q_1} + \|f\|_{p_2}\|D_2^\beta g\|_{q_2}, \tag{3.2}$$

which are valid provided that $1/p_i + 1/q_i = 1/r, 1 < p_i, q_i \leq \infty$, for $i = 1, 2$ and $\max(1/(1+\alpha), 1/(1+\beta)) < r < \infty$.

In general, if h is a Schwartz function of two variables, we denote by $D_1^\alpha h$ and $D_2^\beta h$ its α-derivative with respect to the first variable and its β-derivative with respect to the second variable. These are defined as follows:

$$\widehat{D_1^\alpha h}(\xi) := (2\pi|\xi_1|)^\alpha \widehat{h}(\xi),$$
$$\widehat{D_2^\beta h}(\xi) := (2\pi|\xi_2|)^\beta \widehat{h}(\xi),$$

where $\xi = (\xi_1, \xi_2) \in \mathbb{R}^2$.

Sometimes there are situations when both partial derivatives act on a product of functions, and then it is natural to ask whether expressions such as

$$\|D_1^\alpha D_2^\beta(fg)\|_r$$

can be estimated in a similar way.

The natural candidate, given the previous two inequalities, is the following *biparameter Leibnitz rule*:

$$\begin{aligned}\|D_1^\alpha D_2^\beta (fg)\|_r \lesssim \; & \|D_1^\alpha D_2^\beta f\|_{p_1} \|g\|_{q_1} \\ & + \|f\|_{p_2} \|D_1^\alpha D_2^\beta g\|_{q_2} \\ & + \|D_1^\alpha f\|_{p_3} \|D_2^\beta g\|_{q_3} \\ & + \|D_2^\beta f\|_{p_4} \|D_1^\alpha g\|_{q_4},\end{aligned} \tag{3.3}$$

which one expects to hold true whenever $1/p_i + 1/q_i = 1/r$, $1 < p_i, q_i \le \infty$, for $i = 1, 2, 3, 4$ and $\max(1/(1+\alpha), 1/(1+\beta)) < r < \infty$.

The main goal of this chapter is to prove that this inequality is indeed satisfied under the above assumptions. Notice that if one assumes that both functions f and g have the *tensor product* structure $f(x, y) = f_1 \otimes f_2(x, y) := f_1(x) f_2(y)$ and $g(x, y) = g_1 \otimes g_2(x, y) := g_1(x) g_2(y)$ then (3.3) follows easily from the standard Leibnitz rule of Chapter 2.

Some of the most interesting *nonhomogeneous* PDEs are the so-called Kadomtsev–Petviashvili equations (known as KP-I and KP-II), given by

$$\partial_t u + \partial_x^3 u \mp \partial_x^{-1} \partial_y^2 u + \partial_x \left(\tfrac{1}{2} u^2\right) = 0, \tag{3.4}$$

where $u(t, x, y)$ is a function defined on $\mathbb{R} \times \mathbb{R}^2$. Recent studies of the solutions of these equations by Kenig rely on particular cases of the estimate (3.3).

3.1. Biparameter paraproducts

As in the case of the one-parameter Leibnitz rule of the previous Chapter 2, we will see that this time (3.3) can be reduced to a biparameter generalization of the Coifman–Meyer theorem. To see this, the first task is to decompose the generic product $f(x, y) g(x, y)$ as a sum of several mollified expressions named *biparameter paraproducts*. We start by writing the product of f and g as

$$f(x, y) \cdot g(x, y) = \int_{\mathbb{R}^4} \widehat{f}(\xi_1, \xi_2) \widehat{g}(\eta_1, \eta_2) e^{2\pi i (x,y)\cdot((\xi_1,\xi_2)+(\eta_1,\eta_2))} \, d\xi d\eta. \tag{3.5}$$

Given that the operators D_1^α and D_2^β act on separate variables, it is natural to think of the above implicit symbol $1(\xi_1, \xi_2, \eta_1, \eta_2)$ as being a product:

thus

$$1(\xi_1, \xi_2, \eta_1, \eta_2) = 1(\xi_1, \eta_1) \cdot 1(\xi_2, \eta_2). \tag{3.6}$$

Now, as before, one uses several Littlewood–Paley decompositions and writes $1(\xi_1, \eta_1)$ as follows:[1]

$$\begin{aligned} 1(\xi_1, \eta_1) &= \left(\sum_{k_1} \widehat{\psi}_{k_1}(\xi_1)\right)\left(\sum_{k_2} \widehat{\psi}_{k_2}(\eta_1)\right) = \sum_{k_1 k_2} \widehat{\psi}_{k_1}(\xi_1)\widehat{\psi}_{k_2}(\eta_1) \\ &= \sum_{k_1 \ll k_2} \widehat{\psi}_{k_1}(\xi_1)\widehat{\psi}_{k_2}(\eta_1) + \sum_{k_2 \ll k_1} \widehat{\psi}_{k_1}(\xi_1)\widehat{\psi}_{k_2}(\eta_1) + \sum_{k_1 \simeq k_2} \widehat{\psi}_{k_1}(\xi_1)\widehat{\psi}_{k_2}(\eta_1) \\ &=: \sum_k \widehat{\varphi}_k(\xi_1)\widehat{\psi}_k(\eta_1) + \sum_k \widehat{\psi}_k(\xi_1)\widehat{\psi}_k(\eta_1) + \sum_k \widehat{\psi}_k(\xi_1)\widehat{\varphi}_k(\eta_1). \end{aligned} \tag{3.7}$$

There are finitely many expressions of the type $\sum_k \widehat{\psi}_k(\xi_1)\widehat{\psi}_k(\eta_1)$ above (they correspond to the $k_1 \simeq k_2$ term) but, since then all behave similarly, in an abuse of the notation we have written down only one.

Similarly, we can split $1(\xi_2, \eta_2)$ as

$$1(\xi_2, \eta_2) = \sum_\ell \widehat{\varphi_\ell}(\xi_2)\widehat{\psi_\ell}(\eta_2) + \sum_\ell \widehat{\psi_\ell}(\xi_2)\widehat{\varphi_\ell}(\eta_2) + \sum_\ell \widehat{\psi_\ell}(\xi_2)\widehat{\psi_\ell}(\eta_2)\cdot \tag{3.8}$$

By combining (3.6), (3.7), and (3.8) one obtains a decomposition of $1(\xi_1, \xi_2, \eta_1, \eta_2)$ as a sum of nine terms. One of them is

$$\sum_{k,\ell} \widehat{\varphi_k}(\xi_1)\widehat{\psi_k}(\eta_1)\widehat{\psi_\ell}(\xi_2)\widehat{\varphi_\ell}(\eta_2) := \sum_{k,\ell} (\widehat{\varphi_k} \otimes \widehat{\psi_\ell})(\xi_1, \xi_2) \cdot (\widehat{\psi_k} \otimes \widehat{\varphi_\ell})(\eta_1, \eta_2). \tag{3.9}$$

The part of (3.5) that corresponds to the expression in (3.9) can be also written as

$$\sum_{k,\ell} (f * (\varphi_k \otimes \psi_\ell)) \cdot (g * (\psi_k \otimes \varphi_\ell)),$$

which as before, can be *completed* as

$$\sum_{k,\ell} ((f * (\varphi_k \otimes \psi_\ell)) \cdot (g * (\psi_k \otimes \varphi_\ell)) * \widetilde{\psi}_k \otimes \widetilde{\psi}_\ell. \tag{3.10}$$

Such expressions are called *biparameter paraproducts* and are denoted $\vec{\Pi}(f, g)$. One should observe that, in the simpler case when $f = f_1 \otimes f_2$

[1] As before, $k_a \ll k_b$ means $k_a < k_b - 100$ and $k_a \simeq k_b$ means $k_a - 100 \leq k_b \leq k_a + 100$.

and $g = g_1 \otimes g_2$, one has

$$\overrightarrow{\Pi}(f, g)(x, y) := \Pi_1(f_1, g_1)(x) \cdot \Pi_2(f_2, g_2)(y),$$

where now Π_1 and Π_2 are classical one-parameter paraproducts. Formally, one writes

$$\overrightarrow{\Pi} = \Pi_1 \otimes \Pi_2.$$

Hence, every product of two functions in the plane can be decomposed in a natural way as a finite sum of such biparameter paraproducts.

It is now time to recall why paraproducts were helpful in proving the original Leibnitz rule. Depending on the type of paraproduct Π, the expression $D^\alpha(\Pi(f, g))$ becomes equal to $\widetilde{\Pi}$ or $\widetilde{\Pi}^\alpha$ applied to either $D^\alpha f$ and g or to f and $D^\alpha g$ respectively, and this allowed us to reduce the problem to the corresponding Coifman–Meyer inequality for the bilinear operators $\widetilde{\Pi}$ and $\widetilde{\Pi}^\alpha$ (recall that $\widetilde{\Pi}$ is just another paraproduct while $\widetilde{\Pi}^\alpha$ was been defined in the previous chapter).

In the present case we have a similar situation. Depending on the type of biparameter paraproduct $\overrightarrow{\Pi}$, one can easily see that every expression $D_1^\alpha D_2^\beta(\overrightarrow{\Pi}(f, g))$ becomes equal to a bilinear operator of the type $\Pi_1 \otimes \Pi_2$, $\Pi_1^\alpha \otimes \Pi_2$, $\Pi_1 \otimes \Pi_2^\beta$, or $\Pi_1^\alpha \otimes \Pi_2^\beta$ applied to one of the pairs of functions $(f, D_1^\alpha D_2^\beta g)$, $(D_1^\alpha f, D_2^\beta g)$, $(D_1^\alpha D_2^\beta f, g)$, or $(D_1^\beta f, D_2^\alpha g)$.

This fact reduces our Leibnitz rule (3.3) to the problem of proving Hölder-type estimates for the above biparameter bilinear operators. We will do this for $\Pi_1 \otimes \Pi_2$ and leave the rest of the cases to the reader.

More precisely, we will demonstrate estimates of the form

$$\|\overrightarrow{\Pi}(f, g)\|_r \lesssim \|f\|_p \, \|g\|_q \tag{3.11}$$

for any $1 < p, q \le \infty$ with $1/p + 1/q = 1/r$ and $0 < r < \infty$, where $\overrightarrow{\Pi} = \Pi_1 \otimes \Pi_2$ is a generic biparameter paraproduct. Clearly, inequality (3.11) is the biparameter extension of the Coifman–Meyer theorem studied in the previous chapter. Notice that, as in the one-parameter case, there are no constraints for the index r. They appear only when one studies cases where at least one factor of the type Π^α or Π^β occurs.

Also as before, one can reduce (3.11) to a discrete model. More precisely, consider two discretized classical paraproducts given by

$$\Pi_{1,\mathcal{I}}(f_1, g_1) = \sum_{I \in \mathcal{I}} c_I \frac{1}{|I|^{1/2}} \langle f_1, \varphi_I^1 \rangle \langle g_1, \varphi_I^2 \rangle \varphi_I^3$$

and

$$\Pi_{2,\mathcal{J}}(f_2, g_2) = \sum_{J\in\mathcal{J}} c_J \frac{1}{|J|^{1/2}} \langle f_2, \varphi_J^1\rangle\langle g_2, \varphi_J^2\rangle\varphi_J^3,$$

and define the *biparameter discretized paraproduct* $\vec{\Pi}_{\mathcal{R}}$ by

$$\vec{\Pi}_{\mathcal{R}} = \Pi_{1,\mathcal{I}} \otimes \Pi_{2,\mathcal{J}}$$

or, more generally, by

$$\vec{\Pi}_{\mathcal{R}}(f, g) = \sum_{R\in\mathcal{R}} c_R \frac{1}{|R|^{1/2}} \langle f, \varphi_R^1\rangle\langle g, \varphi_R^2\rangle\varphi_R^3, \tag{3.12}$$

where now the sum is over dyadic rectangles of the form $R = I \times J$ and φ_R^j is defined by $\varphi_R^j := \varphi_I^j \otimes \varphi_J^j$ for $j = 1, 2, 3$. The numbers c_R are all bounded and need not be of the type $c_R = c_I \cdot c_J$.

The following theorem will be proven in the rest of the chapter.

Theorem 3.1 *Any discrete biparameter paraproduct* (3.12) *is bounded from $L^p \times L^q$ into L^r provided that $1 < p, q \leq \infty$, $1/p + 1/q = 1/r$, and* $0 < r < \infty$.

Standard arguments, similar to those in the previous chapter, show that Theorem 3.1 implies (3.11). To be a little more specific let us first remark that, as in the one-parameter case, it is not difficult to see that any biparameter paraproduct $\vec{\Pi}$ can be written as an average of discretized biparameter paraproducts of the type $\vec{\Pi}_{\mathcal{R}}$. As a consequence, at least when $r \geq 1$, the estimates for $\vec{\Pi}_{\mathcal{R}}$ imply the estimates for the original $\vec{\Pi}$ provided that they are independent of the averaging parameter (and they are). To claim the same reduction in the general quasi-Banach case, one has to recall, besides the interpolation procedure, that quasi-norms of the type $L^{r,\infty}$ can be dualized as well (as explained in the first chapter) and everything is fine as long as one chooses the implicit *exceptional sets* independently of the averaging parameter (and as we will see, this is also possible).

Clearly, since there are several types of classical paraproduct, there are as a consequence several types of biparameter paraproduct. To be specific, we will assume that the families $(\varphi_I^1)_I$ and $(\varphi_J^2)_J$ above are nonlacunary (and that all the others are lacunary). However, the argument is completely independent of this particular choice.

3.2. Hybrid square and maximal functions

Let us assume first that we are in the easier, Banach, case, when all the indices p, q, r are all strictly between 1 and ∞, and denote by r' the dual exponent of r; thus $1/r + 1/r' = 1$.

In particular, for a well-chosen $L^{r'}$-normalized function h, one has

$$\begin{aligned}
\left\| \vec{\Pi}(f,g) \right\|_r &= \left| \int_{\mathbb{R}^2} \vec{\Pi}(f,g)(x,y) h(x,y)\, dxdy \right| \\
&\lesssim \sum_R \frac{1}{|R|^{1/2}} |\langle f, \varphi_R^1 \rangle| \, |\langle g, \varphi_R^2 \rangle| \, |\langle h, \varphi_R^3 \rangle| \\
&= \int_{\mathbb{R}^2} \sum_R \frac{|\langle f, \varphi_R^1 \rangle|}{|R|^{1/2}} \frac{|\langle g, \varphi_R^2 \rangle|}{|R|^{1/2}} \frac{|\langle h, \varphi_R^3 \rangle|}{|R|^{1/2}} \chi_R(x,y)\, dxdy \\
&\leq \int_{\mathbb{R}^2} MSf(x,y)\, SM(g)(x,y)\, SS(h)(x,y)\, dxdy, \qquad (3.13)
\end{aligned}$$

where SS is the double square function, defined by

$$SSh(x,y) := \left(\sum_R \frac{|\langle h, \varphi_R^3 \rangle|^2}{|R|} \chi_R(x,y) \right)^{1/2}$$

while the hybrid MS and SM functions are defined by

$$MS(f)(x,y) = \sup_I \frac{1}{|I|^{1/2}} \left(\sum_J \frac{|\langle f, \varphi_I^1 \otimes \varphi_J^1 \rangle|^2}{|J|} \chi_J(y) \right)^{1/2} \chi_I(x)$$

and

$$SM(g)(x,y) = \left(\sum_I \frac{\left(\sup_J (|\langle g, \varphi_I^2 \otimes \varphi_J^2 \rangle| / |J|^{1/2}) \chi_J(y) \right)^2}{|I|} \chi_I(x) \right)^{1/2}.$$

Exercise 3.1 Prove the inequality (3.13).

Another operator, which we have not seen yet but which may appear in a natural way in the analysis of other types of paraproduct, is the double maximal operator, given by

$$MM(h)(x,y) := \sup_{(x,y)\in R} \frac{1}{|R|} \int_R |h(u,v)|\, dudv,$$

where the supremum is taken over all dyadic rectangles containing the given point (x,y).

Lemma 3.2 *The operators MM, SS, MS, and SM are bounded on $L^p(\mathbb{R}^2)$ for every $1 < p < \infty$.*

Proof Let us denote by M_1 and M_2 the Hardy–Littlewood maximal operators with respect to the first and second variable. Then it is not difficult to see that MM is pointwise smaller than $M_1 \circ M_2$ and so its boundedness follows from the classical one-dimensional result and Fubini's theorem.

To understand SM we need to recall the following inequality of Fefferman and Stein (all the functions are now of one variable only; a proof of this can be found in Stein [105] at p. 51):

$$\left\| \left(\sum_k |Mf_k|^2 \right)^{1/2} \right\|_p \lesssim \left\| \left(\sum_k |f_k|^2 \right)^{1/2} \right\|_p .$$

Using it, one can then write

$$\|SM(g)\|_{L^p(\mathbb{R}^2)} = \left\| \left(\sum_I \frac{\left(\sup_J (|\langle\langle g, \varphi_I^2\rangle, \varphi_J^2\rangle|/|J|^{1/2})\chi_J(y) \right)^2}{|I|} \chi_I(x) \right)^{1/2} \right\|_{L^p(\mathbb{R}^2)}$$

$$\lesssim \left\| \left(\sum_I \left(M\left(\frac{\langle g, \varphi_I^2\rangle}{|I|^{1/2}} \right)(y) \right)^2 \chi_I(x) \right)^{1/2} \right\|_{L^p(\mathbb{R}^2)}$$

$$\lesssim \left\| \left(\sum_I \frac{|\langle g, \varphi_I^2\rangle|^2}{|I|}(y)\chi_I(x) \right)^{1/2} \right\|_{L^p(\mathbb{R}^2)} \lesssim \|g\|_{L^p(\mathbb{R}^2)},$$

using Fubini's theorem and the well-known inequality for the one-dimensional square function. The MS function is the simplest to deal with now, since one observes that it is pointwise smaller than the corresponding SM function with the roles of I and J reversed.

We are therefore left with the double square function. As before, in the case of SM its analysis can be reduced to the analysis of a one-dimensional square function, if one can prove the following analogue of the Fefferman–Stein inequality:

$$\left\| \left(\sum_k |Sf_k|^2 \right)^{1/2} \right\|_p \lesssim \left\| \left(\sum_k |f_k|^2 \right)^{1/2} \right\|_p .$$

To prove this estimate, one has to use Khinchine's inequality carefully. First, consider the sequence $(r_I)_I$ of Rademacher functions indexed by the dyadic intervals. Then, consider another sequence of Rademacher functions $(r_k)_k$ indexed by the positive integers. Observe that the *tensor product* sequence $r_{I,k} := r_I \otimes r_k$ becomes a sequence of independent random variables, for which

Khinchine's inequality still holds. Using all these facts one can write

$$\left\| \left(\sum_k |Sf_k|^2 \right)^{1/2} \right\|_p^p$$

$$= \int_{\mathbb{R}} \left(\sum_k \sum_I \frac{|\langle f_k, \varphi_I \rangle|^2}{|I|} \chi_I(x) \right)^{p/2} dx$$

$$\lesssim \int_{\mathbb{R}} \int_0^1 \int_0^1 \left| \sum_k \sum_I r_I(\omega) r_k(\omega') \frac{1}{|I|^{1/2}} \langle f_k, \varphi_I \rangle \chi_I(x) \right|^p d\omega d\omega' dx$$

$$= \int_{\mathbb{R}} \int_0^1 \int_0^1 \left| \sum_I r_I(\omega) \frac{1}{|I|^{1/2}} \left\langle \varphi_I, \sum_k r_k(\omega') f_k \right\rangle \chi_I(x) \right|^p d\omega d\omega' dx.$$

Using Khinchine's inequality again together with the L^p bounds of the one-dimensional square functions, one can bound the last expression as follows:

$$\int_0^1 \int_{\mathbb{R}} \left(\sum_I \frac{1}{|I|} \left| \left\langle \varphi_I, \sum_k r_k(\omega') f_k \right\rangle \right|^2 \chi_I(x) \right)^{p/2} dx d\omega'$$

$$\lesssim \int_0^1 \int_{\mathbb{R}} \left| \sum_k r_k(\omega') f_k(x) \right|^p dx d\omega' \lesssim \left\| \left(\sum_k |f_k|^2 \right)^{1/2} \right\|_p^p,$$

as desired. □

Now the information given by Lemma 3.2, when substituted into (3.13), proves the *Banach case* in Theorem 3.1.

3.3. Biparameter BMO

The general case, when $0 < r \leq 1$, is more difficult and, as we will see, is far from being a routine generalization of the one-parameter case.

In the previous chapter an important role in the proof of the Coifman–Meyer theorem was played by the John–Nirenberg inequality, which enabled us to have good control over the relevant sizes. In what follows we will describe its biparameter analogue. If $(a_R)_R$ is a sequence of complex numbers indexed over dyadic rectangles and if $0 < p < \infty$, one denotes by $\|(a_R)_R\|_{\mathrm{BMO}^{\mathrm{rect}}(p)}$ the expression given by

$$\|(a_R)_R\|_{\mathrm{BMO}^{\mathrm{rect}}(p)} := \sup_{R_0} \frac{1}{|R_0|^{1/p}} \left\| \left(\sum_{R \subseteq R_0} \frac{|a_R|^2}{|R|} \chi_R \right)^{1/2} \right\|_p , \tag{3.14}$$

where the supremum is taken over all possible dyadic rectangles R_0 in the plane.

Similarly, one denotes by $\|(a_R)_R\|_{\mathrm{BMO}(p)}$ the expression given by

$$\|(a_R)_R\|_{\mathrm{BMO}(p)} := \sup_{\Omega} \frac{1}{|\Omega|^{1/p}} \left\| \left(\sum_{R \subseteq \Omega} \frac{|a_R|^2}{|R|} \chi_R \right)^{1/2} \right\|_p , \tag{3.15}$$

where now the supremum is taken over all open subsets Ω in the plane.

Then, the spaces $\mathrm{BMO}^{\mathrm{rect}}(p)$ and $\mathrm{BMO}(p)$ are defined to be the collections of all complex sequences for which the corresponding expressions are finite.

A surprise of the biparameter theory is that the *correct definition* of BMO space is that in (3.15). This fact alone makes it impossible to reproduce many of the standard one-dimensional arguments in a biparameter setting.

The following analogue of the John–Nirenberg inequality holds.

Theorem 3.3 *Let $0 < p < q < \infty$. Then*

$$\|(a_R)_R\|_{\mathrm{BMO}(p)} \simeq \|(a_R)_R\|_{\mathrm{BMO}(q)}. \tag{3.16}$$

Proof One should think of p as being arbitrarily small and of q as being arbitrarily large, since all the intermediate cases follow immediately by Hölder inequality. It is enough to prove the theorem when p is small and q is of the form $q = 2^k$ for some integer $k \geq 1$.

Case 1: $k = 1$. Here the goal is to prove that

$$\|(a_R)_R\|_{\mathrm{BMO}(p)} \simeq \|(a_R)_R\|_{\mathrm{BMO}(2)}. \tag{3.17}$$

Clearly, as before, it is enough to show that

$$\|(a_R)_R\|_{\mathrm{BMO}(2)} \lesssim \|(a_R)_R\|_{\mathrm{BMO}(p,\infty)}.$$

As in the one-parameter case, we denote for simplicity the left-hand side of the above inequality by B and the right-hand side by A. We want to show that

$$B \lesssim A. \tag{3.18}$$

First, choose an open set $\Omega_0 \subseteq \mathbb{R}^2$ such that

$$\frac{1}{|\Omega_0|^{1/2}} \left\| \left(\sum_{R \subseteq \Omega_0} \frac{|a_R|^2}{|R|} \chi_R \right)^{1/2} \right\|_2 = B.$$

However, we also know that

$$\left\| \left(\sum_{R \subseteq \Omega_0} \frac{|a_R|^2}{|R|} \chi_R \right)^{1/2} \right\|_{p,\infty} \leq A|\Omega_o|^{1/p}$$

and, in particular, this means that

$$\left| \left\{ x \in \Omega_0 \left| \left(\sum_{R \subseteq \Omega_0} \frac{|a_R|^2}{|R|} \chi_R(x) \right)^{1/2} \geq CA \right. \right\} \right| \leq \left(\frac{A|\Omega_0|^{1/p}}{CA} \right)^p$$
$$= \frac{|\Omega_0|}{C^p} < \frac{|\Omega_0|}{M}, \tag{3.19}$$

where M is a large constant, provided that the constant $C > 0$ is also sufficiently large. If we denote by E the set that appears in the inequality (3.19), we can write

$$|E| < \frac{|\Omega_0|}{M}. \tag{3.20}$$

Then we have

$$B^2|\Omega_0| = \left\| \left(\sum_{R \subseteq \Omega_0} \frac{|a_R|^2}{|R|} \chi_R \right)^{1/2} \right\|_2^2 = \int_{\mathbb{R}^2} \left(\sum_{R \subseteq \Omega_0} \frac{|a_R|^2}{|R|} \chi_R(x) \right) dx$$
$$= \int_E \cdots + \int_{E^c} \cdots \quad := I + II,$$

where E^c denotes $\Omega_0 \setminus E$. Notice that E^c is an open set. To estimate term II is easy; one can see that

$$|II| \lesssim A^2|E^c| \leq A^2|\Omega_0|,$$

which is a useful bound, given that we want to show (3.18).

To estimate the term I we split it as follows:

$$\int_E \left(\sum_{R \subseteq \Omega_0} \frac{|a_R|^2}{|R|} \chi_R \right) dx = \int_E \left(\sum_{\substack{R \subseteq \Omega_0 \\ R \in \Re_1}} \frac{|a_R|^2}{|R|} \chi_R \right) dx + \int_E \left(\sum_{\substack{R \subseteq \Omega_0 \\ R \in \Re_2}} \frac{|a_R|^2}{|R|} \chi_R \right) dx, \tag{3.21}$$

where $\Re_1$ and $\Re_2$ are defined by

$$\Re_1 := \left\{ R \,\middle|\, |R \cap E| > \tfrac{1}{2}|R| \right\}$$

and

$$\Re_2 := \left\{ R \,\middle|\, |R \cap E| \le \tfrac{1}{2}|R| \right\}. \tag{3.22}$$

To estimate the second term in (3.21) we write

$$\begin{aligned}
\int_E \left(\sum_{\substack{R \subseteq \Omega_0 \\ R \in \Re_2}} \frac{|a_R|^2}{|R|} \chi_R \right) dx &= \sum_{\substack{R \in \Re_2 \\ R \subseteq \Omega_0}} \frac{|a_R|^2}{|R|} |R \cap E| \\
&\le \frac{1}{2} \sum_{\substack{R \subseteq \Omega_0 \\ R \in \Re_2}} \frac{|a_R|^2}{|R|} |R| \le \sum_{\substack{R \subseteq \Omega_0 \\ R \in \Re_2}} \frac{|a_R|^2}{|R|} |R \cap E^c| \\
&= \int_{E^c} \left(\sum_{\substack{R \subseteq \Omega_0 \\ R \in \Re_2}} \frac{|a_R|^2}{|R|} \chi_R \right) dx \le A^2 |E^c| \le A^2 |\Omega_0|,
\end{aligned}$$

which is again acceptable, given that we want to prove (3.18). Finally, to estimate the first term in (3.21) we observe that, since $R \in \Re_1$, one has

$$R \subseteq \widetilde{E},$$

where $\widetilde{E}$ is defined as $\{MM(\chi_E) > \frac{1}{2}\}$.

Since $\widetilde{E}$ is clearly an open set, one has

$$\int_E \left(\sum_{\substack{R\subseteq\Omega_0 \\ R\in\Re_1}} \frac{|a_R|^2}{|R|}\chi_R \right) dx \leq \int_{\widetilde{E}} \left(\sum_R \frac{|a_R|^2}{|R|}\chi_R \right) dx$$

$$\leq B^2|\widetilde{E}| \leq CB^2|E| \leq B^2C\frac{1}{M}|\Omega_0|. \qquad (3.23)$$

Putting everything together, we obtain

$$B^2|\Omega_0| \leq C_1A^2|\Omega_0| + C_2\frac{1}{M}|\Omega_0|B^2$$

and if we take M large enough, this implies that

$$B \lesssim A,$$

as desired.

Case 2: $k \geq 2$. Here one follows the same argument but some adjustment is needed. Since the essence of our approch is captured in the $k = 2$ case, we will assume for simplicity that $q = 4$ and leave the details of the general case to the reader. The only difference is that now one has to estimate the expression

$$\int_{\mathbb{R}^2} \left(\sum_{R\subseteq\Omega_0} \frac{|a_R|^2}{|R|}\chi_R(x) \right)^2 dx = \int_E \cdots + \int_{E^c} \cdots \quad := I + II.$$

Observe that the integrand is the square of that in case 1. The second term, II, can be treated as before, so we only need to consider the first. We write

$$\int_E \left(\sum_{R\subseteq\Omega_0} \frac{|a_R|^2}{|R|}\chi_R \right)^2 dx \leq 2\int_E \left(\sum_{\substack{R\subseteq\Omega_0 \\ R\in\Re_1}} \frac{|a_R|^2}{|R|}\chi_R \right)^2 dx$$

$$+2\int_E \left(\sum_{\substack{R\subseteq\Omega_0 \\ R\in\Re_2}} \frac{|a_R|^2}{|R|}\chi_R \right)^2 dx, \qquad (3.24)$$

where this time (using a factor $1/100$ to be safe) we define $\Re_1$ and $\Re_2$ by

$$\Re_1 := \left\{ R \,\middle|\, |R\cap E| > \frac{1}{100}|R| \right\}$$

and

$$\mathfrak{R}_2 := \left\{ R \,\middle|\, |R \cap E| \leq \frac{1}{100}|R| \right\}. \tag{3.25}$$

The first term in (3.24) can clearly be estimated as before. To estimate the second we write

$$\int_E \left(\sum_{\substack{R \subseteq \Omega_0 \\ R \in \mathfrak{R}_2}} \frac{|a_R|^2}{|R|} \chi_R \right)^2 dx = \int_E \left(\sum_{\substack{R, R' \subseteq \Omega_0 \\ R, R' \in \mathfrak{R}_2}} \frac{|a_R|^2}{|R|} \frac{|a_{R'}|^2}{|R'|} \chi_R \chi_{R'} \right) dx$$

$$\leq \sum_{\substack{R, R' \subseteq \Omega_0 \\ R, R' \in \mathfrak{R}_2}} \frac{|a_R|^2}{|R|} \frac{|a_{R'}|^2}{|R'|} |R \cap R'|. \tag{3.26}$$

Since now R and R' belong to $\mathfrak{R}_2$ we know that $|R \cap E^c| > (99/100)|R|$ and similarly $|R' \cap E^c| > (99/100)|R'|$. In particular, for any such pair one has

$$|R \cap R' \cap E^c| > \frac{98}{100}|R \cap R'|. \tag{3.27}$$

Using this in (3.26) one can estimate the right-hand side by

$$\sum_{\substack{R, R' \subseteq \Omega_0 \\ R, R' \in \mathfrak{R}_2}} \frac{|a_R|^2}{|R|} \frac{|a_{R'}|^2}{|R'|} |R \cap R' \cap E^c|$$

and it is not difficult to see that this is equal to

$$\int_{E^c} \left(\sum_{\substack{R \subseteq \Omega_0 \\ R \in \mathfrak{R}_2}} \frac{|a_R|^2}{|R|} \chi_R \right)^2 dx.$$

Now using the definition of A and the fact that E^c is open, we deduce that this is smaller than $A^4|E^c| \leq A^4|\Omega_0|$ as desired. The proof is now complete. □

3.4. Carleson's counterexample

Now that we have seen that the John–Nirenberg inequality holds for the more complicated BMO norms in (3.15), one may wonder if the two spaces (the *rectangular* and the *general* BMO) coincide.

The following well-known counterexample, due to Carleson, proves that this is not the case.

Theorem 3.4 *One has*

$$\mathrm{BMO}^{\mathrm{rect}}(2) \neq \mathrm{BMO}(2).$$

Proof Clearly, by definition, one has the inclusion

$$\mathrm{BMO}(2) \subseteq \mathrm{BMO}^{\mathrm{rect}}(2),$$

and so it is enough to construct a sequence $(a_R)_R$ that belongs to the *rectangular* BMO but not to the *general* BMO. We start with the following definition.

A collection of dyadic rectangles $\mathfrak{R}$ is said to have property $(*)$ if and only if the following hold:

(1) $R \subseteq [0,1] \times [0,1]$, for every $R \in \mathfrak{R}$;

(2) $\sum_{R \in \mathfrak{R}} |R| = 1$;

(3) $\sum_{R \subseteq Q} |R| \leq |Q|$ for every dyadic rectangle $Q \subseteq [0,1] \times [0,1]$.

The claim now is that there exist collections $\mathfrak{R}$ with property $(*)$ having arbitrarily small area (i.e., such that $|\cup_{R \in \mathfrak{R}} R|$ is arbitrarily small). Let us first observe that if we assume the claim we can easily construct such a sequence.

Indeed, we first define $a_R := |R|^{1/2}$ for $R \in \mathfrak{R}$ and $a_R := 0$ otherwise. If $R_0 \subseteq [0,1] \times [0,1]$ is an arbitrary rectangle then we obtain on the one hand

$$\frac{1}{|R_0|^{1/2}} \left\| \left(\sum_{R \subseteq R_0} \frac{|a_R|^2}{|R|} \chi_R \right)^{1/2} \right\|_2 = \frac{1}{|R_0|^{1/2}} \left(\sum_{R \subseteq R_0} |R| \right)^{1/2} \leq 1,$$

using the third property defining $(*)$. This proves that the sequence has rectangular BMO norm less than 1.

On the other hand, if we let $\Omega := \cup_{R \in \mathfrak{R}} R$ then we can also write

$$\frac{1}{|\Omega|^{1/2}} \left\| \left(\sum_{R \subseteq \Omega} \frac{|a_R|^2}{|R|} \chi_R \right)^{1/2} \right\|_2 = \frac{1}{|\Omega|^{1/2}} \left(\sum_{R \in \mathfrak{R}} |R| \right)^{1/2} = \frac{1}{|\Omega|^{1/2}}$$

and this, as we said, can be arbitrarily large, which shows that the *general* BMO norm of the sequence can be arbitrarily large.

It is therefore enough to show the above claim. The idea is to prove that, given a collection $\mathfrak{R}$ with property $(*)$ and area σ, one can construct from it another collection $\widetilde{\mathfrak{R}}$ also having property $(*)$ but whose area is $\sigma - \frac{1}{4}\sigma^2$. Assuming that such a construction is possible, one can then iterate it, starting with $\mathfrak{R} = [0,1] \times [0,1]$ and area $\sigma = 1$ and obtaining a sequence $(\mathfrak{R}_n)_n$ and corresponding areas $(\sigma_n)_n$ satisfying the recursive relation

$$\sigma_{n+1} = \sigma_n - \tfrac{1}{4}\sigma_n^2. \tag{3.28}$$

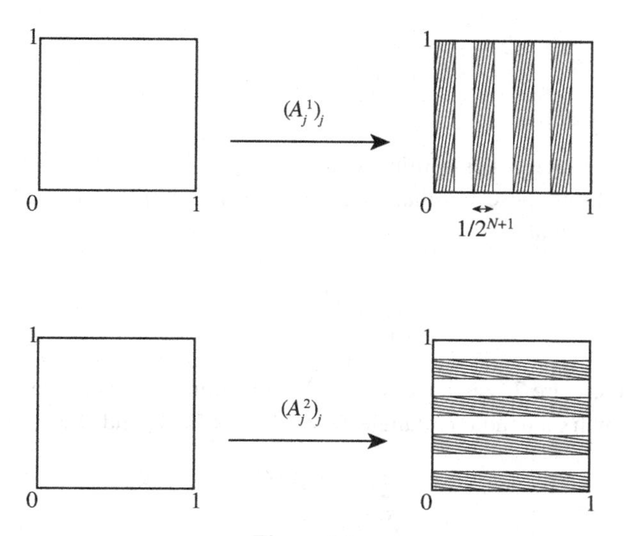

Figure 3.1

Clearly, since $(\sigma_n)_n$ is a positive and decreasing sequence, it must have a limit ℓ, which has to be zero because of (3.28). This shows that collections with arbitrarily small areas can indeed be constructed, provided that collections $\widetilde{\mathfrak{R}}$ with area $\sigma - \frac{1}{4}\sigma^2$ exist.

Fix $\mathfrak{R}$ with property $(*)$ and assume that its area is σ. Let N be a positive integer, large enough that the side lengths of every rectangle in $\mathfrak{R}$ are greater than or equal to $1/2^N$.

For every $1 \leq j < 2^N$ define the transformations A_j^1 and A_j^2 by

$$A_j^1(x, y) = \left(\frac{j}{2^N} + \frac{x}{2^{N+1}}, y\right)$$

and

$$A_j^2(x, y) = \left(x, \frac{j}{2^N} + \frac{y}{2^{N+1}}\right).$$

Then define a new collection $\widetilde{\mathfrak{R}}$ by

$$\widetilde{\mathfrak{R}} := \bigcup_{i=1}^{2} \bigcup_{j=0}^{2^N-1} A_j^i(\mathfrak{R}). \tag{3.29}$$

The transformations $(A_j^1)_j$ and $(A_j^2)_j$ are clearly geometrical and they are described in Figure 3.1. To prove that $\widetilde{\mathfrak{R}}$ has the property $(*)$ let us first show that

$$\sum_{R\in\widetilde{\mathfrak{R}}} |R| = 1. \tag{3.30}$$

(Clearly, every such R lies within the unit cube $[0, 1] \times [0, 1]$ by construction). To see (3.30), one observes that, for every $i = 1, 2$ and $0 \leq j < 2^N$, one has by construction that

$$\sum_{R\in A^i_j(\mathfrak{R})} |R| = \frac{1}{2^{N+1}}$$

and, since there are 2^N such rows and 2^N such columns, this proves (3.30).

Let us now fix a dyadic rectangle $Q \subseteq [0, 1] \times [0, 1]$ and check that

$$\sum_{\substack{R\in\widetilde{\mathfrak{R}} \\ R\subseteq Q}} |R| \leq |Q|. \tag{3.31}$$

Assume that $Q = I \times J$ with $|I| = 1/2^{k_1}$ and $|J| = 1/2^{k_2}$ and such that both k_1 and k_2 are positive integers.

If $k_1 > N$ this means that Q is very thin in the horizontal direction and as a consequence at most one column $A^1_j(\mathfrak{R})$ and no row $A^2_j(\mathfrak{R})$ contributes to (3.31) (to see this recall that N was chosen to be sufficiently large that every side length of $R \in \mathfrak{R}$ is greater than $1/2^N$). In this case (3.31) follows from the corresponding inequality for the collection $\mathfrak{R}$, by rescalling. One can treat the case $k_2 > N$ similarly.

If instead we have that both $k_1, k_2 \leq N$ then, in principle, many rows and columns can contribute. One observes that $2^N/2^{k_1}$ columns $A^1_j(R)$ contribute to the summation in (3.31) and we deduce that, each gives a contribution of at most $(1/2^{N+1})|J|$, again by using the corresponding inequality for $\mathfrak{R}$ and rescaling. Hence the total contribution of the columns is at most

$$\frac{2^N}{2^{k_1}} \frac{1}{2^{N+1}} |J| = \frac{|I|\,|J|}{2} = \frac{|Q|}{2}.$$

Since the contribution of the rows $A^2_j(\mathfrak{R})$ can be estimated in the same way, this gives (3.31).

Finally, we are left with proving that

$$\left| \bigcup_{R\in\widetilde{\mathfrak{R}}} R \right| = \sigma - \tfrac{1}{4}\sigma^2. \tag{3.32}$$

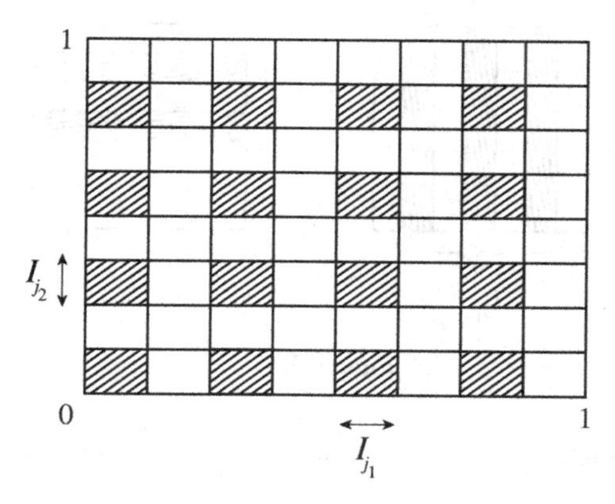

Figure 3.2

Clearly by construction the area of $\bigcup_j A_j^1(\mathfrak{R})$ is $\frac{1}{2}\sigma$ and that of $\bigcup_j A_j^2(\mathfrak{R})$ is also $\frac{1}{2}\sigma$. Then, by the inclusion–exclusion principle, we just need to show that

$$\left|\left(\bigcup_j A_j^1(\mathfrak{R})\right) \cap \left(\bigcup_j A_j^2(\mathfrak{R})\right)\right| = \tfrac{1}{4}\sigma^2. \tag{3.33}$$

If one writes $I_j := \left[j/2^N, j/2^N + 1/2^{N+1}\right]$ for $0 \le j < 2^N - 1$ then the left-hand side of (3.33) can be decomposed as

$$\sum_{j_1=0}^{2^N-1}\sum_{j_2=0}^{2^N-1}\left|\left(I_{j_1} \times I_{j_2} \cap \bigcup_{R \in A_{j_1}^1(\mathfrak{R})} R\right) \cap \left(I_{j_1} \times I_{j_2} \cap \bigcup_{R \in A_{j_2}^2(\mathfrak{R})} R\right)\right|. \tag{3.34}$$

Fix $0 \le j_1, j_2 \le 2^N - 1$ and consider the corresponding intersection in (3.34). If we inspect at the square $I_{j_1} \times I_{j_2}$ (see Figure 3.2) we see that there exist one-dimensional sets E_{j_1,j_2} and F_{j_1,j_2} such that

$$I_{j_1} \times I_{j_2} \cap \bigcup_{R \in A_{j_1}^1(\mathfrak{R})} R = E_{j_1,j_2} \times I_{j_2}$$

and

$$I_{j_1} \times I_{j_2} \cap \bigcup_{R \in A_{j_2}^2(\mathfrak{R})} R = I_{j_1} \times F_{j_1,j_2}.$$

By Fubini's theorem, we have that

$$\left|\left(E_{j_1,j_2} \times I_{j_2}\right) \cap \left(I_{j_1} \times F_{j_1,j_2}\right)\right| = |E_{j_1,j_2}|\,|F_{j_1,j_2}|$$

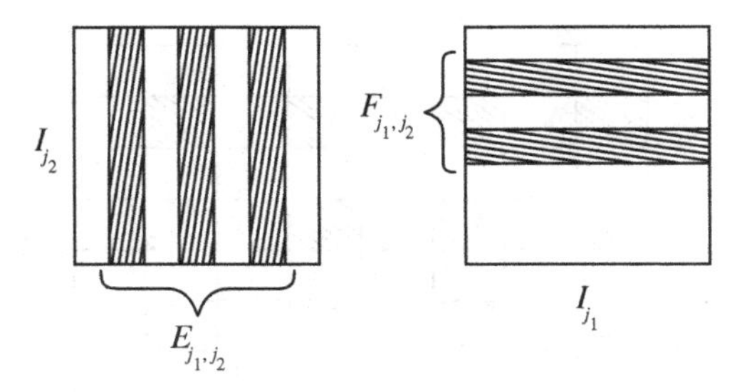

Figure 3.3

and as a consequence, the left-hand side of (3.34) becomes (see Figure 3.3)

$$\sum_{j_1=0}^{2^N-1}\sum_{j_2=0}^{2^N-1} |E_{j_1,j_2}|\,|F_{j_1,j_2}|. \tag{3.35}$$

Since all the rows and columns are identical it follows that the lengths $\left|E_{j_1,j_2}\right|$ do not depend on j_1, so we will denote them by $|E_{j_2}|$. Similarly, the lengths $\left|F_{j_1,j_2}\right|$ do not depend on j_2 and we can denote them by $|F_{j_1}|$. In particular (3.35) splits as follows:

$$\left(\sum_{j_1=0}^{2^N-1} |F_{j_1}|\right)\left(\sum_{j_2=0}^{2^N-1} |E_{j_2}|\right). \tag{3.36}$$

We claim now that

$$\sum_{j_2=0}^{2^N-1} |E_{j_2}| = \tfrac{1}{2}\sigma, \tag{3.37}$$

which (together with the similar expression $\sum_{j_1=0}^{2^N-1} |F_{j_1}| = \frac{1}{2}\sigma$) is enough to complete (3.33). To see this, one simply has to observe that for every given $A_j^1(\mathfrak{R})$ we have

$$\sum_{j_2=0}^{2^N-1} |E_{j_2}|\,\frac{1}{2^N} = \left|\bigcup_{R\in A_j^1(\mathfrak{R})} R\right| = \frac{\sigma}{2^{N+1}},$$

which is clearly equivalent to (3.37). See Figure 3.4. □

3.5. Proof of Theorem 3.1; part 1

We can now start the proof of the general case of Theorem 3.1. Let us consider $1 < p, q < \infty$ and $0 < r \leq 1$ such that $1/p + 1/q = 1/r$. Assume also that

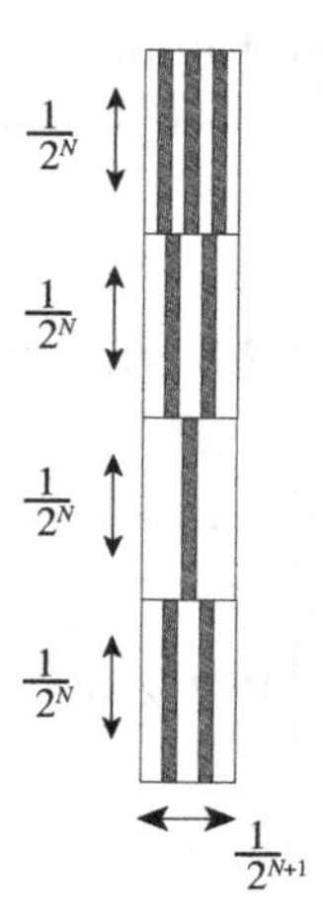

Figure 3.4

p, q are close to 1. We would like to show that discrete paraproducts such as (3.12) satisfy

$$\|\vec{\Pi}_{\mathfrak{R}}(f,g)\|_{r,\infty} \lesssim \|f\|_p \, \|g\|_q. \tag{3.38}$$

If we can prove (3.38) then Theorem 3.1 will be completely proven by using the symmetry of $\vec{\Pi}_{\mathfrak{R}}$ and standard interpolation arguments. Fix f and g with $\|f\|_p = \|g\|_q = 1$. Using the scaling invariance of $\vec{\Pi}_{\mathfrak{R}}$ and the duality lemma of the previous chapter, Lemma 2.6, it is enough to show that, for any measurable subset E of $\mathbb{R}^2$ of measure 1, there exists $E' \subseteq E$ with $|E'| \simeq 1$ such that

$$\left| \int_{\mathbb{R}^2} \vec{\Pi}_{\mathfrak{R}}(f,g)(x)h(x)\,dx \right| \lesssim 1, \tag{3.39}$$

where $h := \chi_{E'}$. Equivalently, using (3.12) we need to show that

$$\left| \sum_{R\in\mathfrak{R}} \frac{1}{|R|^{1/2}} \langle f, \varphi_R^1\rangle \langle g, \varphi_R^2\rangle \langle h, \varphi_R^3\rangle \right| \lesssim 1. \tag{3.40}$$

First, we define Ω_0 by

$$\begin{aligned}\Omega_0 := &\left\{x \in \mathbb{R}^2 \middle| MS(f)(x) > C\right\}\\ &\cup \left\{x \in \mathbb{R}^2 \middle| SM(g)(x) > C\right\}\\ &\cup \left\{x \in \mathbb{R}^2 \middle| MM(f)(x) > C\right\}\\ &\cup \left\{x \in \mathbb{R}^2 \middle| MM(g)(x) > C\right\},\end{aligned}$$

where the *MS* and *SM* functions are defined, as before, with respect to the finite collection of rectangles $\mathfrak{R}$. Then, we define successively the sets $\Omega_0 \subseteq \Omega \subseteq \widetilde{\Omega} \subseteq \widetilde{\widetilde{\Omega}} \subseteq \widetilde{\widetilde{\widetilde{\Omega}}}$ as follows:

$$\Omega := \left\{ x \in \mathbb{R}^2 \middle| MM(\chi_{\Omega_0})(x) > \frac{1}{100} \right\},$$

$$\widetilde{\Omega} := \left\{ x \in \mathbb{R}^2 \middle| MM(\chi_{\Omega})(x) > \frac{1}{2} \right\},$$

$$\widetilde{\widetilde{\Omega}} := \left\{ x \in \mathbb{R}^2 \middle| MM(\chi_{\widetilde{\Omega}})(x) > \frac{1}{2} \right\},$$

$$\widetilde{\widetilde{\widetilde{\Omega}}} := \left\{ x \in \mathbb{R}^2 \middle| MM(\chi_{\widetilde{\widetilde{\Omega}}})(x) > \frac{1}{10} \right\}.$$

Clearly, since all the functions *MS*, *SM*, *MM* are bounded on L^s for every $1 < s < \infty$, one has that

$$|\widetilde{\widetilde{\widetilde{\Omega}}}| < 1/10$$

if C in the definition of Ω_0 is a large enough constant.

Then, we set $E' := E \setminus \widetilde{\widetilde{\widetilde{\Omega}}}$; the goal is to show (3.40) for this subset $E' \subseteq E$. At this point the reader may wonder why it is necessary to consider this *tower* of *omega sets*. Let us just say that this fact will not be used explicitly yet but will be needed later on.

Theorems 3.3 and 3.4 show that a strategy based on sizes and energies as in the one-parameter case is hard to realize, mainly owing to the fact that natural biparameter sizes need to be defined using arbitrary open sets in $\mathbb{R}^2$ and these are clearly difficult to handle. Therefore, a new approach needs to be developed. First, we split the left-hand side of (3.40) as follows:

$$\sum_R \frac{1}{|R|^{1/2}} \langle f, \varphi_R^1 \rangle \langle g, \varphi_R^2 \rangle \langle h, \varphi_R^3 \rangle$$

$$= \sum_R \sum_{\substack{R: \\ R \cap \Omega^c \neq \emptyset}} \frac{1}{|R|^{1/2}} \langle f, \varphi_R^1 \rangle \langle g, \varphi_R^2 \rangle \langle h, \varphi_R^3 \rangle$$

$$+ \sum_{\substack{R: \\ R \cap \Omega^c = \emptyset}} \frac{1}{|R|^{1/2}} \langle f, \varphi_R^1 \rangle \langle g, \varphi_R^2 \rangle \langle h, \varphi_R^3 \rangle := I + II. \qquad (3.41)$$

Estimates for term I in (3.41)

Clearly, since $R \cap \Omega^c \neq \emptyset$ there exists $x_0 \in R \cap \Omega^c$. In particular one has $MM(\chi_{\Omega_0})(x_0) \leq \frac{1}{100}$, which implies that

$$\frac{1}{|R|} \int_R \chi_{\Omega_0} \leq \frac{1}{100}$$

and, as a consequence,

$$|R \cap \Omega_0^c| > \frac{99}{100}|R|. \tag{3.42}$$

In other words, the rectangles that appear in I have the property that they are 99% within the set Ω_0^c where the corresponding functions are bounded.

Then, we define

$$\Omega_1 := \left\{x \in \mathbb{R}^2 \middle| MS(f)(x) > \frac{C}{2^1}\right\}$$

and set

$$\mathfrak{R}_1 := \left\{R \in \mathfrak{R} \middle| |R \cap \Omega_1| > \frac{1}{100}|R|\right\};$$

likewise, we define

$$\Omega_2 := \left\{x \in \mathbb{R}^2 \middle| MS(f)(x) > \frac{C}{2^2}\right\}$$

and set

$$\mathfrak{R}_2 := \left\{R \in \mathfrak{R} \setminus \mathfrak{R}_1 \middle| |R \cap \Omega_2| > \frac{1}{100}|R|\right\}.$$

If this process is continued it produces the sets $(\Omega_n)_n$ and $(\mathfrak{R}_n)_n$. Independently, define

$$\Omega_1' := \left\{x \in \mathbb{R}^2 \middle| SM(g)(x) > \frac{C}{2^1}\right\}$$

and set

$$\mathfrak{R}_1' := \left\{R \in \mathfrak{R} \middle| |R \cap \Omega_1'| > \frac{1}{100}|R|\right\};$$

define

$$\Omega_2' := \left\{x \in \mathbb{R}^2 \middle| MS(g)(x) > \frac{C}{2^2}\right\}$$

and set

$$\mathfrak{R}_2' := \left\{R \in \mathfrak{R} \setminus \mathfrak{R}_1' \middle| |R \cap \Omega_2'| > \frac{1}{100}|R|\right\}.$$

This process produces the sets $(\Omega_n')_n$ and $(\mathfrak{R}_n')_n$.

We would now like to obtain a similar decomposition for the function h. To achieve this, we first need to define the analogue of the set Ω_0 for it.

Since $\mathfrak{R}$ is a finite collection of rectangles, one can choose N large enough that for every $R \in \mathfrak{R}$ one has

$$|R \cap \Omega''^{c}_{-N}| > \frac{99}{100}|R|,$$

where

$$\Omega''_{-N} := \{x \in \mathbb{R}^2 | SS(h)(x) > C2^N\}.$$

Then, in a similar way to the previous algorithms, we define

$$\Omega''_{-N+1} := \left\{x \in \mathbb{R}^2 \middle| SSh(x) > \frac{C2^N}{2}\right\}$$

and set

$$\mathfrak{R}''_{-N+1} := \left\{R \in \mathfrak{R} \middle| |R \cap \Omega''_{-N+1}| > \frac{1}{100}|R|\right\},$$

likewise defining

$$\Omega''_{-N+2} = \left\{x \in \mathbb{R}^2 \middle| SSh(x) > \frac{C2^N}{2^2}\right\}$$

and setting

$$\mathfrak{R}''_{-N+2} = \left\{R \in \mathfrak{R}\backslash\mathfrak{R}''_{-N+1} \middle| |R \cap \Omega''_{-N+2}| > \frac{1}{100}|R|\right\}$$

and so on, producing the sets $(\Omega''_n)_n$ and $\{\mathfrak{R}''_n\}_n$.

Then, using these decompositions one can estimate the left-hand side of term I in (3.41) by

$$\sum_{\substack{n_1,n_2>0 \\ n_3>-N}} \sum_{\substack{R\in \\ \mathfrak{R}_{n_1,n_2,n_3}}} \frac{1}{|R|^{1/2}}|\langle f, \varphi_R^1\rangle||\langle g, \varphi_R^2\rangle||\langle h, \varphi_R^3\rangle|, \tag{3.43}$$

where

$$\mathfrak{R}_{n_1,n_2,n_3} = \mathfrak{R}_{n_1} \cap \mathfrak{R}'_{n_2} \cap \mathfrak{R}''_{n_3}.$$

Now, since $R \in \mathfrak{R}_{n_1,n_2,n_3}$, this means in particular that R has not been selected at the previous, $(n_1 - 1)$th, step during the first stopping-time argument, which means that

$$\left|R \cap \Omega_{n_1-1}\right| \le \frac{1}{100}|R|.$$

This implies that

$$|R \cap \Omega^c_{n_1-1}| > \frac{99}{100}|R|.$$

Similarly, one has that

$$|R \cap \Omega'^c_{n_2-1}| > \frac{99}{100}|R|$$

and also that

$$|R \cap \Omega''^c_{n_3-1}| > \frac{99}{100}|R|.$$

These three inequalities imply that

$$\left|R \cap \Omega^c_{n_1-1} \cap \Omega'^c_{n_2-1} \cap \Omega''^c_{n_3-1}\right| > \frac{97}{100}\,|R|\,. \tag{3.44}$$

In particular, (3.43) can be estimated as follows:

$$\sum_{\substack{n_1,n_2>0\\ n_3>-N}} \sum_{R\in\mathfrak{R}_{n_1,n_2,n_3}} \frac{\left|\langle f,\varphi^1_R\rangle\right|}{|R|^{1/2}} \frac{\left|\langle g,\varphi^2_R\rangle\right|}{|R|^{1/2}} \frac{\left|\langle h,\varphi^3_R\rangle\right|}{|R|^{1/2}} |R|$$

$$\lesssim \sum_{\substack{n_1,n_2>0\\ n_3>-N}} \sum_{R\in\mathfrak{R}_{n_1,n_2,n_3}} \frac{|\langle f,\varphi^1_R\rangle|}{|R|^{1/2}} \frac{|\langle g,\varphi^2_R\rangle|}{|R|^{1/2}} \frac{|\langle h,\varphi^3_R\rangle|}{|R|^{1/2}}$$

$$\times \left|R \cap \Omega^c_{n_1-1} \cap \Omega'^c_{n_2-1} \cap \Omega''^c_{n_3-1}\right|$$

$$= \sum_{\substack{n_1,n_2>0\\ n_3>-N}} \int_{\Omega^c_{n_1-1}\cap\Omega'^c_{n_2-1}\cap\Omega''^c_{n_3-1}} \left(\sum_{R\in\mathfrak{R}_{n_1,n_2,n_3}} \frac{|\langle f,\varphi^1_R\rangle|}{|R|^{1/2}} \frac{|\langle g,\varphi^2_R\rangle|}{|R|^{1/2}} \frac{|\langle h,\varphi^3_R\rangle|}{|R|^{1/2}} \chi_R(x) \right) dx$$

$$\lesssim \sum_{\substack{n_1,n_2>0\\ n_3>-N}} \int_{\Omega^c_{n_1-1}\cap\Omega'^c_{n_2-1}\cap\Omega''^c_{n_3-1}} MSf(x)\, SM(g)(x)\, SSh(x)\, dx, \tag{3.45}$$

since the respective functions MS, SM, SS each become larger when their implicit sums run over larger collections of dyadic rectangles. Using the definitions of $\Omega^c_{n_1-1}$, $\Omega'^c_{n_2-1}$, and $\Omega''^c_{n_3-1}$ one can further estimate (3.45) by

$$\sum_{\substack{n_1,n_2>0\\ n_3>-N}} 2^{-n_1}2^{-n_2}2^{-n_3}|\Omega_{n_1,n_2,n_3}|, \tag{3.46}$$

where

$$\Omega_{n_1,n_2,n_3} = \bigcup_{R\in\mathfrak{R}_{n_1,n_2,n_3}} R.$$

However, we have that

$$|\Omega_{n_1,n_2,n_3}| \leq \left| \bigcup_{R \in \Re_{n_1}} R \right| \leq \left| \left\{ x \middle| MM\left(\chi_{\Omega_{n_1}}\right)(x) > \frac{1}{100} \right\} \right|$$
$$\lesssim |\Omega_{n_1}| = \left| \left\{ x \middle| MS(f)(x) > \frac{C}{2^{n_1}} \right\} \right| \lesssim 2^{n_1 p},$$

since $\|f\|_p = 1$.

Similarly, we can estimate $|\Omega_{n_1,n_2,n_3}|$ by $2^{n_2 q}$ and by $2^{n_3 \alpha}$ for every $\alpha > 1$. In particular, we have

$$|\Omega_{n_1,n_2,n_3}| \lesssim 2^{n_1 p \theta_1} 2^{n_2 q \theta_2} 2^{n_3 \alpha \theta_3}$$

for any $\theta_1, \theta_2, \theta_3 \in [0, 1)$ with the property that $\theta_1 + \theta_2 + \theta_3 = 1$. Now we can split our sum in (3.46) as

$$\sum_{\substack{n_1,n_2>0 \\ n_3>0}} 2^{-n_1} 2^{-n_2} 2^{-n_3} |\Omega_{n_1,n_2,n_3}| + \sum_{\substack{n_1,n_2>0 \\ 0 \geq n_3 > -N}} 2^{-n_1} 2^{-n_2} 2^{-n_3} |\Omega_{n_1,n_2,n_3}| . \tag{3.47}$$

In the first case we can simply take $\theta_1 = \theta_2 = \theta_3 = \frac{1}{3}$ to make the sum convergent, while in the second situation one has to pick $\theta_1, \theta_2, \theta_3$ so that $\alpha\theta_3 > 1$, $p\theta_1 < 1$ and $q\theta_2 < 1$ which is possible since p, q are close to 1 while α can be arbitrarily big. This ends the discussion of term I in (3.41).

Estimates for term II in **3.41**

In order to understand how to deal with term *II* in (3.41), let us first address a similar question in the classical, already understood, one-parameter case. In other words let us estimate an expression of the type

$$\sum_I \frac{1}{|I|^{1/2}} |\langle f, \varphi_I^1 \rangle| \, |\langle g, \varphi_I^2 \rangle| \, |\langle h, \varphi_I^3 \rangle| , \tag{3.48}$$

when the sum runs over dyadic intervals within a set Ω that itself is part of an inclusion

$$\Omega_0 \subseteq \Omega \subseteq \widetilde{\Omega} \subseteq \widetilde{\widetilde{\Omega}} \subseteq \widetilde{\widetilde{\widetilde{\Omega}}}.$$

Now we assume that Ω_0 is given by

$$\Omega_0 = \{x | M(f)(x) > C\} \cup \{x | M(g)(x) > C\},$$

while the other sets are defined in precisely the same way. Let $I \subseteq \Omega$ and denoted by $9I$ the interval having the same center as I but nine times as long. One observes that

$$\frac{1}{9|I|}\int_{9I} \chi_{\widetilde{\widetilde{\Omega}}}(x)\,dx \geq \frac{1}{9|I|}\int_{I} \chi_{\widetilde{\widetilde{\Omega}}}(x)\,dx = \frac{|I|}{9|I|} = \frac{1}{9},$$

which implies that $9I \subseteq \widetilde{\widetilde{\Omega}}$.

Denote by $\mathbb{J}$ the set of all dyadic intervals in Ω and, for any $d \geq 2$, denote by $\mathbb{J}_d$ the set of all dyadic intervals I in Ω such that

$$2^d \leq \frac{\operatorname{dist}(I, \widetilde{\widetilde{\Omega}}^c)}{|I|} < 2^{d+1}.$$

Clearly, because of the previous observation one has that

$$J = \bigcup_{d \geq 2} J_d$$

and, as a consequence, (3.48) splits as follows:

$$\sum_{d\geq 2}\sum_{I\in\mathbb{J}_d} \frac{|\langle f, \varphi_I^1\rangle|}{|I|^{1/2}} \frac{|\langle g, \varphi_I^2\rangle|}{|I|^{1/2}} \frac{|\langle h, \varphi_I^3\rangle|}{|I|^{1/2}} |I|.$$

Given the fact that the support of h lies within $\widetilde{\widetilde{\Omega}}^c$, one can estimate the above expression by

$$\sum_{d\geq 2} 2^d 2^d 2^{-10d} \sum_{I\in\mathbb{J}_d} |I|, \tag{3.49}$$

taking into account the definition of Ω_0. Also, it is not difficult to observe that the intervals in $\mathbb{J}_d$ have bounded overlaps and this means that

$$\sum_{I\in\mathbb{J}_d} |I| \lesssim |\Omega| \lesssim 1, \tag{3.50}$$

which makes the whole expression in (3.49) $O(1)$ as desired.

If one tries to apply a similar argument in the biparameter case, one realizes that the inequality analogous to (3.50) is simply false, since this time the rectangles in the corresponding set $\mathfrak{R}_d$ may overlap considerably, as in Figure 3.5. There is, however, a way to get around this difficulty, by using Journé's lemma; this will be described in the next section.

Before doing that, two comments are in order. The first concerns the fact that (as mentioned earlier) the present proof can be adjusted to handle not only discretized paraproducts but also generic biparameter paraproducts. To see this,

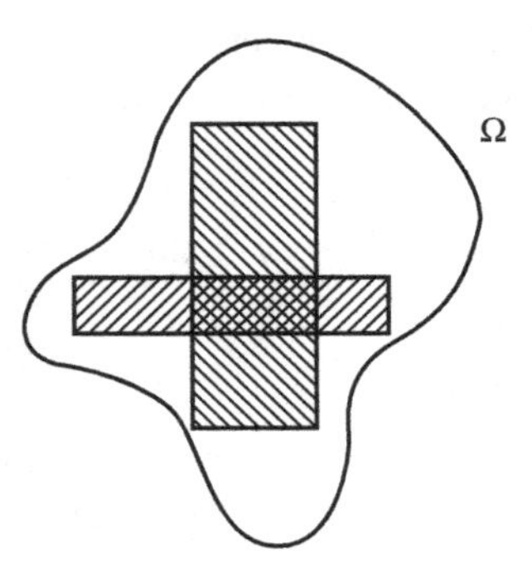

Figure 3.5. Maximal dyadic rectangles in an open set.

recall that every biparameter paraproduct $\vec{\Pi}$ can be written as an average of discretized paraproducts:

$$\vec{\Pi} = \int_{[0,1]^2} \vec{\Pi}^{\gamma}_{\mathfrak{R}} \, d\gamma.$$

To prove (3.38) for $\vec{\Pi}$ one proceeds in the same way. The only difference is in the definition of the exceptional set Ω_0. For two fixed functions f and g as before, one has to define Ω_0 not using the $MS(f)$ and $SM(g)$ functions (now they depend on the averaging parameter γ) but instead by using the more natural $\sup_\gamma MS_\gamma(f)$ and $\sup_\gamma SM_\gamma(g)$ functions, which are still bounded operators in L^p for every $1 < p < \infty$.

The second comment concerns the proof of Theorem 3.1 itself. We will learn later (in the last two sections of the chapter) that, modulo some technical adjustments, one can organize the proof in such a way that the term II in (3.41) simply disappears. The reason why we have decided to take this longer route first is partly due to its *mathematical scenery*, which we hope the reader will enjoy.

3.6. Journé's lemma

As we pointed out before, the next statement, *Journé's lemma*, will allow us to obtain a weaker (but still very useful) variant of the inequality (3.50).

Lemma 3.5 *Let $\Omega \subseteq R^2$ be an open set and k a fixed positive integer. Write $\widetilde{\Omega} := \{x | MM(\chi_\Omega)(x) > \frac{1}{2}\}$. Let $\mathfrak{R}$ be a collection of dyadic rectangles $R = I \times J \subseteq \Omega$ that are maximal with respect to inclusion in the Oy direction* (see Figure 3.5). *Assume that for each $R = I \times J \in \mathfrak{R}$ one has $2^k I \times J \subseteq \widetilde{\Omega}$ and that k is maximal with this property (here $2^k I$ is the unique dyadic interval of*

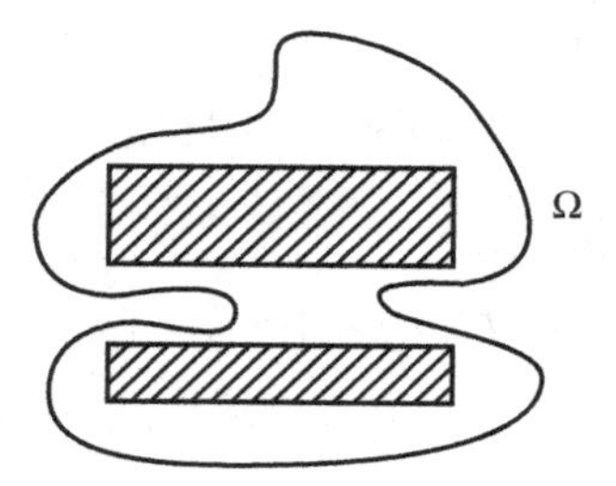

Figure 3.6. An example of a set of the type $K \times E_K$.

length $2^k|I|$ containing I). Then one has

$$\sum_{R \in \mathfrak{R}} |R| \lesssim 2^{k\varepsilon} |\Omega| \tag{3.51}$$

for every $\varepsilon > 0$, the implicit constant depending on $\varepsilon > 0$.

Proof If K is any dyadic interval, we write

$$E_K := \bigcup_{K \times J \subseteq \Omega} J,$$

where the intervals J are dyadic. In particular, one always has $K \times E_K \subseteq \Omega$ (see Figure 3.6).

Fix $I \times J \in \mathfrak{R}$. Since $2^{k+1}I \times J \not\subseteq \widetilde{\Omega}$ it follows that there exists $x_0 \in \widetilde{\Omega}^c$ such that $x_0 \in 2^{k+1}I \times J$. One has

$$\frac{1}{\left|2^{k+1}I \times J\right|} \int_{2^{k+1}I \times J} \chi_\Omega(x)dx \leq \tfrac{1}{2},$$

which implies that

$$\left|\left(2^{k+1}I \times J\right) \cap \Omega\right| \leq \tfrac{1}{2}\left|2^{k+1}I \times J\right|$$

and in particular

$$|(2^{k+1}I \times J) \cap (2^{k+1}I \times E_{2^{k+1}I})| \leq \tfrac{1}{2}|2^{k+1}I \times J| \tag{3.52}$$

since $2^{k+1}I \times E_{2^{k+1}I} \subseteq \Omega$, as pointed out earlier.

In particular (3.52) implies that

$$\left|2^{k+1}I \times (J \cap E_{2^{k+1}I})\right| \leq \tfrac{1}{2}\left|2^{k+1}I \times J\right|$$

and so

$$\left|2^{k+1}I\right| |J \cap E_{2^{k+1}I}| \leq \tfrac{1}{2}\left|2^{k+1}I\right| |J|,$$

from which we deduce that

$$|J \cap E_{2^{k+1}I}| \le \tfrac{1}{2}|J|.$$

This implies that

$$|J \setminus E_{2^{k+1}I}| > \tfrac{1}{2}|J|,$$

in other words,

$$|J| < 2|J \setminus E_{2^{k+1}I}|. \tag{3.53}$$

Using this information, one can write

$$\sum_{R\in\mathfrak{R}} |R| = \sum_{I\times J\in\mathfrak{R}} |I|\,|J| \lesssim \sum_{I\times J\in\mathfrak{R}} |I|\,|J\setminus E_{2^{k+1}I}| = \sum_{I} |I| \sum_{J:I\times J\in\mathfrak{R}} |J\setminus E_{2^{k+1}I}|. \tag{3.54}$$

Since for every I the corresponding rectangles $I \times J$ are all disjoint (by their maximality in their Oy direction), we can estimate (3.54) further by

$$\sum_{I} |I|\,|E_I\setminus E_{2^{k+1}I}|. \tag{3.55}$$

Then, since $I \subseteq 2I \subseteq 2^2 I \subseteq \cdots \subseteq 2^k I \subseteq 2^{k+1} I$, it follows that $E_I \supseteq E_{2I} \supseteq \cdots \supseteq E_{2^{k+1}I}$ and, as a consequence, (3.55) can be estimated by

$$\begin{aligned}
&\sum_{I} |I|(|E_I\setminus E_{2I}| + |E_{2I}\setminus E_{2^2 I}| + \cdots + |E_{2^k I}\setminus E_{2^{k+1}I}|) \\
&= \sum_{j=0}^{k}\sum_{I} |I|\,|E_{2^j I}\setminus E_{2^{j+1}I}| \\
&= \sum_{j=0}^{k} 2^{-j}\sum_{I} |2^j I|\,|E_{2^j I}\setminus E_{2^{j+1}I}|.
\end{aligned} \tag{3.56}$$

Now, it is not difficult to observe that in general

$$\sum_{I} |I|\,|E_I\setminus E_{2I}| \le |\Omega| \tag{3.57}$$

since

$$\bigcup_{I} I \times (E_I\setminus E_{2I}) \subseteq \Omega$$

and all the sets $I \times (E_I\setminus E_{2I})$ are disjoint. Since in (3.56) we are summing over I and not over $2^j I$, if one takes into account the fact that there are 2^j intervals

K for which $2^j K = 2^j I$ in the same length of I, one can estimate (3.56) by

$$\sum_{j=0}^{k} 2^{-j} 2^j \sum_{2^j I} |2^j I|\,|E_{2^j I} \backslash E_{2^{j \times 1} I}| \lesssim \sum_{j=0}^{k} |\Omega| = (k+1)|\Omega| \lesssim 2^{k\varepsilon}|\Omega|,$$

as desired. □

Corollary 3.6 *Let $\Omega \subseteq \mathbb{R}^2$ be an open set as before and $d \geq 0$ a fixed integer. Assume also that $\mathfrak{R}$ is a collection of dyadic rectangles which are maximal with respect to inclusion and which all lie in Ω.*

Suppose that for each $R \in \mathfrak{R}$ one has $2^d R \subseteq \widetilde{\widetilde{\Omega}}$ and that d is maximal with this property ($2^d R := 2^d I \times 2^d J$ if $R = I \times J$). Then

$$\sum_{R \varepsilon \mathfrak{R}} |R| \lesssim 2^{d\varepsilon}|\Omega|, \tag{3.58}$$

where, as before, the implicit constant depends on ε.

Proof Let us start recalling that $\widetilde{\widetilde{\Omega}}$ is obtained from Ω by performing the tilde operation twice, see Lemma 3.5. The idea is to apply Journé's lemma twice, taking advantage first of the maximality in the Oy direction and then of the maximality in the Ox direction. Fix $R \in \mathfrak{R}$, $R = I \times J$. Clearly, there exists a maximal integer $k \geq 0$ such that $2^k I \times J \subseteq \widetilde{\Omega}$. Denote by $\mathfrak{R}_k$ the collection of all such $R \in \mathfrak{R}$.

In this way one obtains a natural decomposition of $\mathfrak{R}$:

$$\mathfrak{R} = \bigcup_{k=0}^{\infty} \mathfrak{R}_k = \left(\bigcup_{k=0}^{d} \mathfrak{R}_k \right) \cup \left(\bigcup_{k=d+1}^{\infty} \mathfrak{R}_k \right). \tag{3.59}$$

Using Journé's lemma, we obtain

$$\sum_{R \in \bigcup_{k=0}^{d} \mathfrak{R}_k} |R| \leq \sum_{k=0}^{d} \sum_{R \in \mathfrak{R}_k} |R| \lesssim \sum_{k=0}^{d} 2^{k\varepsilon}\,|\Omega| \lesssim 2^{\varepsilon d}\,|\Omega|,$$

which satisfies (3.58).

We are left with estimating the contribution of $\bigcup_{k=d+1}^{\infty} \mathfrak{R}_k$. If $R = I \times J \in \bigcup_{k=d+1}^{\infty} \mathfrak{R}_k$ then there is a unique maximal k such that $2^k I \times J \subseteq \widetilde{\Omega}$. Denote by $\mathfrak{R}'$ the collection of all such *dilated* rectangles $2^k I \times J$. Clearly, there are in principle at most 2^k rectangles $I \times J$ having the same dilation, but they are all disjoint. If we denote by $\mathfrak{R}'_{\text{distinct}}$ the collection of all distinct rectangles in

$\mathfrak{R}'$ then it is easy to see that on the one hand we have

$$\sum_{R \in \bigcup_{k=d+1}^{\infty} \mathfrak{R}_k} |R| \leq \sum_{R \in \mathfrak{R}'_{\text{distinct}}} |R|. \tag{3.60}$$

On the other hand, each $2^k I \times J$ in $\mathfrak{R}'_{\text{distinct}}$ is maximal in $\widetilde{\Omega}$ in the Ox direction and also there must be an integer $\alpha \geq 0$ for which $2^k I \times 2^\alpha J \subseteq \widetilde{\widetilde{\Omega}}$ and such that α is maximal with this property. Obviously α must be smaller than d, given the assumption of the corollary and the fact that k is at least as large as $d+1$. This allows us to split $\mathfrak{R}'_{\text{distinct}}$ as follows:

$$\mathfrak{R}'_{\text{distinct}} = \bigcup_{\alpha=0}^{d} \mathfrak{R}'^{\alpha}_{\text{distinct}},$$

where $\mathfrak{R}'^{\alpha}_{\text{distinct}}$ is defined in a natural way, as before. As a consequence of Journé's lemma we have

$$\sum_{R \in \mathfrak{R}'_{\text{distinct}}} |R| \leq \sum_{\alpha=0}^{d} \sum_{R \in \mathfrak{R}'^{\alpha}_{\text{distinct}}} |R| \lesssim \sum_{\alpha=0}^{d} 2^{\alpha\varepsilon} |\widetilde{\Omega}| \lesssim 2^{d\varepsilon} |\Omega|,$$

using the boundedness of the maximal operator MM. □

3.7. Proof of Theorem 3.1; part 2

We now return to the problem of estimating term II in (3.41). Recall that

$$II = \sum_{R \subseteq \Omega} \frac{1}{|R|^{1/2}} |\langle f, \varphi_R^1 \rangle| \, |\langle g, \varphi_R^2 \rangle| \, |\langle h, \varphi_R^3 \rangle|.$$

Recall also the tower of inclusions $\Omega \subseteq \widetilde{\Omega} \subseteq \widetilde{\widetilde{\Omega}} \subseteq \widetilde{\widetilde{\widetilde{\Omega}}}$, which now plays an important role. For every rectangle $R \subseteq \Omega$ there exists $R \subseteq \widetilde{R} \subseteq \Omega$ such that $\widetilde{R}$ is maximal with this property. Since we are working with dyadic rectangles, note that for a given R there could exist more than one maximal $\widetilde{R}$ having the above property. Also, it might well happen that distinct rectangles R could generate the same maximal rectangle $\widetilde{R}$. Regardless of these facts, we will consider the family of these maximal rectangles $\widetilde{R}$ and we collect all those that are distinct in a set $\mathfrak{R}_{\max}$. For any integer $d \geq 0$ we denote by $\mathfrak{R}^d_{\max}$ the set of all $\widetilde{R} \in \mathfrak{R}_{\max}$ for which $2^d \widetilde{R} \subseteq \widetilde{\widetilde{\Omega}}$, where d is maximal with this property. By Corollary 3.6, we know that

$$\sum_{\widetilde{R} \subset \mathfrak{R}^d_{\max}} |\widetilde{R}| \lesssim 2^{\varepsilon d} |\Omega|. \tag{3.61}$$

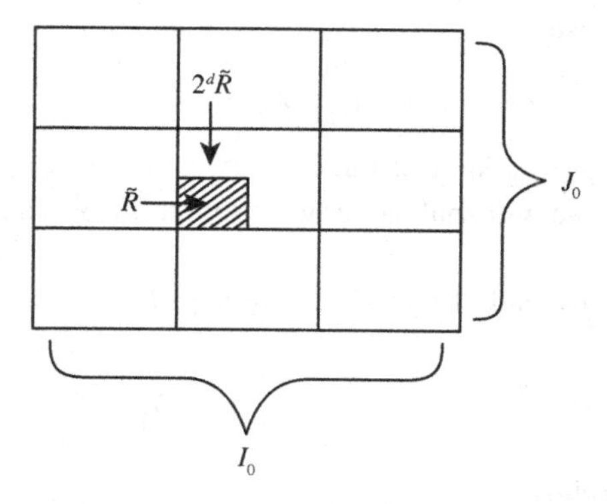

Figure 3.7

Fix such a $d \geq 0$ and $\tilde{R} \in \Re^d_{\max}$. We claim that one has the estimate

$$\sum_{R\subseteq\tilde{R}} \frac{1}{|R|^{1/2}} |\langle f, \varphi_R^1\rangle| \, |\langle g, \varphi_R^2\rangle| \, |\langle h, \varphi_R^3\rangle| \lesssim 2^{-Nd}|\tilde{R}|, \tag{3.62}$$

for a large constant $N > 0$.

Clearly, if we assume this claim then we can estimate term II by

$$\sum_{d=0}^{\infty} \sum_{\tilde{R}\in\Re^d_{\max}} \sum_{R\subseteq\tilde{R}} \frac{1}{|R|^{1/2}} |\langle f, \varphi_R^1\rangle| \, |\langle g, \varphi_R^2\rangle| \, |\langle h, \varphi_R^3\rangle|$$

$$\lesssim \sum_{d=0}^{\infty} \sum_{\tilde{R}\in\Re^d_{\max}} 2^{-Nd}|\tilde{R}| \lesssim \sum_{d=0}^{\infty} 2^{-Nd}2^{\varepsilon d}|\Omega| \lesssim |\Omega| \lesssim 1,$$

as desired. It is therefore enough to treat (3.62).

Fix $\tilde{R} := \tilde{I} \times \tilde{J} \subseteq \Omega$, $\tilde{R} \in \Re^d_{\max}$. Consider also the rectangle $I_0 \times J_0$, as in Figure 3.7, whose area is nine times that of $2^d\tilde{R}$. Clearly, since $2^d\tilde{R} \subseteq \tilde{\tilde{\Omega}}$ we have that

$$\frac{1}{|I_0 \times J_0|} \int_{I_0\times J_0} \chi_{\tilde{\tilde{\Omega}}}(z)\,dz \geq \frac{1}{|I_0 \times J_0|} \int_{2^d\tilde{R}} \chi_{\tilde{\tilde{\Omega}}}(z)\,dz = \frac{1}{|I_0 \times J_0|} |2^d\tilde{R}| = \tfrac{1}{9}$$

and, as a consequence, $I_0 \times J_0 \subseteq \tilde{\tilde{\tilde{\Omega}}}$. This is important, since now we know that

$$(I_0 \times J_0) \cap E' = \emptyset. \tag{3.63}$$

In particular, we have that

$$\chi_{E'} = \chi_{E'}\chi_{(I_0\times J_0)^c}.$$

Since we can also write

$$\chi_{(I_0\times J_0)^c} = \chi_{I_0^c} + \chi_{J_0^c} - \chi_{I_0^c}\chi_{J_0^c},$$

we can split $h = \chi_{E'}$ as a sum of three terms, and this allows us to estimate (3.62) accordingly. We will analyze only the first term, $\chi_{I_0^c}$, since the other two are similar.

So from now on, instead of h in (3.62) we have $h\chi_{I_0^c} = \chi_{E'}\chi_{I_0^c}$.

Next, write

$$\mathbb{L} := \{I | R = I \times J \subseteq \widetilde{R}\}.$$

Then split $\mathbb{L}$ as follows:

$$\mathbb{L} = \bigcup_{d_1\geq 0} \mathbb{L}_{d_1},$$

where

$$\mathbb{L}_{d_1} = \left\{ K \in \mathbb{L} \,\middle|\, \frac{|\widetilde{I}|}{|K|} \simeq 2^{d_1} \right\}.$$

Then, we can decompose the left-hand side of (3.62) as

$$\begin{aligned}
&\sum_{d_1\geq 0}\sum_{K\in\mathbb{L}_{d_1}}\sum_{\substack{R=I\times J:\\ I=K}} \frac{1}{|R|^{1/2}} |\langle f,\varphi_R^1\rangle|\,|\langle g,\varphi_R^2\rangle|\,|\langle h,\varphi_R^3\rangle| \\
&= \sum_{d_1\geq 0}\sum_{K\in\mathbb{L}_{d_1}}\sum_{R=K\times J\subseteq\widetilde{R}} |K| \frac{1}{|J|^{1/2}} \left|\left\langle \frac{\langle f,\varphi_K^1\rangle}{|K|^{1/2}},\varphi_J^1\right\rangle\right| \\
&\quad\times \left|\left\langle \frac{\langle g,\varphi_K^2\rangle}{|K|^{1/2}},\varphi_J^2\right\rangle\right| \left|\left\langle \frac{\langle h,\varphi_K^3\rangle}{|K|^{1/2}},\varphi_J^3\right\rangle\right| \\
&= \sum_{d_1\geq 0}\sum_{K\in\mathbb{L}_{d_1}}\sum_{J\in\mathbb{J}_K} |K| \frac{1}{|J|^{1/2}} \left|\left\langle \frac{\langle f,\varphi_K^1\rangle}{|K|^{1/2}},\varphi_J^1\right\rangle\right| \\
&\quad\times \left|\left\langle \frac{\langle g,\varphi_K^2\rangle}{|K|^{1/2}},\varphi_J^2\right\rangle\right| \left|\left\langle \frac{\langle h,\varphi_K^3\rangle}{|K|^{1/2}},\varphi_J^3\right\rangle\right|,
\end{aligned}$$

where

$$\mathbb{J}_K := \{J | K \times J \subseteq \widetilde{R}\}.$$

Now split $\mathbb{J}_K$:

$$\mathbb{J}_K = \bigcup_{d_2\geq 0} \mathbb{J}_K^{d_2},$$

with each $\mathbb{J}_K^{d_2}$ given by

$$\mathbb{J}_K^{d_2} := \left\{ J \in \mathbb{J}_K \,\middle|\, \frac{|\widetilde{J}|}{|J|} \simeq 2^{d_2} \right\}.$$

In particular, our expression can be rewritten as

$$\begin{aligned}
&\sum_{d_1 \geq 0} \sum_{K \in \mathbb{L}_{d_1}} |K| \sum_{d_2 \geq 0} \sum_{J \in \mathbb{J}_K^{d_2}} \frac{1}{|J|^{1/2}} \left| \left\langle \frac{\langle f, \varphi_K^1 \rangle}{|K|^{1/2}}, \varphi_J^1 \right\rangle \right| \left| \left\langle \frac{\langle g, \varphi_K^2 \rangle}{|K|^{1/2}}, \varphi_J^2 \right\rangle \right| \\
&\quad \times \left| \left\langle \frac{\langle h, \varphi_K^3 \rangle}{|K|^{1/2}}, \varphi_J^3 \right\rangle \right| \\
&\quad = \sum_{d_1 \geq 0} \sum_{K \in \mathbb{L}_{d_1}} |K| \sum_{J \in \bigcup_{d_2 < d_1} \mathbb{J}_K^{d_2}} \frac{1}{|J|^{1/2}} \left| \left\langle \frac{\langle f, \varphi_K^1 \rangle}{|K|^{1/2}}, \varphi_J^1 \right\rangle \right| \left| \left\langle \frac{\langle g, \varphi_K^2 \rangle}{|K|^{1/2}}, \varphi_J^2 \right\rangle \right| \\
&\quad \times \left| \left\langle \frac{\langle h, \varphi_K^3 \rangle}{|K|^{1/2}}, \varphi_J^3 \right\rangle \right| + \sum_{d_1 \geq 0} \sum_{K \in \mathbb{L}_{d_1}} |K| \sum_{J \in \bigcup_{d_2 \geq d_1} \mathbb{J}_K^{d_2}} \frac{1}{|J|^{1/2}} \\
&\quad \times \left| \left\langle \frac{\langle f, \varphi_K^1 \rangle}{|K|^{1/2}}, \varphi_J^1 \right\rangle \right| \left| \left\langle \frac{\langle g, \varphi_K^2 \rangle}{|K|^{1/2}}, \varphi_J^2 \right\rangle \right| \left| \left\langle \frac{\langle h, \varphi_K^3 \rangle}{|K|^{1/2}}, \varphi_J^3 \right\rangle \right| := A + B.
\end{aligned} \tag{3.64}$$

Observe now that, for each K, the *inner* sums are one-parameter discretized paraproducts, for which the generic Proposition 2.12 can be applied. To estimate A we note that

$$\operatorname{size}_{\bigcup_{d_2<d_1} \mathbb{J}_K^{d_2}} \left(\left(\left\langle \frac{\langle f, \varphi_K^1 \rangle}{|K|^{1/2}}, \varphi_J^1 \right\rangle \right)_J \right) \lesssim 2^{2d_1+2d},$$

$$\operatorname{size}_{\bigcup_{d_2<d_1} \mathbb{J}_K^{d_2}} \left(\left(\left\langle \frac{\langle g, \varphi_K^2 \rangle}{|K|^{1/2}}, \varphi_J^2 \right\rangle \right)_J \right) \lesssim 2^{2d_1+2d},$$

and

$$\operatorname{size}_{\bigcup_{d_2<d_1} \mathbb{J}_K^{d_2}} \left(\left(\left\langle \frac{\langle h, \varphi_K^3 \rangle}{|K|^{1/2}}, \varphi_J^3 \right\rangle \right)_J \right) \lesssim 2^{-N(d_1+d)},$$

for a large constant $N > 0$. Similarly, we have

$$\text{energy}_{\bigcup_{d_2<d_1}\mathbb{J}_K^{d_2}}\left(\left(\left\langle \frac{\langle f, \varphi_K^1\rangle}{|K|^{1/2}}, \varphi_J^1\right\rangle\right)_J\right) \lesssim 2^{2d_1+2d}|\widetilde{J}|,$$

$$\text{energy}_{\bigcup_{d_2<d_1}\mathbb{J}_K^{d_2}}\left(\left(\left\langle \frac{\langle g, \varphi_K^2\rangle}{|K|^{1/2}}, \varphi_J^2\right\rangle\right)_J\right) \lesssim 2^{2d_1+2d}|\widetilde{J}|,$$

and also

$$\text{energy}_{\bigcup_{d_2<d_1}\mathbb{J}_K^{d_2}}\left(\left(\left\langle \frac{\langle h, \varphi_K^3\rangle}{|K|^{1/2}}, \varphi_J^3\right\rangle\right)_J\right) \lesssim 2^{-N(d_1+d)}|\widetilde{J}|.$$

These estimates can then be used along with Proposition 2.12; they allow us, since N is an arbitrarily large constant, to estimate A by

$$\sum_{d_1\geq 0}\sum_{K\in\mathbb{L}_{d_1}} |K|\, 2^{2d_1+2d}2^{2d_1+2d}2^{-N(d_1+d)}|\widetilde{J}| \lesssim 2^{-(N-4)d}|\widetilde{J}|\,|\widetilde{I}| = 2^{-(N-4)d}|\widetilde{R}|$$

as required by the claim (3.62).

To estimate B, the argument above does not work directly. Instead, one has first to decompose $\bigcup_{d_2\geq d_1}\mathbb{J}_K^{d_2}$ further, as follows:

$$\bigcup_{d_2\geq d_1}\mathbb{J}_K^{d_2} = \bigcup_{d_3}\mathbb{J}_{K,d_3},$$

where

$$\mathbb{J}_{K,d_3} := \left\{ J \in \bigcup_{d_2\geq d_1}\mathbb{J}_K^{d_2} \,\middle|\, \mathrm{Dil}_{2^{d_3}}(\mathrm{K}\times\mathrm{J}) \subseteq \widetilde{\widetilde{\widetilde{\Omega}}} \right\}$$

and d_3 is maximal with this property. By $\mathrm{Dil}_{2^{d_3}}(K\times J)$ we denote the rectangle having the same center as the original $K\times J$ but whose sidelength is 2^{d_3} times greater.

It is then not difficult to observe that one has the constraint

$$d_1 + d \leq d_3. \tag{3.65}$$

To see this, let us first observe that $2^{d_1}(K\times J)\subseteq \widetilde{I}\times\widetilde{J} = \widetilde{R}$. In particular, this implies that on the one hand $2^{d_1+d}(K\times J)\subseteq 2^d\widetilde{R}$. On the other hand, $2^d\widetilde{R}$ is a subset of $I_0\times J_0$ and this implies that $\mathrm{Dil}_{2^{d_1+d}}(K\times J)\subseteq I_0\times J_0\subseteq \widetilde{\widetilde{\widetilde{\Omega}}}$, which is enough to conclude (3.65).

Now we rewrite B as follows:

$$\sum_{d_1\geq 0}\sum_{K\in\mathbb{L}_{d_1}}|K|\sum_{d_3}\sum_{J\in\mathbb{J}_{K,d_3}}$$
$$\times\frac{1}{|J|^{1/2}}\left|\left\langle\frac{\langle f,\varphi_K^1\rangle}{|K|^{1/2}},\varphi_J^1\right\rangle\right|\left|\left\langle\frac{\langle g,\varphi_K^2\rangle}{|K|^{1/2}},\varphi_J^2\right\rangle\right|\left|\left\langle\frac{\langle h,\varphi_K^3\rangle}{|K|^{1/2}},\varphi_J^3\right\rangle\right|. \quad (3.66)$$

To estimate the factors in (3.66) we use Proposition 2.12 again. The corresponding sizes and energies can now be estimated by

$$\text{size}_{\mathbb{J}_{K,d_3}}\left(\left(\left\langle\frac{\langle f,\varphi_K^1\rangle}{|K|^{1/2}},\varphi_J^1\right\rangle\right)_J\right)\lesssim 2^{2d_3},$$
$$\text{size}_{\mathbb{J}_{K,d_3}}\left(\left(\left\langle\frac{\langle g,\varphi_K^2\rangle}{|K|^{1/2}},\varphi_J^2\right\rangle\right)_J\right)\lesssim 2^{2d_3},$$

and

$$\text{size}_{\mathbb{J}_{K,d_3}}\left(\left(\left\langle\frac{\langle h,\varphi_K^3\rangle}{|K|^{1/2}},\varphi_J^3\right\rangle\right)_J\right)\lesssim 2^{-Nd_3},$$

again for an arbitrarily large N, this time using the fact that E' lies in the complement of $\widetilde{\widetilde{\widetilde{\Omega}}}$. Similarly, we have

$$\text{energy}_{\mathbb{J}_{K,d_3}}\left(\left(\left\langle\frac{\langle f,\varphi_K^1\rangle}{|K|^{1/2}},\varphi_J^1\right\rangle\right)_J\right)\lesssim 2^{2d_3}|\widetilde{J}|,$$
$$\text{energy}_{\mathbb{J}_{K,d_3}}\left(\left(\left\langle\frac{\langle g,\varphi_K^2\rangle}{|K|^{1/2}},\varphi_J^2\right\rangle\right)_J\right)\lesssim 2^{2d_3}|\widetilde{J}|,$$

and

$$\text{energy}_{\mathbb{J}_{K,d_3}}\left(\left(\left\langle\frac{\langle h,\varphi_K^3\rangle}{|K|^{1/2}},\varphi_J^3\right\rangle\right)_J\right)\lesssim 2^{-Nd_3}|\widetilde{J}|.$$

Using all these inequalities, one can estimate (3.66) further by

$$\sum_{d_1\geq 0}\sum_{K\in\mathbb{L}_{d_1}}|K|\sum_{d_3\geq d_1+d}2^{2d_3}2^{2d_3}2^{-Nd_3}|\widetilde{J}|$$
$$\lesssim\sum_{d_1\geq 0}2^{-(N-4)(d_1+d)}|\widetilde{I}|\,|\widetilde{J}|\lesssim 2^{-(N-4)d}|\widetilde{R}|,$$

and this ends our proof.

Exercise 3.2 Check carefully all the size and energy estimates needed in the above proof.

3.8. Multiparameter paraproducts

As in the classical one-parameter case, there is also a generic multiplier variant of Theorem 3.1.

Let $m(\xi, \eta)$ be a symbol in $\mathbb{R}^2 \times \mathbb{R}^2$ that is smooth away from the planes $(\xi_1, \eta_1) = (0, 0)$ and $(\xi_2, \eta_2) = (0, 0)$ and satisfying the *Marcinkiewicz* condition

$$\left|\partial_{\xi_1}^{\alpha_1} \partial_{\xi_2}^{\alpha_2} \partial_{\eta_1}^{\beta_1} \partial_{\eta_2}^{\beta_2} m(\xi, \eta)\right| \lesssim \frac{1}{|(\xi_1, \eta_1)|^{\alpha_1+\beta_1}} \frac{1}{|(\xi_2, \eta_2)|^{\alpha_2+\beta_2}} \tag{3.67}$$

for sufficiently many multi-indices α, β. Define the bilinear operator T_m by

$$T_m^{(2)}(f, g)(x) = \int_{\mathbb{R}^4} m(\xi, \eta)\widehat{f}(\xi)\widehat{g}(\eta)e^{2\pi i x(\xi+\eta)}\, d\xi d\eta. \tag{3.68}$$

Theorem 3.7 *The operator $T_m^{(2)}$ extends to a bounded bilinear operator from $L^p(\mathbb{R}^2) \times L^q(\mathbb{R}^2)$ into $L^r(\mathbb{R}^2)$ provided that $1 < p, q \leq \infty$, $1/p + 1/q = 1/r$, and $0 < r < \infty$.*

Theorem 3.7 can be reduced to Theorem 3.1 in precisely the way in which the one-parameter Coifman–Meyer theorem was reduced to its corresponding discrete variant, in Chapter 2.

Exercise 3.3 Check carefully the details of this reduction.

Note that since every biparameter paraproduct is a tensor product of two one-parameter paraproducts, its corresponding symbol, see (3.68), is of the form

$$m(\xi, \eta) = \widetilde{m}(\xi_1, \eta_1)\widetilde{\widetilde{m}}(\xi_2, \eta_2),$$

where both $\widetilde{m}$ and $\widetilde{\widetilde{m}}$ are classical symbols, and clearly satisfies the product structure condition (3.67). It is then important to point out the fact that operators such as $T_m^{(2)}$ commute with biparameter dilations. More precisely, let us first observe that if one denotes by m_{λ_1,λ_2} the *dilated* symbol $m_{\lambda_1,\lambda_2}(\xi, \eta) := m(\xi/\lambda_1, \eta/\lambda_2)$ then m_{λ_1,λ_2} satisfies (3.67) as well, uniformly in $\lambda_1, \lambda_2 > 0$. A simple calculation shows that if one dilates both functions f and g in $T_m^{(2)}(f, g)$ by the same parameters (λ_1, λ_2) then, modulo replacing m by m_{λ_1,λ_2}, this is the same as dilating $T_m^{(2)}(f, g)$ by (λ_1, λ_2).

Exercise 3.4 Check that if m is given by $m(\xi,\eta) = \widetilde{m}(\xi_1,\eta_1)\widetilde{\widetilde{m}}(\xi_2,\eta_2)$ as before then $T_m^{(2)}(f,g)$ can be written as

$$T_m^{(2)}(f,g)(x_1,x_2) = p.v. \int_{\mathbb{R}^4} f(x_1-t_1,x_2-t_2)g(x_1-s_1,x_2-s_2) \times \widetilde{K}(t_1,s_1)\widetilde{\widetilde{K}}(t_2,s_2)\,dt_1dt_2ds_1ds_2$$

where the two Calderón–Zygmund kernels in the above expression are the inverse Fourier transforms of $\widetilde{m}$ and $\widetilde{\widetilde{m}}$ respectively.

It is also natural to ask whether there is a *d-parameter* n-linear generalization of Theorem 3.7. To be more specific, suppose that $d \geq 1$ is a fixed integer. In general, any collection of n vectors in $\mathbb{R}^d$, $\xi_1 = (\xi_1^i)_{i=1}^d, \xi_2 = (\xi_2^i)_{i=1}^d, \ldots, \xi_n = (\xi_n^i)_{i=1}^d$, generates the following collection of d vectors in $\mathbb{R}^n$:

$$\bar{\xi}_1 = (\xi_j^1)_{j=1}^n, \quad \bar{\xi}_2 = (\xi_j^2)_{j=1}^n, \quad \ldots, \quad \bar{\xi}_d = (\xi_j^d)_{j=1}^n.$$

Assume now that $m = m(\xi) = m(\bar{\xi})$ is a bounded symbol in $L^\infty(\mathbb{R}^{dn})$ that is smooth away from the subspaces $\{\bar{\xi}_1 = 0\} \cup \cdots \cup \{\bar{\xi}_d = 0\}$ and satisfying

$$\left|\partial_{\bar{\xi}_1}^{\alpha_1} \cdots \partial_{\bar{\xi}_d}^{\alpha_d} m(\bar{\xi})\right| \lesssim \prod_{i=1}^{d} \frac{1}{|\bar{\xi}_i|^{|\alpha_i|}} \tag{3.69}$$

for sufficiently many multi-indices $\alpha_1, \ldots, \alpha_d$. Denote by $T_m^{(d)}$ the n-linear multiplier defined by

$$T_m^d(f_1,\ldots,f_n)(x) = \int_{\mathbb{R}^{dn}} m(\xi)\widehat{f_1}(\xi_1)\cdots\widehat{f_n}(\xi_n)e^{2\pi i x(\xi_1+\cdots+\xi_n)}\,d\xi. \tag{3.70}$$

The following theorem holds.

Theorem 3.8 *The multiplier $T_m^{(d)}$ extends to a bounded n-linear operator from $L^{p_1}(\mathbb{R}^d) \times \cdots \times L^{p_n}(\mathbb{R}^d)$ into $L^p(\mathbb{R}^d)$ provided that $1/p_1 + \cdots + 1/p_n = 1/p$, $1 < p_1, \ldots, p_n \leq \infty$ and $0 < p < \infty$.*

Arguments that should be by now standard allow one to reduce this theorem to its discrete variant, which is a generalization of Theorem 3.1. The *discretized d-parameter paraproducts* are now expressions of the form

$$\sum_R \frac{1}{|R|^{(n-1)/2}} \langle f_1, \varphi_R^1\rangle \cdots \langle f_n, \varphi_R^n\rangle \varphi_R^{n+1},$$

where the sum runs over the dyadic parallelepipeds $R = I_1 \times \cdots \times I_d \subseteq \mathbb{R}^d$.

Clearly, in order to understand these paraproducts, one first needs to understand the L^p-boundedness properties of the various d-parameter square and maximal functions together with their hybrids that generalize in a natural way the operators MM, SS, MS, and SM appearing in the biparameter bilinear case. This turns out to be not too difficult, since a similar argument based on Fubini's theorem and the Fefferman–Stein inequality reduces the problem to a one-parameter situation. These are the only observations that one needs to make to be able to prove the Banach case of the above theorem.

The reason why the previous biparameter proof (of the general case) does not generalize easily to the d-parameter setting is partly related to the Journé's lemma part of the argument. The d-parameter generalization of the estimation of term *II* is not a routine task, as the reader can check. Our goal for the rest of the chapter is to present a simplified version of the whole biparameter proof that can be generalized in a straightforward manner to the d-parameter situation, for any $d \geq 2$. As we will see, one can avoid the use of Journé's lemma completely.

3.9. Proof of Theorem 3.1; a simplification

Let us therefore consider once more the discrete expression

$$\sum_R \frac{1}{|R|^{1/2}} \langle f, \varphi_R^1 \rangle \langle g, \varphi_R^2 \rangle \varphi_R^3$$

and reprove Theorem 3.1 for it.

Fix as before p, q greater than, but close to, 1; f, g such that $\|f\|_p = \|g\|_q = 1$; and $E \subseteq \mathbb{R}^2$ a measurable set with $|E| = 1$. Our task is to show the existence of a subset $E' \subseteq E$ of comparable measure $|E'| \simeq 1$ such that

$$\sum_R \frac{1}{|R|^{1/2}} |\langle f, \varphi_R^1 \rangle| \, |\langle g, \varphi_R^2 \rangle| \, |\langle h, \varphi_R^3 \rangle| \lesssim 1, \tag{3.71}$$

where $h := \chi_{E'}$.

We would like to pause for a moment to motivate and describe what is coming up next. Suppose that one considers the Haar analogue of inequality (3.71). In other words, assume that all the implicit functions $(\varphi_I)_I$ are replaced with L^2-normalized Haar functions $(h_I)_I$ in the *lacunary* case and with L^2-normalized characteristic functions $(\chi_I/|I|^{1/2})_I$ in the *nonlacunary* case. The reader may recall the Haar system from the first volume of the book. Then, if one applies the method of proving Theorem 3.1 to this new Haar trilinear form, one imediately observes that term *II* simply disappears since all the functions $(\varphi_R^3)_R$ are now supported on the rectangles R while the support of h is in the complement of the set Ω. Since part 1 of the proof (see Section 3.5) is easily extendable to the general d-parameter case, this shows that Haar d-parameter paraproducts satisfy the generic Theorem 3.8. The main new observation is that there is a similar argument that works even in the *continuous* case, where all the bump functions have tails. There are two ingredients. First, one rewrites the functions φ_R^3 as series of similar but compactly supported functions. Second, one defines the exceptional sets more carefully, to make sure that the analogous term II still vanishes.

The following lemma explains how it is possible to decompose a bump function as a series of compactly supported functions without changing its oscillation properties in an essential way. It is instructive to compare it with the atomic decomposition in Lemma I.8.10.

Lemma 3.9 *Let $J \subseteq \mathbb{R}$ be a fixed interval. Then every smooth bump function ϕ_J adapted to J can be naturally decomposed as follows:*

$$\phi_J = \sum_{k \in \mathbb{N}} 2^{-1000k} \phi_J^k,$$

where for every $k \in \mathbb{N}$, ϕ_J^k is also a bump function adapted to J but having the additional property that $\operatorname{supp}(\varphi_J^k) \subseteq 2^k J$. If in addition we assume that $\int_{\mathbb{R}} \phi_J(x)\, dx = 0$ then the functions ϕ_J^k can be chosen such that $\int_{\mathbb{R}} \phi_J^k(x)\, dx = 0$ for every $k \in \mathbb{N}$.

Here and below, by $2^k J$ we mean the interval having the same center as J but 2^k times larger.

We postpone the proof of Lemma 3.9 until the next section. Using it, one can estimate the left-hand side of (3.71) by

$$\sum_{\vec{k} \in \mathbb{N}^2} 2^{-1000|\vec{k}|} \sum_R \frac{1}{|R|^{1/2}} |\langle f, \varphi_R^1 \rangle| |\langle g, \varphi_R^2 \rangle| |\langle h, \varphi_R^{3,\vec{k}} \rangle| \tag{3.72}$$

where the new functions $\varphi_R^{3,\vec{k}} := \varphi_I^{3,k_1} \otimes \varphi_J^{3,k_2}$, $R = I \times J$, are similar to the old ones, $\varphi_R^3 = \varphi_I^3 \otimes \varphi_J^3$, but have the additional property that $\operatorname{supp}(\varphi_R^{3,\vec{k}}) \subseteq 2^{\vec{k}} R$, where we have set $2^{\vec{k}} R := 2^{k_1} I \times 2^{k_2} J$ for $\vec{k} = (k_1, k_2) \in \mathbb{N}^2$; $|\vec{k}| = \max(k_1, k_2)$. To obtain our desired set E' we proceed as follows.

For each $\vec{k} \in \mathbb{N}^2$ we first define

$$\Omega_{-10|\vec{k}|} := \{x \in \mathbb{R}^2 | MS(f)(x) > C2^{10|\vec{k}|}\}$$
$$\cup \{x \in \mathbb{R}^2 | SM(g)(x) > C2^{10|\vec{k}|}\}.$$

Then, we define

$$\widetilde{\Omega}_{-10|\vec{k}|} := \left\{ x \in \mathbb{R}^2 \middle| MM\left(\chi_{\Omega_{-10|\vec{k}|}}\right)(x) > \frac{1}{100} \right\}$$

and then

$$\widetilde{\widetilde{\Omega}}_{-10|\vec{k}|} := \left\{ x \in \mathbb{R}^2 \middle| MM\left(\chi_{\widetilde{\Omega}_{-10|\vec{k}|}}\right)(x) > \frac{1}{2^{|\vec{k}|}} \right\}.$$

Finally, we set

$$\Omega := \bigcup_{\vec{k} \in \mathbb{N}^2} \tilde{\tilde{\Omega}}_{-10|\vec{k}|}.$$

It is clear that $|\Omega| \ll 1$ if C is a large enough constant and so, as before, we can then define E' to be $E \setminus \Omega$.

Now we return to (3.72). Fix $\vec{k} \in \mathbb{N}^2$ and consider the corresponding inner sum. Split the sum into two parts. Part I sums over the rectangles R with the property that

$$R \cap \tilde{\Omega}^c_{-10|\vec{k}|} \neq \emptyset$$

and part II sums over those rectangles R with the property that

$$R \cap \tilde{\Omega}^c_{-10|\vec{k}|} = \emptyset.$$

Now we observe that part II is in fact identically equal to zero. Indeed, if $R \subseteq \tilde{\Omega}_{-10|\vec{k}|}$ then $2^{\vec{k}} R \subseteq \tilde{\tilde{\Omega}}_{-10|\vec{k}|}$, which means that the support of the corresponding function $\varphi_R^{3,\vec{k}}$ does not intersect E' since E' lies outside the set $\tilde{\tilde{\Omega}}_{-10|\vec{k}|}$, in particular. It is therefore enough to estimate part I of the sum. This can be achieved by using the technique developed before. Since $R \cap \tilde{\Omega}^c_{-10|\vec{k}|} \neq \emptyset$, it follows that

$$\frac{|R \cap \Omega_{-10|\vec{k}|}|}{|R|} \leq \frac{1}{100}|R| \quad \text{or, equivalently,} \quad |R \cap \Omega^c_{-10|\vec{k}|}| > \frac{99}{100}|R|.$$

First, we describe a decomposition procedure for our function f from (3.71).

Define

$$\Omega_{-10|\vec{k}|+1} := \left\{ x \in \mathbb{R}^2 \middle| MS(f)(x) > \frac{C2^{10|\vec{k}|}}{2^1} \right\}$$

and set

$$\mathfrak{R}_{-10|\vec{k}|+1} := \left\{ R \middle| \left| R \cap \Omega_{-10|\vec{k}|+1} \right| > \frac{1}{100}|R| \right\}.$$

Now define

$$\Omega_{-10|\vec{k}|+2} = \left\{ x \in \mathbb{R}^2 \middle| MSf(x) > \frac{C2^{10|\vec{k}|}}{2^2} \right\}$$

and set

$$\mathfrak{R}_{-10|\vec{k}|+2} := \left\{ R \in \mathfrak{R} \setminus \mathfrak{R}_{-10|\vec{k}|+1} \middle| \left| R \cap \Omega_{-10|\vec{k}|+2} \right| > \frac{1}{100}|R| \right\},$$

and so on. The constant $C > 0$ above is the one in the definition of E'. Since there are finitely many rectangles in our collection, the algorithm ends after finitely many steps, producing the sets $(\Omega)_n$ and $(\mathfrak{R}_n)_n$.

Independently and similarly, define

$$\Omega'_{-10|\vec{k}|+1} := \left\{ x \in \mathbb{R}^2 \,\middle|\, SM(g)(x) > \frac{C2^{10|\vec{k}|}}{2^1} \right\}$$

and set

$$\mathfrak{R}'_{-10|\vec{k}|+1} := \left\{ R \,\middle|\, \left| R \cap \Omega'_{-10|\vec{k}|+1} \right| > \frac{1}{100}|R| \right\}.$$

Then define

$$\Omega'_{-10|\vec{k}|+2} := \left\{ x \in \mathbb{R}^2 \,\middle|\, SM(g)(x) > \frac{C2^{10|\vec{k}|}}{2^2} \right\}$$

and set

$$\mathfrak{R}'_{-10|\vec{k}|+2} := \left\{ R \in \mathfrak{R} \backslash \mathfrak{R}'_{-10|\vec{k}|+1} \,\middle|\, \left| R \cap \Omega'_{-10|\vec{k}|+1} \right| > \frac{1}{100}|R| \right\},$$

and so on, producing the sets $(\Omega'_n)_n$ and $(\mathfrak{R}'_n)_n$.

To produce a similar decomposition for h we first choose $N > 0$ large enough that for every R one has

$$\left| R \cap \Omega''^{c}_{-N} \right| > \frac{99}{100}|R|,$$

where

$$\Omega''_{-N} := \left\{ x \in \mathbb{R}^2 \,\middle|\, SS^{\vec{k}}(h)(x) > C2^N \right\}.$$

By $SS^{\vec{k}}$ we denote the usual double square function, but defined in terms of the functions $\varphi_R^{3,\vec{k}}$ instead of φ_R^3. Since these functions still have zero integral, $SS^{\vec{k}}$ will be bounded on any L^s space as well, for any $1 < s < \infty$. Then, in the same way as in the previous two stopping-time algorithms, we define

$$\Omega''_{-N+1} := \left\{ x \in \mathbb{R}^2 \,\middle|\, SS^{\vec{k}}(h)(x) > \frac{C2^N}{2^1} \right\}$$

and set

$$\mathfrak{R}''_{-N+1} := \left\{ R \,\middle|\, \left| R \cap \Omega''_{-N+1} \right| > \frac{1}{100}|R| \right\}.$$

As before we define

$$\Omega''_{-N+2} := \left\{ x \in \mathbb{R}^2 \,\middle|\, SS^{\vec{k}}(h)(x) > \frac{C2^N}{2^2} \right\}$$

and set

$$\mathfrak{R}''_{-N+2} := \left\{ R \in \mathfrak{R} \backslash \mathfrak{R}''_{-N+1} \,\middle|\, \left| R \cap \Omega''_{-N+2} \right| > \frac{1}{100} |R| \right\},$$

and so on, producing the sets $(\Omega''_n)_n$ and $(\mathfrak{R}''_n)_n$. As a consequence, the corresponding inner sum in (3.72) can be estimated by

$$\sum_{\substack{n_1, n_2 > -10|\vec{k}| \\ n_3 > -N}} \sum_{R \in \mathfrak{R}_{n_1, n_2, n_3}} \frac{|\langle f, \varphi_R{}^1 \rangle|}{|R|^{1/2}} \frac{|\langle g, \varphi_R{}^2 \rangle|}{|R|^{1/2}} \frac{\left|\langle h, \varphi_R{}^{3, \vec{k}} \rangle\right|}{|R|^{1/2}} |R|, \tag{3.73}$$

where $\mathfrak{R}_{n_1, n_2, n_3}$ stands for $\mathfrak{R}_{n_1} \cap \mathfrak{R}'_{n_2} \cap \mathfrak{R}''_{n_3}$. Now, since R belongs to $\mathfrak{R}_{n_1} \cap \mathfrak{R}'_{n_2} \cap \mathfrak{R}''_{n_3}$ this means in particular that it has not been selected at any of the previous $n_1 - 1, n_2 - 1$, and $n_3 - 1$ steps respectively. Thus

$$\left| R \cap \Omega_{n_1 - 1} \right| < \frac{1}{100} |R|, \; \left| R \cap \Omega'_{n_2 - 1} \right| < \frac{1}{100} |R| \text{ and } \left| R \cap \Omega''_{n_3 - 1} \right| < \frac{1}{100} |R|$$

or, equivalently,

$$\left| R \cap \Omega^c_{n_1 - 1} \right| \geq \frac{99}{100} |R|, \; \left| R \cap \Omega'^c_{n_2 - 1} \right| \geq \frac{99}{100} |R| \text{ and } \left| R \cap \Omega''^c_{n_3 - 1} \right| \geq \frac{99}{100} |R|.$$

As a consequence, one has that

$$\left| R \cap \Omega^c_{n_1 - 1} \cap \Omega'^c_{n_2 - 1} \cap \Omega''^c_{n_3 - 1} \right| \geq \frac{97}{100} |R|$$

and so (3.73) can be estimated by

$$\begin{aligned}
&\sum_{\substack{n_1, n_2 > -10|\vec{k}| \\ n_3 > -N}} \sum_{R \in \mathfrak{R}_{n_1, n_2, n_3}} \frac{|\langle f, \varphi_R^1 \rangle|}{|R|^{1/2}} \frac{|\langle g, \varphi_R^2 \rangle|}{|R|^{1/2}} \frac{|\langle h, \varphi_R^{3, \vec{k}} \rangle|}{|R|^{1/2}} \\
&\quad \times \left| R \cap \Omega^c_{n_1 - 1} \cap \Omega'^c_{n_2 - 1} \cap \Omega''^c_{n_3 - 1} \right| \\
&= \sum_{\substack{n_1, n_2 > -10|\vec{k}| \\ n_3 > -N}} \int_{\substack{\Omega^c_{n_1 - 1} \cap \Omega'^c_{n_2 - 1} \\ \cap \Omega''^c_{n_3 - 1}}} \sum_{R \in \mathfrak{R}_{n_1, n_2, n_3}} \frac{|\langle f, \varphi_R^1 \rangle|}{|R|^{1/2}} \frac{|\langle g, \varphi_R^2 \rangle|}{|R|^{1/2}} \frac{|\langle h, \varphi_R^{3, \vec{k}} \rangle|}{|R|^{1/2}} \chi_R(x) \, dx \\
&\lesssim \sum_{\substack{n_1, n_2 > -10|\vec{k}| \\ n_3 > -N}} \int_{\substack{\Omega^c_{n_1 - 1} \cap \Omega'^c_{n_2 - 1} \\ \cap \Omega''^c_{n_3 - 1} \cap \Omega_{\mathfrak{R}_{n_1, n_2, n_3}}}} MS(f)(x) \, SM(g)(x) \, SS^{\vec{k}}(h)(x) \, dx \\
&\lesssim \sum_{\substack{n_1, n_2 > -10|\vec{k}| \\ n_3 > -N}} 2^{-n_1} 2^{-n_2} 2^{-n_3} \left| \Omega_{\mathfrak{R}_{n_1, n_2, n_3}} \right|,
\end{aligned} \tag{3.74}$$

where

$$\Omega_{\mathfrak{R}_{n_1,n_2,n_3}} := \bigcup_{R \in \mathfrak{R}_{n_1,n_2,n_3}} R.$$

However, we also have that

$$\begin{aligned} \left|\Omega_{\mathfrak{R}_{n_1,n_2,n_3}}\right| \leq \left|\Omega_{\mathfrak{R}_{n_1}}\right| &\leq \left|\left\{x \in \mathbb{R}^2 \middle| MM(\chi_{\Omega_{n_1}})(x) > \frac{1}{100}\right\}\right| \\ &\lesssim \left|\Omega_{n_1}\right| = \left|\left\{x \in \mathbb{R}^2 \middle| MS(f)(x) > \frac{C}{2^{n_1}}\right\}\right| \lesssim 2^{n_1 p}. \end{aligned}$$

Similarly,

$$\left|\Omega_{\mathfrak{R}_{n_1,n_2,n_3}}\right| \lesssim 2^{n_2 q}$$

and also

$$\left|\Omega_{\mathfrak{R}_{n_1,n_2,n_3}}\right| \lesssim 2^{n_3 \alpha}$$

for any $\alpha > 1$. We have used here the fact that all the operators SM, MS, $SS^{\vec{k}}$, and MM are bounded (independently of $\vec{k}$) and also the fact that $|E'| \simeq 1$.

As a consequence, we deduce that

$$\left|\Omega_{\mathfrak{R}_{n_1,n_2,n_3}}\right| \lesssim 2^{n_1 p \theta_1} 2^{n_2 q \theta_2} 2^{n_3 \alpha \theta_3}, \tag{3.75}$$

for any $0 \leq \theta_1, \theta_2, \theta_3 < 1$ such that $\theta_1 + \theta_2 + \theta_3 = 1$. Then we split the sum in (3.74) as

$$\begin{aligned} &\sum_{\substack{n_1,n_2>-10|\vec{k}| \\ n_3>0}} 2^{-n_1} 2^{-n_2} 2^{-n_3} |\Omega_{\mathfrak{R}n_1,n_2,n_3}| \\ &+ \sum_{\substack{n_1,n_2>-10|\vec{k}| \\ 0>n_3>-N}} 2^{-n_1} 2^{-n_2} 2^{-n_3} |\Omega_{\mathfrak{R}n_1,n_2,n_3}|. \end{aligned} \tag{3.76}$$

To estimate the first term in (3.76) we use inequality (3.75) in the particular case $\theta_1 = \theta_2 = \frac{1}{2}$ and $\theta_3 = 0$, and to estimate the second term in (3.76) we use (3.75) for $(\theta_j)_j$ such that $p\theta_1 < 1$, $q\theta_2 < 1$, and $\alpha\theta_3 > 1$. Given these the sum in (3.76) becomes $O(2^{20|\vec{k}|})$ at most, and this makes the initial expression in (3.72) $O(1)$ after summing over the indices $\vec{k} \in \mathbb{N}^2$.

3.10. Proof of the generic decomposition

We are therefore left with proving Lemma 3.9 in order to complete this new general proof.

Consider first the case where there is no *oscillation* of our bump function. This case is easier. Fix an interval $J \subseteq \mathbb{R}$ and let ϕ_J be a smooth bump function adapted to J. Pick another smooth function ψ such that $\mathrm{supp}(\psi) \subseteq [\frac{-1}{2}, \frac{1}{2}]$ and such that $\psi = 1$ on $[\frac{-1}{4}, \frac{1}{4}]$, say. If $I \subseteq \mathbb{R}$ is any interval with center x_I, set

$$\psi_I(x) := \psi\left(\frac{x - x_I}{|I|}\right).$$

We then observe that we can write

$$1 = \psi_J + (\psi_{2J} - \psi_J) + (\psi_{2^2J} - \psi_{2J}) + \cdots$$

and this allows us to conclude that

$$\begin{aligned}
\phi_J &= \phi_J\psi_J + \sum_{k=1}^{\infty} \phi_J(\psi_{2^kJ} - \psi_{2^{k-1}J}) \\
&= \phi_J\psi_J + \sum_{k=1}^{\infty} 2^{-1000k}(2^{1000k}\phi_J(\psi_{2^kJ} - \psi_{2^{k-1}J})) \\
&:= \sum_{k=0}^{\infty} 2^{-1000k}\phi_J^k.
\end{aligned}$$

It is not difficult to see that all the functions ϕ_J^k are bumps adapted to J, with the important additional property that $\mathrm{supp}(\phi_J^k) \subseteq 2^kJ$ as desired.

Assume now that $\int_{\mathbb{R}} \phi_J(x)\,dx = 0$. Now we can write

$$\begin{aligned}
\phi_J &= \phi_J\psi_J + \phi_J(1 - \psi_J) \\
&= \left(\phi_J\psi_J - \left(\frac{1}{\int_{\mathbb{R}} \psi_J(x)\,dx} \int_{\mathbb{R}} \phi_J(x)\psi_J(x)\,dx\right)\psi_J\right) \\
&\quad + \left(\left(\frac{1}{\int_{\mathbb{R}} \psi_J(x)\,dx} \int_{\mathbb{R}} \phi_J(x)\psi_J(x)\,dx\right)\psi_J + \phi_J(1 - \psi_J)\right) := \phi_J^0 + E_J^0.
\end{aligned}$$

Clearly, by construction, we have that $\int_{\mathbb{R}} \phi_J^0(x)\,dx = 0$ and, as a consequence, we also have that

$$\int_{\mathbb{R}} E_J^0(x)\,dx = 0.$$

Furthermore, ϕ_J^0 is a bump adapted to the interval J and with the property that $\mathrm{supp}\,\phi_J^0 \subseteq J$.

However, given that

$$\left| \frac{1}{\int_{\mathbb{R}} \psi_J(x)\,dx} \int_{\mathbb{R}} \phi_J(x)\psi_J(x)\,dx \right|$$

$$= \left| \frac{1}{\int_{\mathbb{R}} \psi_J(x)\,dx} \int_{\mathbb{R}} \phi_J(x)(1-\psi_J(x))\,dx \right| \lesssim 2^{-1000},$$

we can deduce that $\|E_J^0\|_\infty \lesssim 2^{-1000}$. Here we have used the fact that ψ_J is identically equal to 1 on an interval of length $|J|/2$ and so one is only integrating the tail of ϕ_J in the integral above. Now we perform a similar decomposition for the error function E_J^0, but this time we localize it on the longer interval $2J$. We can write

$$\begin{aligned}
E_J^0 &= E_J^0\psi_{2J} + E_J^0(1-\psi_{2J}) \\
&= \left(E_J^0\psi_{2J} - \left(\frac{1}{\int_{\mathbb{R}} \psi_{2J}(x)\,dx} \int_{\mathbb{R}} E_J^0(x)\psi_{2J}(x)\,dx \right) \psi_{2J} \right) \\
&\quad + \left(\left(\frac{1}{\int_{\mathbb{R}} \psi_{2J}(x)\,dx} \int_{\mathbb{R}} E_J^0(x)\psi_{2J}(x)\,dx \right) \psi_{2J} + E_J^0(1-\psi_{2J}) \right) \\
&=: 2^{-1000}\phi_J^1 + E_J^1.
\end{aligned}$$

In the same way as before, we observe that $\int_{\mathbb{R}} \phi_J^1(x)\,dx = 0$ and so $\int_{\mathbb{R}} E_J^1(x)\,dx = 0$ also. Moreover, notice that ϕ_J^1 is a bump adapted to J whose support lies in $2J$ while $\|E_J^1\|_\infty \lesssim 2^{-1000\times 2}$. Iterating this procedure N times, we obtain the splitting

$$\phi_J = \sum_{k=0}^{N} 2^{-1000\times k}\phi_J^k + E_J^N,$$

where the ϕ_J^k are bump functions adapted to J with the property that $\int_{\mathbb{R}} \phi_J^k(x)\,dx = 0$ and $\mathrm{supp}(\phi_J^k) \subseteq 2^k J$, while

$$\|E_J^N\|_\infty \lesssim 2^{-1000\times N}.$$

Sending N to infinity completes the proof of Lemma 3.9.

As we come to the end of this chapter, let us emphasize that the techniques that we have learned here work equally well in the original framework of the Coifman–Meyer theorem, described in Chapter 2. In fact, in that classical setting, one works with intervals rather than with rectangles, which is easier. This technique provides an alternative proof of Theorem 2.15 that is simple, as it requires no knowledge of Carleson measures or BMO space. As we will see in Chapter 4, this approach (along with various other nontrivial ideas, of course) can be used to prove the usual L^p estimates for the so-called Calderón commutators and the Cauchy integral on Lipschitz curves.

Notes

Essentially, the content of the chapter comes from the two papers Muscalu, Pipher, Tao *et al.* [93, 96]. Particular cases of the polydisk Coifman–Meyer theorem were proven earlier by Journé [61] by different methods. Endpoint estimates of the $L \log L$ type were proven by Workman [119]. Applications to PDEs can be found in Kenig [64]. Journé's lemma appeared in [60]. The proof that we have described here is from Pipher [101]. A generalization of it to higher dimensions can be found in Pipher [100]. The corollary of Journé's lemma is from Ferguson and Lacey [43]. The presentation of Carleson's counterexample follows some notes of Tao [109]. For other results on multiparameter commutators, see Lacey, Petermichl, Pipher *et al.* [73].

For expositions on *linear multiparameter* harmonic analysis see Chang and Fefferman [13] and Fefferman and Stein [42].

Problems

Problem 3.1 Complete the results of this chapter by proving natural L^p estimates for the operators $\Pi_1^\alpha \otimes \Pi_2$, $\Pi_1 \otimes \Pi_2^\beta$, and $\Pi_1^\alpha \otimes \Pi_2^\beta$.

Problem 3.2 Extend the biparameter Leibnitz rule and the biparameter Coifman–Meyer theorem in a natural way to an arbitrary number of Cartesian products of Euclidean spaces of arbitrary dimensions.

Problem 3.3 Reprove the original Coifman–Meyer theorem, Theorem 2.15, using the methods of this chapter.

Problem 3.4 Show that Theorem 3.7 can be *periodized*. More precisely, assume that m is a symbol satisfying (3.68). Define the bilinear operator

$$T_m^2(f, g)(\vec{x}) := \sum_{\vec{n}_1, \vec{n}_2 \in \mathbb{Z}^2} m(\vec{n}_1, \vec{n}_2)\widehat{f}(\vec{n}_1)\widehat{g}(\vec{n}_2)e^{2\pi i \vec{x}(\vec{n}_1+\vec{n}_2)},$$

where now $\vec{x} \in \mathbb{T}^2$, $\mathbb{T}^2$ being the two-dimensional torus, and the functions f, g are also defined on $\mathbb{T}^2$. Show that the same L^p estimates as in Theorem 3.7 hold true for the operator T_m^2.

Problem 3.5 Prove the following *pseudodifferential* variant of Theorem 3.7, using ideas from the previous chapter. Suppose that $a(x, \xi, \eta)$ is a symbol satisfying

$$\left|\partial_{x_1}^{\gamma_1}\partial_{x_2}^{\gamma_2}\partial_{\xi_1}^{\alpha_1}\partial_{\xi_2}^{\alpha_2}\partial_{\eta_1}^{\beta_1}\partial_{\eta_2}^{\beta_2}a(x,\xi,\eta)\right| \lesssim \frac{1}{(1+|(\xi_1,\eta_1)|)^{\alpha_1+\beta_1}}\,\frac{1}{(1+|(\xi_2,\eta_2)|)^{\alpha_2+\beta_2}}$$

and denote by T_a^2 the bilinear operator given by

$$T_a^2(f,g)(x) := \int_{\mathbb{R}^4} a(x,\xi,\eta)\widehat{f}(\xi)\widehat{g}(\eta)e^{2\pi i x(\xi+\eta)}\,d\xi d\eta.$$

Show that the same L^p estimates as in Theorem 3.7 hold true for the operator T_a^2.

4

Calderón commutators and the Cauchy integral on Lipschitz curves

The goal of this chapter is to describe the theory of Calderón commutators and the Cauchy integral on Lipschitz curves. These objects were introduced by Calderón in the early 1960s and since then have played a prominent role in analysis. *Multilinear harmonic analysis* started essentially with the study of these operators. The reader may then naturally ask why we did not begin this second volume of the book with them, preferring to describe the theory of paraproducts? The short answer would be that commutators require a much more careful analysis than paraproducts even though they are deeply related, as we will see later.

4.1. History

4.1.1. Calderón commutators

Let us start by recaling the classical linear differential equation

$$Lu = f \tag{4.1}$$

where f is a given smooth function on the real line $\mathbb{R}$ and L is a differential operator with variable coefficients:

$$L = \sum_{\alpha=1}^{m} a_\alpha(x) \left(\frac{\partial}{\partial x} \right)^\alpha . \tag{4.2}$$

If all the coefficients a_α are smooth functions then one can clearly compose operators such as L and get another of a similar kind. This fact can be used to construct a *calculus with pseudodifferential operators*, which eventually would allow one to understand when one can solve an equation of the type (4.1), to study the properties of its solutions, etc. The reader will find discussions of some of these fundamental issues in the first volume of the book.

In the early 1960s Calderón proposed a very general method for the study of such equations even when the coefficients a_α are far from being smooth functions. More precisely, he assumed that all the coefficients are nearly L^∞, while that corresponding to the highest derivative, a_m, is also a Lipschitz function, in other words it satisfies $a'_m(x) \in L^\infty$. This Lipschitz condition is appropriate, since if those coefficients had lower regularity then there would be pathological counterexamples to some natural uniqueness properties for the corresponding PDEs.

We will describe Caldéron's method in what follows. First, consider a smooth and strictly positive function $\varphi(\xi)$ having also the property that $\varphi(\xi) = |\xi|$ for $|\xi| \geq 1$. Rewrite L as a pseudodifferential operator defined by

$$Lu(x) = \int_{\mathbb{R}} \left(\sum_{\alpha=1}^{m} a_\alpha(x)\xi^\alpha \right) \hat{u}(\xi) e^{2\pi i x \xi} \, d\xi. \tag{4.3}$$

Its symbol then becomes

$$\sum_{\alpha=1}^{m} a_\alpha(x)\xi^\alpha = \left(\frac{\sum_{\alpha=1}^{m} a_\alpha(x)\xi^\alpha}{\varphi(\xi)^m} \right) \varphi(\xi)^m.$$

Split the contents of the large parentheses as follows:

$$\begin{aligned} &\sum_{\alpha=1}^{m-1} \frac{a_\alpha(x)\xi^\alpha}{\varphi(\xi)^m} + \frac{a_m(x)\xi^m}{\varphi(\xi)^m} \\ &= \left(\frac{a_m(x)\xi^m}{|\xi|^m} \right) + \left(a_m(x)\xi^m \left(\frac{1}{\varphi(\xi)^m} - \frac{1}{|\xi|^m} \right) + \sum_{\alpha=1}^{m-1} \frac{a_\alpha(x)\xi^\alpha}{\varphi(\xi)^m} \right) \\ &=: q(x, \xi) + r(x, \xi). \end{aligned}$$

As a consequence, L itself decomposes; we obtain

$$L = (Q + R)\Lambda^m,$$

where Q, R, and Λ are the Fourier integral operators defined naturally by the symbols q, r, and φ respectively. Observe that the symbol $q(x, \xi)$ is homogeneous of degree zero in ξ and bounded in x, while R is smoothing of order 1; in other words both R and $R \circ \partial/\partial x$ are bounded as linear operators on every L^p space, for $1 < p < \infty$.

Exercise 4.1 Check that the operator R is indeed smoothing of order 1.

Since Λ is invertible (that is why the function φ has been introduced), equation (4.1) becomes

$$(R + Q)u = \Lambda^{-m} f \tag{4.4}$$

and the problem reduces to that of understanding the family of operators of the type $Q + R$. *Are they closed under composition? If $Q + R$ is invertible, is its inverse an operator of the same kind?* These are the questions that one would like to understand.

Let us consider now a particular case when the operator $Q + R$ is indeed invertible. Assume that $R = 0$ and that $Q = I - AH$, where A is the operator giving multiplication with the bounded Lipschitz function $A(x)$ and H is the classical Hilbert transform. The reader may recall its definition and significance from the first volume of the book. The symbol for Q is clearly given by $q(x, \xi) = 1 + i\pi A(x)\,\mathrm{sgn}(\xi)$ and is indeed bounded in x and homogeneous of degree zero in ξ. Assume also that $\|A\|_\infty \ll 1$. Then the inverse of Q can be written as

$$(I - AH)^{-1} = \sum_{n=0}^{\infty} (AH)^n. \tag{4.5}$$

Notice that the first two operators in the series are Q type, while the third, $(AH)^2$, is not. However, it can be rewritten as follows:

$$(AH)(AH) = A^2H^2 + A(HA - AH)H = -\pi^2 A^2 + A[H, A]H.$$

We will prove later on that the commutator $[H, A]$ is an operator that is smoothing of order 1, and this will allow us to conclude that $(AH)^2$ is also $Q + R$ type. This important result was proved by Calderón in 1965. Using it one can show that all the other terms and in fact the whole series are operators of the type $Q + R$.

Exercise 4.2 Assume Calderón's theorem and prove that $(I - AH)^{-1}$ is an operator of the type $Q + R$.

To check that the commutator $[H, A]$ is smoothing of order 1, one has to establish whether $(HA - AH) \circ \partial/\partial x$ is a bounded operator on $L^p(\mathbb{R})$ for $1 < p < \infty$. One can write

$$\begin{aligned}
(HA - AH) \circ \frac{\partial}{\partial x}(f)(x) &= (HA - AH)\left(f'\right)(x) \\
&= p.v. \int_{\mathbb{R}} \frac{A(y) f'(y)}{x - y}\, dy - p.v. \int_{\mathbb{R}} \frac{A(x) f'(y)}{x - y}\, dy \\
&= -p.v. \int_{\mathbb{R}} \left(\frac{A(x) - A(y)}{x - y}\right) f'(y)\, dy.
\end{aligned}$$

Assuming now that f is a smooth function that vanishes at infinity, one can integrate by parts in the above formula and obtain

$$p.v. \int_{\mathbb{R}} \left(\frac{A(x) - A(y)}{x - y} \right)' f(y)\,dy$$

$$= -p.v. \int_{\mathbb{R}} \frac{A'(y) f(y)}{x - y}\,dy + p.v. \int_{\mathbb{R}} \frac{A(x) - A(y)}{(x - y)^2} f(y)\,dy. \tag{4.6}$$

The first term in (4.6) is $-H(A'f)$, which is clearly bounded on every $L^p(\mathbb{R})$ since $A' \in L^\infty(\mathbb{R})$, while the second term, Calderón's first commutator C_1, is given by

$$C_1 f(x) = p.v. \int_{\mathbb{R}} \frac{A(x) - A(y)}{(x - y)^2} f(y)\,dy. \tag{4.7}$$

The fact that C_1 is a bounded operator on $L^p(\mathbb{R})$ constitutes the theorem of Calderón mentioned earlier. We will return to it and describe its proof in the next section of the chapter.

There is an alternative and interesting way to arrive at the operator C_1. Start with the Leibnitz formula

$$(Af)' = A'f + Af'$$

and solve for $A'f$:

$$A'f = (Af)' - Af' = D(Af) - ADf = [D, A]f,$$

where D is a convenient notation for the derivative operator. Obviously the operator $[D, A]$ is bounded on $L^p(\mathbb{R})$. *Does this property still hold for the operator* $[|D|, A]$? A simple and direct calculation shows that this last operator coincides with Calderón's first commutator C_1, as in (4.7). More complex calculations, but of a similar kind, give rise to operators of the type

$$C_k f(x) = p.v. \int_{\mathbb{R}} \frac{(A(x) - A(y))^k}{(x - y)^{k+1}} f(y)\,dy$$

for $k \geq 1$, which are called Calderón's kth commutators, or, more generally, to operators of the type given by

$$p.v. \int_{\mathbb{R}} \prod_{j=1}^{k} \left(\frac{A_j(x) - A_j(y)}{x - y} \right) \frac{1}{x - y} f(y)\,dy \tag{4.8}$$

where the functions $(A_j(x))_j$ are Lipschitz functions. The L^p-boundedness of these operators for $k \geq 2$ was proved by Coifman and Meyer in 1975. We will present a proof of their theorem later in the chapter.

Exercise 4.3 Show directly that the operator in (4.8) is equal to the iterated commutator

$$\frac{\pi}{k!i^k}[[\ldots[[D^k H, A_1], A_2], \ldots], A_k].$$

Using such arguments based on commutators, Calderón was eventually able to prove that the family of operators of the type $Q + R$ forms an *algebra*.

4.1.2. Cauchy integral on Lipschitz curves

The Cauchy integral on Lipschitz curves is an operator which appears naturally in complex analysis. Let Γ be a simple, closed, oriented, and rectifiable curve in the complex plane $\mathbb{C}$. Take $f \in L^2(\Gamma)$ (with respect to arclength measure) and consider the Cauchy integral

$$Cf(z) = \frac{1}{2\pi i}\int_\Gamma \frac{f(w)}{w - z}\,dw.$$

Clearly, $Cf(z)$ is well defined for $z \notin \Gamma$ and is a holomorphic function in the complement of the curve. It is a natural to ask whether the limit of $Cf(z)$ as z approaches a point on Γ exists and, if so, whether the limit function remains in $L^2(\Gamma)$. In order to understand this question let us write

$$C_\Gamma f(z_0) := \lim_{\varepsilon\to 0}\frac{1}{2\pi i}\int_{w\in\Gamma:\,|w-z_0|>\varepsilon}\frac{f(w)}{w - z_0}\,dw, \quad z_0 \in \Gamma,$$

whenever the limit exists. Also, let us write

$$C^+ f(z_0) := \lim_{\substack{z\to z_0\\ z\in D^+}} Cf(z) \quad \text{and} \quad C^- f(z_0) := \lim_{\substack{z\to z_0\\ z\in D^-}} Cf(z),$$

whenever the limits exist. Here D^+ and D^- are the interior and exterior domains determined by Γ. Denote by γ_ε and Γ_ε the curves

$$\gamma_\varepsilon := \{w \,|\, |w - z_0| = \varepsilon\} \cap D^+$$

and

$$\Gamma_\varepsilon := \{w \in \Gamma \,|\, |w - z_0| > \varepsilon\}$$

(see Figure 4.1). Assume in addition that f is smooth.

It is then not difficult to observe that

$$\begin{aligned}\int_{\Gamma_\varepsilon}\frac{f(w)}{w - z_0}\,dw &= \int_{\Gamma_\varepsilon}\frac{f(w) - f(z_0)}{w - z_0}\,dw + \int_{\Gamma_\varepsilon}\frac{f(z_0)}{w - z_0}\,dw \\ &= \int_{\Gamma_\varepsilon}\frac{f(w) - f(z_0)}{w - z_0}\,dw + \int_{\gamma_\varepsilon}\frac{f(z_0)}{w - z_0}\,dw, \qquad (4.9)\end{aligned}$$

using Cauchy's theorem.

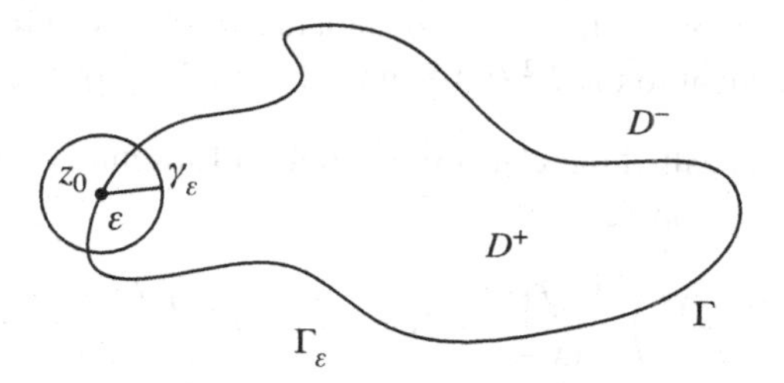

Figure 4.1

Letting ε tend to 0 and dividing by $2\pi i$ one obtains

$$C_\Gamma f(z_0) = \frac{1}{2\pi i} \int_\Gamma \frac{f(w) - f(z_0)}{w - z_0} \, dw + \frac{f(z_0)}{2}, \tag{4.10}$$

provided that Γ has a tangent at z_0, which is true almost everywhere. Similarly, one can conclude that $C^+ f(z_0)$ also exists and satisfies

$$C^+ f(z_0) = \frac{1}{2\pi i} \int_\Gamma \frac{f(w) - f(z_0)}{w - z_0} \, dw + f(z_0). \tag{4.11}$$

By combining (4.10) and (4.11) one sees that

$$C^+ f(z_0) = C_\Gamma f(z_0) + \tfrac{1}{2} f(z_0). \tag{4.12}$$

One can deduce a similar formula for $C^- f(z_0)$, namely

$$C^- f(z_0) = C_\Gamma f(z_0) - \tfrac{1}{2} f(z_0). \tag{4.13}$$

These two relations are Plemelj's formulae. From them we see that $C^+ f$ and $C^- f$ are $L^2(\Gamma)$ functions if and only if $C_\Gamma f$ is an $L^2(\Gamma)$ function. In other words, one needs to know whether this linear operator C_Γ is bounded on $L^2(\Gamma)$. For general functions f in $L^2(\Gamma)$ that are not necessarily smooth it turns out that analogues to formulae (4.11) and (4.13) hold if, in the definitions of $C^+ f$ and $C^- f$, the limits are understood in a certain nontangential sense. Suppose now that Γ is a Lipschitz graph given by a parametrization $x \mapsto x + iA(x)$ with $A' \in L^\infty(\mathbb{R})$. A simple change of variables show that the boundedness of C_Γ on $L^2(\Gamma)$ is equivalent to the boundedness on $L^2(\mathbb{R})$ of another linear operator (also denoted C_Γ, for convenience) given by

$$C_\Gamma f(x) = p.v. \int_{\mathbb{R}} \frac{f(y)}{x - y + i(A(x) - A(y))} \, dy. \tag{4.14}$$

Exercise 4.4 Show that the $L^2(\Gamma)$-boundedness of the original operator C_Γ is indeed equivalent to the L^2 $(\mathbb{R})$-boundedness of this operator.

The operator C_Γ is called the Cauchy integral on Lipschitz curves. It is interesting to note that it can be written as

$$\int_{\mathbb{R}} \left(1 + i\frac{A(x) - A(y)}{x - y}\right)^{-1} \frac{f(y)}{(x - y)} dy = \sum_{k=0}^{\infty} (-i)^k \int_{\mathbb{R}} \frac{(A(x) - A(y))^k}{(x - y)^{k+1}} f(y)\, dy$$
$$= \sum_{k=0}^{\infty} (-i)^k C_k f(x), \tag{4.15}$$

where the operators C_k are the previous Calderón commutators. It is remarkable that these two problems (that is, the PDE problem that generated the commutators and the complex analysis problem that gave rise to the Cauchy integral on Lipschitz curves), so distinct at first glance, are so closely related to one another.

The following simple but important observation allows us to reduce the problem of the $L^2(\mathbb{R})$-boundedness of C_Γ to a similar problem in which the Lipschitz constant of the curve is strictly smaller than 1.

Indeed, let as assume that A satisfies the estimate

$$|A(x) - A(y)| \leq M|x - y|$$

for every $x, y \in \mathbb{R}$; here M is the Lipschitz constant of the curve. Then one can write

$$\begin{aligned}(x - y) + i(A(x) - A(y)) = (x - y) + \mu(x - y) + i(A(x) - A(y))\\ - \mu(x - y),\end{aligned} \tag{4.16}$$

where μ will be determined later. Then (4.16) can be continued as follows:

$$\begin{aligned}&(1 + \mu)(x - y) + (iA(x) - \mu x) - (iA(y) - \mu y)\\ &= (1 + \mu)\left((x - y) + \frac{iA(x) - \mu x}{\mu + 1} - \frac{iA(y) - \mu y}{\mu + 1}\right)\\ &= (1 + \mu)\left((x - y) + \left(\widetilde{A}(x) - \widetilde{A}(y)\right)\right),\end{aligned}$$

where $\widetilde{A}(x) := (iA(x) - \mu x)/(\mu + 1)$.

It is then easy to see that $\widetilde{A}'(x) = (iA'(x) - \mu)/(\mu + 1)$ and so $\|\widetilde{A}'\|_\infty \leq (\mu^2 + M^2)^{1/2}/(\mu + 1)$.

Now, if one picks $\mu := M^2$ one sees that the above expression is indeed strictly smaller than 1.

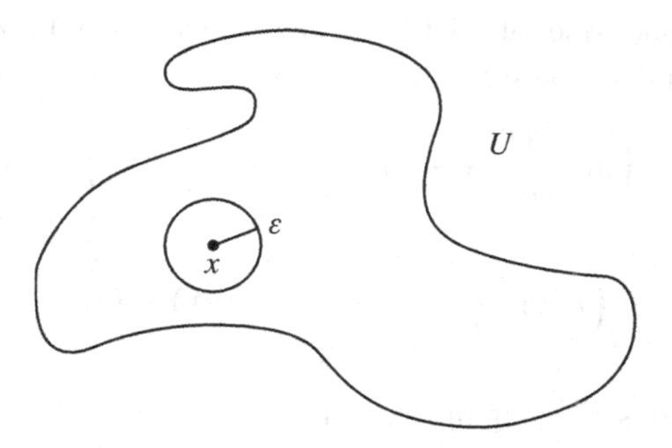

Figure 4.2

Taking into account formula (4.15), we will show later that C_Γ is bounded on $L^2(\mathbb{R})$ as long as one can prove that every C_k is bounded on $L^2(\mathbb{R})$ with a bound that grows at most polynomially in k. This is precisely what Coifman, McIntosh, and Meyer did when they proved the boundedness of the Cauchy integral in 1982. These polynomial bounds will be proven carefully in the third section of this chapter.

4.1.3. Dirichlet problem on Lipschitz domains

Finally, in this section, we will show that operators very similar to Cauchy integrals on Lipschitz curves appear in PDEs as well.

Assume that U is a bounded Lipschitz domain in the complex plane and consider the following classical Dirichlet problem for the Laplace operator:

$$\begin{cases} \Delta u = 0 & \text{in } U, \\ u = f & \text{on } \partial U. \end{cases} \tag{4.17}$$

Let us denote by

$$F(x) = -\frac{1}{2\pi} \log |x|$$

the fundamental solution for the Laplacian in the complex plane. In other words $\Delta F = \delta_0$, the Dirac delta distribution centered at the origin. Let $x \in U$ and let $\varepsilon > 0$ be a small number such that $B(x, \varepsilon) \subseteq U$ (see Figure 4.2). Write $U_\varepsilon := U \backslash B(x, \varepsilon)$. By applying Green's formula to the functions $y \mapsto u(y)$ and $y \mapsto F(y - x)$ in the domain U_ε, one obtains

$$\int_{U_\varepsilon} \big(u(y)\Delta F(y-x) - F(y-x)\Delta u(y)\big)\, dy$$

$$= \int_{\partial U_\varepsilon} \left(u(y)\frac{\partial F}{\partial \nu}(y-x) - F(y-x)\frac{\partial u}{\partial \nu}(y)\right) d\sigma(y). \tag{4.18}$$

Since the left-hand side of (4.18) is identically equal to zero and since $\partial U_\varepsilon = \partial U \cup \partial B(x, \varepsilon)$, we see that

$$\int_{\partial B(x,\varepsilon)} \left(u(y)\frac{\partial F}{\partial \nu}(y-x) - F(y-x)\frac{\partial u}{\partial \nu}(y) \right) d\sigma(y)$$

$$= \int_{\partial U} \left(u(y)\frac{\partial F}{\partial \nu}(y-x) - F(y-x)\frac{\partial u}{\partial \nu}(y) \right) d\sigma(y). \tag{4.19}$$

It is not difficult to see (using the explicit formula for F) that the second term on the left-hand side of (4.19) tends to zero as $\varepsilon \to 0$ while the first term is an average of u that tends to $u(x)$ as $\varepsilon \to 0$. Thus one obtains the representation formula

$$u(x) = \int_{\partial U} \left(u(y)\frac{\partial F}{\partial \nu}(y-x) - F(y-x)\frac{\partial u}{\partial \nu}(y) \right) d\sigma(y) \tag{4.20}$$

for the solution of (4.17). Since one does not know $\partial u/\partial \nu$ on ∂U, (4.20) is not a useful formula and instead one needs to look for a solution $u(x)$ of the form

$$u(x) = \int_{\partial U} \frac{\partial F}{\partial \nu}(y-x) g(y)\, d\sigma(y) \tag{4.21}$$

for a certain function g defined on ∂U. Clearly, the function $u(x)$ is harmonic in U, so one wonders about the limit of the expression in (4.21) as x approaches a generic point $x_0 \in \partial U$. A calculation almost identical to that performed earlier in the case of the Cauchy integral shows that this limit exists (if g is sufficiently well behaved) and is equal to

$$Kg(x_0) + \tfrac{1}{2} g(x_0),$$

where K is the *double-layer potential* operator, given by

$$Kg(x) = p.v. \int_{\partial U} \frac{\partial F}{\partial \nu}(y-x) g(y)\, d\sigma(y) \tag{4.22}$$

for $x \in \partial U$. To solve (4.17) one needs to find a function g such that

$$Kg + \tfrac{1}{2} g = f,$$

and this amounts to inverting the linear operator $\frac{1}{2}I + K$ on the boundary of U. An implicit question is whether this operator K is bounded on $L^2(\partial U)$. Assuming as before that ∂U is given by a Lipschitz graph whose parametrization is $x \mapsto (x, A(x))$, the boundedness of K on $L^2(\partial U)$ becomes equivalent to the boundedness of the operator T on $L^2(\mathbb{R})$, given by

$$Tf(x) = p.v. \int_{\mathbb{R}} \frac{(A(x) - A(y)) - (x - y)A'(y)}{(x - y)^2 + (A(x) - A(y))^2} f(y)\,dy. \tag{4.23}$$

Exercise 4.5 Check that the boundedness of K in (4.22) is indeed equivalent to the boundedness of T in (4.23).

We claim that the operator T is similar to the Cauchy integral on Lipschitz curves. In particular it splits as a infinite sum of commutators. Indeed, it is easy to see that for the first term in (4.23) we have

$$\begin{aligned} p.v. &\int_{\mathbb{R}} \frac{A(x) - A(y)}{(x - y)^2 + (A(x) - A(y))^2} f(y)\,dy \\ &= p.v. \int_{\mathbb{R}} \frac{A(x) - A(y)}{(x - y)^2} \left(1 + \left(\frac{A(x) - A(y)}{x - y}\right)^2\right)^{-1} f(y)\,dy \\ &= \sum_{n=0}^{\infty} (-1)^n p.v. \int_{\mathbb{R}} \frac{(A(x) - A(y))^{2n+1}}{(x - y)^{2n+2}} f(y)\,dy \end{aligned}$$

while for the second term we have

$$\begin{aligned} p.v. &\int_{\mathbb{R}} \frac{(x - y)}{(x - y)^2 + (A(x) - A(y))^2} A'(y) f(y)\,dy \\ &= \sum_{n=0}^{\infty} (-1)^n p.v. \int_{\mathbb{R}} \frac{(A(x) - A(y))^{2n}}{(x - y)^{2n+1}} A'(y) f(y)\,dy. \end{aligned}$$

As mentioned earlier, this shows that the analysis of the operator T in (4.23) can be reduced to the analysis of the Calderón commutators.

4.2. The first Calderón commutator

Now we consider the first Calderón commutator. For any Lipschitz function A on the real line (so that $A' \in L^\infty(\mathbb{R})$) one defines a linear operator $C_1(f)$ by

$$C_1(f)(x) = p.v. \int_{\mathbb{R}} \frac{A(x) - A(y)}{(x - y)^2} f(y)\,dy, \tag{4.24}$$

where the principal-value of the integral is given by

$$\lim_{\epsilon \to 0} \int_{\epsilon < |x-y| < 1/\epsilon} \frac{A(x) - A(y)}{(x - y)^2} f(y)\,dy, \tag{4.25}$$

whenever the limit exists. We emphasize that it is not necessary to assume that A is also a bounded function, as was the case in the previous section. Note that the simplest particular case is thus obtained when $A(x) = x$ and $C_1(f)$ becomes the classical Hilbert transform. The reader is invited to recall the theory of this operator from Chapter I.3.

Theorem 4.1 *The first Calderón commutator C_1 extends naturally as a bounded linear operator on every $L^p(\mathbb{R})$ space for $1 < p < \infty$ and satisfies the estimate*

$$\|C_1(f)\|_p \lesssim \|A'\|_\infty \|f\|_p. \tag{4.26}$$

It should be pointed out at the beginning that since C_1 is not an operator of the convolution type there are no easy estimates that can be proved for it. Recall from Chapter I.7 that for classical Calderón–Zygmund operators one immediately concludes their L^2-boundedness; however, a new method is needed for Calderón commutators.

Of course, as is usual with singular integrals, it is not at all clear at first glance how one should define $C_1(f)$ for an arbitrary $L^p(\mathbb{R})$ function f, and we need to clarify this issue first.

Write $a := A'$ and observe that, at least, when a and f are both Schwartz functions (4.25) makes perfect sense.

Indeed, for a fixed $\epsilon > 0$, rewrite (4.25) as

$$\begin{aligned}
&-\int_{\epsilon<|t|<1/\epsilon} \frac{A(x+t) - A(x)}{t^2} f(x+t)\,dt \\
&= -\int_{\epsilon<|t|<1/\epsilon} \left(\frac{A(x+t) - A(x)}{t}\right) f(x+t)\frac{dt}{t} \\
&= -\int_{\epsilon<|t|<1/\epsilon} \left(\int_0^1 a(x+\alpha t)\,d\alpha\right) f(x+t)\frac{dt}{t}.
\end{aligned} \tag{4.27}$$

Then, write a and f using the inverse Fourier transform as

$$a(x+\alpha t) = \int_{\mathbb{R}} \widehat{a}(\xi_1) e^{2\pi i (x+\alpha t)\xi_1}\, d\xi_1$$

and

$$f(x+t) = \int_{\mathbb{R}} \widehat{f}(\xi) e^{2\pi i (x+t)\xi}\, d\xi.$$

Using these formulae, (4.27) becomes

$$-\int_{\mathbb{R}^2} m^\epsilon(\xi, \xi_1)\widehat{f}(\xi)\widehat{a}(\xi_1) e^{2\pi i x(\xi+\xi_1)}\, d\xi d\xi_1, \tag{4.28}$$

where

$$m^{\epsilon}(\xi, \xi_1) = \int_0^1 \int_{\epsilon<|t|<1/\epsilon} \frac{1}{t} e^{2\pi i t(\xi+\alpha\xi_1)} \, dt d\alpha,$$

which converges pointwise to

$$-\int_0^1 \operatorname{sgn}(\xi + \alpha\xi_1) \, d\alpha.$$

Using the dominated convergence theorem one can see that the limit as $\epsilon \to 0$ exists in (4.28) and is equal to

$$\int_{\mathbb{R}^2} \left(\int_0^1 \operatorname{sgn}(\xi + \alpha\xi_1) \, d\alpha \right) \widehat{f}(\xi)\widehat{a}(\xi_1) e^{2\pi i x(\xi+\xi_1)} \, d\xi d\xi_1. \tag{4.29}$$

Note that the above integral is absolutely integrable, since both f and a are Schwartz functions. Because of (4.29) one can think of the linear operator C_1 as being a *bilinear operator* in f and a, and we will denote it from now on as $C_1(f, a)$. In particular, as we have seen before, C_1 has two natural *bilinear adjoints*, which we shall denoted by C_1^{*1} and C_1^{*2}. To define $C_1(f, a)$ for any $a \in L^\infty$ and any $f \in L^p$ we proceed as follows.

First, if a and f are Schwartz functions then on the one hand (4.26) is clearly equivalent to

$$\left| \int_{\mathbb{R}} C_1(f, a)(x) g(x) \, dx \right| \lesssim \|a\|_\infty \|f\|_p \|g\|_{p'}$$

for any Schwartz function g in the dual space $L^{p'}$. On the other hand we also know from the definition of the adjoint of a bilinear operator that

$$\int_{\mathbb{R}} C_1(f, a)(x) g(x) \, dx = \int_{\mathbb{R}} C_1^{*2}(f, g)(x) a(x) \, dx. \tag{4.30}$$

We will prove in the rest of the section that

$$\|C_1^{*2}(f, g)\|_1 \lesssim \|f\|_p \|g\|_{p'} \tag{4.31}$$

for any Schwartz functions f, g and this shows that one can extend C_1^{*2} naturally, by density, to the whole product space $L^p \times L^{p'}$. This means, however, that the right-hand side of (4.30) makes sense for any $a \in L^\infty$, not only for bounded Schwartz functions, which suggests that we extend $C_1(f, a)$ by duality. More precisely, for generic $f \in L^p$ and $a \in L^\infty$ we define $C_1(f, a)$ to be the unique L^p function satisfying (4.30) for any $g \in L^{p'}$.

This discussion shows that in order to demonstrate the main result, Theorem 4.1, it is enough to prove (4.31) for Schwartz functions.

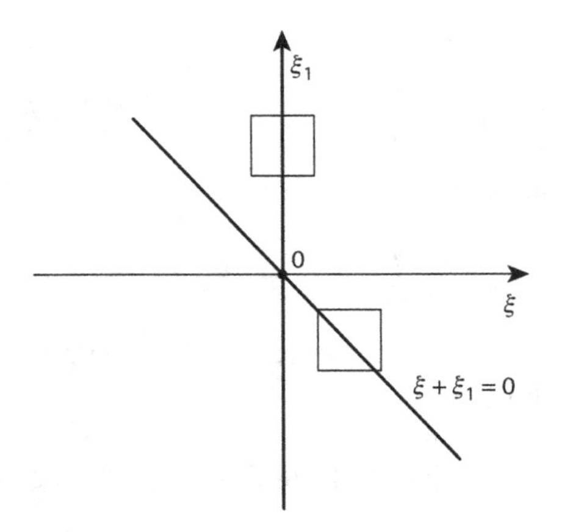

Figure 4.3. Singularities of the symbol $m_1(\xi, \xi_1)$; see (4.32).

At this point, it is important to observe that the symbol of the operator (4.29) given by

$$m_1(\xi, \xi_1) := \int_0^1 \operatorname{sgn}(\xi + \alpha\xi_1)\, d\alpha \tag{4.32}$$

is not a classical Marcinkiewicz–Mikhlin–Hörmander symbol and, as a consequence, the Coifman–Meyer theorem on paraproducts proved in Section 2.12 cannot be applied. More precisely, one can see that in the complement of the lines $\xi = 0$ and $\xi + \xi_1 = 0$ the symbol (4.32) is smooth and behaves like a classical symbol but along these lines it is just continuous (see Figure 4.3). However, in spite of this bad news we would still like to develop for this symbol a method similar to that for paraproducts.

There are two new ideas. First, one has to observe that, in spite of this lack of differentiability, when one smoothly restricts the symbol (4.32) to an arbitrary Whitney square with respect to the origin (recall that these are squares whose sides are parallel to the coordinate axes and whose distance to the origin is comparable with their side lengths), the Fourier coefficients of the corresponding restricted functions decay at least quadratically. This fact will be proved in detail later and will reduce our problem to that of proving estimates for the associated shifted discretized bilinear operators, which do not grow too fast with respect to the indices of the Fourier coefficients. Second, we will see that these upper bounds grow at most logarithmically, which is clearly more than enough to complete our analysis.

This constitutes a short outline of the upcoming proof.

Before starting the actual proof let us also mention that if one freezes the variable $\alpha \in (0, 1)$ in the formula (4.32) and considers the bilinear operator having the symbol $\operatorname{sgn}(\xi + \alpha\xi_1)$ then one obtains a *bilinear Hilbert transform.* This operator was introduced by Calderón at a time when only the first commutator had been understood, in the hope that a general method would be found that would eventually lead to the understanding of all Calderón commutators and of the Cauchy integral on Lipschitz curves. As it turned out, in the end the bilinear Hilbert transform was not needed for the completion of the whole Calderón program; however, the question whether this operator satisfies any L^p estimates of the Hölder type remained open. It was solved in a series of two papers by Lacey and Thiele. This will be discussed in Chapters 5 and 6.

4.2.1. Quadratic decay estimates for the symbol of C_1

First let us remark that we can analyze the symbol $\int_0^1 1_{\mathbb{R}_+}(\xi + \alpha\xi_1)\,d\alpha$ instead of $\int_0^1 \operatorname{sgn}(\xi + \alpha\xi_1)\,d\alpha$ since the difference between the second and twice the first is a multiple of the identity, for which the corresponding bilinear operator becomes a simple product. Let us also observe that, since this symbol is homogeneous of degree zero, to study the decay properties of its Fourier coefficients at scale 1 only is sufficient.

Therefore, consider now two smooth bump functions $\widehat{\varphi}(\xi)$ and $\widehat{\varphi}(\xi_1)$ such that the support of

$$(\xi, \xi_1) \mapsto \widehat{\varphi}(\xi)\widehat{\varphi}(\xi_1) \tag{4.33}$$

lies inside a Whitney square Q, with respect to the origin, of size 1. The double Fourier coefficient of our symbol smoothly restricted to Q is the following expression

$$\int_{\mathbb{R}^2} \left(\int_0^1 1_{\mathbb{R}_+}(\xi + \alpha\xi_1)\,d\alpha \right) \widehat{\varphi}(\xi)\widehat{\varphi}(\xi_1) e^{-2\pi i n\xi} e^{-2\pi i n_1 \xi_1}\, d\xi d\xi_1, \tag{4.34}$$

where n and n_1 are two arbitrary integers. To be able to estimate it (4.34) we need to consider several cases. If Q does not intersect either of the lines $\xi = 0$ and $\xi + \xi_1 = 0$ then one can integrate by parts multiple times and obtain estimates for (4.34) that decay faster than any polynomial. This situation is identical to the *paraproduct case* studied in Chapter 2 in relation to the Coifman–Meyer theorem. In the cases where Q intersects $\xi = 0$ or $\xi + \xi_1 = 0$, let us denote by C_{n,n_1} and $\widetilde{C}_{n,n_1}$ the corresponding Fourier coefficients (4.34).

The following lemma will play an important role later.

Lemma 4.2 *One has the estimates*

$$|C_{n,n_1}| \lesssim \frac{1}{<n>^2} \frac{1}{<n_1>^{\mu}}$$

and similarly

$$|\widetilde{C}_{n,n_1}| \lesssim \frac{1}{<n>^2} \frac{1}{<n-n_1>^{\mu}} + \frac{1}{<n>^{\mu}} \frac{1}{<n_1>^{\mu}},$$

for a fixed large integer μ *where* $<n>$ *is a notation for* $2+|n|$.

Proof We should observe that in neither case does the support of $\widehat{\varphi}(\xi_1)$ contain the origin. Thus we can record the following simple but crucial observation.

Lemma 4.3 *If* δ_0 *is the Dirac distribution with respect to the origin, one has the following identities:*

(i) $\partial_{\xi}^2 \left(\int_0^{\xi_1} 1_{\mathbb{R}_+}(\xi+\alpha)\, d\alpha \right) = \delta_0(\xi+\xi_1) - \delta_0(\xi)$;

(ii) $\partial_{\xi_1}^2 \left(\int_0^{\xi_1} 1_{\mathbb{R}_+}(\xi+\alpha)\, d\alpha \right) = \delta_0(\xi+\xi_1)$;

(iii) $\partial_{\xi}\partial_{\xi_1} \left(\int_0^{\xi_1} 1_{\mathbb{R}_+}(\xi+\alpha)\, d\alpha \right) = \partial_{\xi_1}\partial_{\xi} \left(\int_0^{\xi_1} 1_{\mathbb{R}_+}(\xi+\alpha)\, d\alpha \right) = \delta_0(\xi+\xi_1)$.

Proof This is a straightforward verification. Since (ii) and (iii) can be seen directly from the formulae, we will check only (i). One has

$$\partial_{\xi}^2 \left(\int_0^{\xi_1} 1_{\mathbb{R}_+}(\xi+\alpha)\, d\alpha \right) = \partial_{\xi} \left(\int_0^{\xi_1} \delta_0(\xi+\alpha)\, d\alpha \right)$$
$$= \partial_{\xi} \left(\int_{\xi}^{\xi+\xi_1} \delta_0(\alpha)\, d\alpha \right) = \delta_0(\xi+\xi_1) - \delta_0(\xi),$$

as desired. □

Returning to the estimates in Lemma 4.2, we rewrite $\widetilde{C}_{n,n_1}$, for instance, as

$$\int_{\mathbb{R}^2} \left(\frac{1}{\xi_1} \int_0^{\xi_1} 1_{\mathbb{R}_+}(\xi+\alpha)\, d\alpha \right) \widehat{\varphi}(\xi)\widehat{\varphi}(\xi_1) e^{-2\pi i n \xi} e^{-2\pi i n_1 \xi_1}\, d\xi d\xi_1$$
$$=: \int_{\mathbb{R}^2} \left(\int_0^{\xi_1} 1_{\mathbb{R}_+}(\xi+\alpha\xi_1)\, d\alpha \right) \widehat{\varphi}(\xi)\widehat{\widehat{\varphi}}(\xi_1) e^{-2\pi i n \xi} e^{-2\pi i n_1 \xi_1}\, d\xi d\xi_1. \quad (4.35)$$

The idea is to integrate by parts as much as we can, in (4.35). We start by integrating by parts in ξ as much as we can. Since both the factors $\int_0^{\xi_1} 1_{\mathbb{R}_+}(\xi+\alpha)\, d\alpha$ and $\widehat{\varphi}(\xi)$ depend on ξ, the ξ derivative operator $d/d\xi$ can act upon either factor. If $d/d\xi$ acts twice upon the first factor, $\int_0^{\xi_1} 1_{\mathbb{R}_+}(\xi+\alpha)\, d\alpha$, then, because of Lemma 4.3 the ξ variable becomes $-\xi_1$ (observe that ξ cannot be zero in this case), at which point (4.35) becomes an expression of the type

$$\int_{\mathbb{R}} \widehat{\varphi}(-\xi_1)\widehat{\widehat{\varphi}}(\xi_1)e^{-2\pi i \xi_1(n-n_1)}\,d\xi_1.$$

This term can clearly be integrated by parts an arbitrary number of times, and this explains the first upper bound for $|\widetilde{C}_{n,n_1}|$. If, however, the ξ derivative operator has not yet acted twice upon the factor $\int_0^{\xi_1} 1_{\mathbb{R}_+}(\xi+\alpha)\,d\alpha$, even after we have integrated by parts many times, this means that we must have already obtained a factor of the type $1/{<}n{>}^{\mu}$. So, we stop integrating by parts in ξ and start integrating by parts in ξ_1 instead. Exactly as before, if the ξ_1 derivative operator acts upon the awkward term $\int_0^{\xi_1} 1_{\mathbb{R}_+}(\xi+\alpha)\,d\alpha$ until one arrives at $\delta_0(\xi+\xi_1)$ then ξ_1 becomes $-\xi$ and, after that, one is simply integrating by parts a smooth function and one obtains an upper bound of the type $(1/{<}n{>}^{\mu})(1/{<}n-n_1{>}^{\mu})$, which is smaller than the previously discussed bound.

Finally, if it happens that the ξ_1 derivative operator does not act upon the factor $\int_0^{\xi_1} 1_{\mathbb{R}_+}(\xi+\alpha)\,d\alpha$ until it becomes $\delta_0(\xi+\xi_1)$ then this means that it keeps acting upon the smooth bump function of ξ_1, in which case we obtain an upper bound of the type $(1/{<}n{>}^{\mu})(1/{<}n_1{>}^{\mu})$, which is more than sufficient.

The term C_{n,n_1} in Lemma 4.2 can be treated in a similar way. One should just observe, however, that in this situation the equality $\xi_1 = \xi$ is impossible and only $\delta_0(\xi)$ survives after the integration by parts; this explains the minor difference between the two upper bounds. □

4.2.2. A discrete theorem

As in the case of paraproducts, our goal here is to reduce (4.31) to a natural estimate for some discrete operators. If $I = [2^k n, 2^k(n+1)]$ is a generic dyadic interval and $\mathfrak{n} \in \mathbb{Z}$ is an arbitrary integer, we denote by $I_{\mathfrak{n}} := [2^k(n-\mathfrak{n}), 2^k(n+1-\mathfrak{n})]$ the new dyadic interval lying $\mathfrak{n}$ units each of length $|I|$ away from I.

Consider now $\mathfrak{n}_1, \mathfrak{n}_2$ two fixed integers and a finite collection $\mathbb{I}$ of dyadic intervals. Consider also three sequences of L^2-normalized bump functions $(\Phi^1_{I_{\mathfrak{n}_1}})_I$, $(\Phi^2_{I_{\mathfrak{n}_2}})_I$, and $(\Phi^3_I)_I$ adapted to $I_{\mathfrak{n}_1}$, $I_{\mathfrak{n}_2}$, and I respectively and such that at least two are lacunary in the sense of Definition 2.2. The following theorem holds.

Theorem 4.4 *The discrete bilinear operator defined by*

$$T_{\mathbb{I}}(f,g) = \sum_{I\in\mathbb{I}} \frac{1}{|I|^{1/2}} \langle f, \Phi^1_{I_{\mathfrak{n}_1}}\rangle\langle g, \Phi^2_{I_{\mathfrak{n}_2}}\rangle \Phi^3_I$$

is bounded from $L^p \times L^q$ into L^r for any $1 < p, q < \infty$ and $0 < r < \infty$ such that $1/p + 1/q = 1/r$, with an operatorial bound of the type

$$O(\log \langle\mathfrak{n}_1\rangle \log \langle\mathfrak{n}_2\rangle) \tag{4.36}$$

and depending implicitly on p, q but otherwise independent of the cardinality of $\mathbb{I}$ and of the families of bump functions considered.

We claim now that it is enough to prove this theorem to understand completely Calderón's theorem 4.1. Indeed, using arguments by now standard, one can show exactly as in Section 2.13 that, modulo some paraproducts (which correspond to those regions of the symbol away from the problematic lines), C_1^{*2} splits naturally as

$$C_1^{*2}(f, g) = \sum_{\mathfrak{n}_1, \mathfrak{n}_2 \in \mathbb{Z}} C(\mathfrak{n}_1, \mathfrak{n}_2) T_{\mathfrak{n}_1, \mathfrak{n}_2}(f, g), \tag{4.37}$$

where $T_{\mathfrak{n}_1, \mathfrak{n}_2}(f, g)$ is an average of discrete operators similar to those in Theorem 4.4 and the coefficients $C(\mathfrak{n}_1, \mathfrak{n}_2)$ have the property that

$$\sum_{\mathfrak{n}_1, \mathfrak{n}_2 \in \mathbb{Z}} |C(\mathfrak{n}_1, \mathfrak{n}_2)| \log \langle\mathfrak{n}_1\rangle \log \langle\mathfrak{n}_2\rangle < \infty. \tag{4.38}$$

More precisely, these coefficients are clearly related to the Fourier coefficients of the symbol of C_1, discussed earlier, and we know from Lemma 4.2 that they decay at least quadratically. Notice that (4.37) is an analogue of (2.85).

Exercise 4.6 Review the discretization procedure from Section 2.13 and prove the details of the decomposition in (4.37).

Now standard arguments similar to those in Chapter 2, based on Theorem 4.4, the decomposition (4.37), Fatou's lemma, and the triangle inequality, allow one to prove the desired (4.31) immediately.

Before going any further let us emphasize again the difference between Carlderón's first commutator and the paraproduct case studied earlier. In the paraproduct case the corresponding coefficients $C(\mathfrak{n}_1, \mathfrak{n}_2)$ decay as fast as we like, as a consequence of the smoothness (away from the origin) of their symbol, while in the first-commutator case they decay only quadratically.

4.2.3. Proof of the discrete theorem

We now describe the proof of Theorem 4.4. The proof is similar to that used in Chapter 3 for biparameter paraproducts. However, this time we have to be particularly careful about the upper bounds since we need them to grow at most logarithmically.

Let us assume without loss of generality that the families $(\Phi^2_{I_{\mathfrak{n}_2}})_I$ and $(\Phi^3_I)_I$ are lacunary since, as usual, the other possibilities can all be treated in a similar way. Fix the indices $1 < p, q < \infty$ and $0 < r < \infty$ with the property $1/p + 1/q = 1/r$ and functions f, g normalized in L^p and L^q respectively. We will prove that $T_{\mathbb{I}}$ maps $L^p \times L^q$ to $L^{r,\infty}$, since then, as we have seen, the theorem follows easily by standard interpolation arguments.

Because of the duality lemma of Chapter 2, it is clearly enough to show that, given a measurable set $E \subseteq \mathbb{R}$ with $|E| = 1$, we can find $E' \subseteq E$ with $|E'| \simeq 1$ and such that

$$\sum_{I \in \mathbb{I}} \frac{1}{|I|^{1/2}} \left|\left\langle f, \Phi^1_{I_{\mathfrak{n}_1}} \right\rangle\right| \left|\left\langle g, \Phi^2_{I_{\mathfrak{n}_2}} \right\rangle\right| \left|\langle h, \Phi^3_I \rangle\right| \lesssim \log{<}\mathfrak{n}_1{>} \log{<}\mathfrak{n}_2{>}, \tag{4.39}$$

where we have written $h := \chi_{E'}$.

We need to define now the *shifted maximal* operator $M^{\mathfrak{n}_1}$ and the *shifted square* function $S^{\mathfrak{n}_2}$, which will appear in a natural way later. First,

$$M^{\mathfrak{n}_1} f(x) := \sup_{x \in I} \frac{1}{|I|} \int_{\mathbb{R}} |f(y)| \widetilde{\chi}_{I_{\mathfrak{n}_1}}(y)\, dy$$

where recall that $\widetilde{\chi}_{I_{\mathfrak{n}_1}}(y)$ denotes the function

$$\widetilde{\chi}_{I_{\mathfrak{n}_1}}(x) = \left(1 + \frac{\operatorname{dist}(x, I_{\mathfrak{n}_1})}{|I_{\mathfrak{n}_1}|}\right)^{-100}$$

while $S^{\mathfrak{n}_2}$ is defined by

$$S^{\mathfrak{n}_2} g(x) := \left(\sum_I \frac{|\langle g, \Phi^2_{I_{\mathfrak{n}_2}} \rangle|^2}{|I|} 1_I(x)\right)^{1/2}.$$

We will prove later on that both operators are bounded on every L^p space for $1 < p < \infty$, with bounds of the type $O(\log{<}\mathfrak{n}_1{>})$ and $O(\log{<}\mathfrak{n}_2{>})$ respectively. Until then we will use these two facts and define an exceptional set as follows.

We start by defining the set $\widetilde{\Omega_0}$:

$$\widetilde{\Omega_0} := \left\{x \,\middle|\, M^{\mathfrak{n}_1} f(x) > C \log{<}\mathfrak{n}_1{>} \right\} \cup \left\{x \,\middle|\, S^{\mathfrak{n}_2} g(x) > C \log{<}\mathfrak{n}_2{>} \right\}.$$

Let d be a nonnegative integer and ℓ another integer with the property $2^d \leq |\ell| < 2^{d+1}$. For any such d and ℓ define the set Ω^d_ℓ by

$$\Omega^d_\ell := \left\{x \,\middle|\, M^{\mathfrak{n}_1 - \ell} f(x) > C \log{<}\mathfrak{n}_1 - \ell{>}\, 2^{5d} \right\}$$

and then consider $\widetilde{\widetilde{\Omega_0}}$, defined by

$$\widetilde{\widetilde{\Omega_0}} := \bigcup_{d \geq 0} \bigcup_{2^d \leq |\ell| < 2^{d+1}} \Omega^d_\ell.$$

Then define the set $\widetilde{\widetilde{\widetilde{\Omega}}}_0$ in a similar way to $\widetilde{\widetilde{\Omega}}_0$ but using the function g and the index $\mathfrak{n}_2$ instead. Next, define

$$\Omega_0 := \widetilde{\Omega}_0 \cup \widetilde{\widetilde{\Omega}}_0 \cup \widetilde{\widetilde{\widetilde{\Omega}}}_0 \tag{4.40}$$

and, finally, the exceptional set

$$\Omega := \left\{ x \,\middle|\, M(1_{\Omega_0})(x) > \frac{1}{100} \right\}.$$

Given that $|\Omega| < 1/10$ for large enough C, one can define the set E' by $E' := E \setminus \Omega$ and observe that $|E'| \simeq 1$ as intended. To estimate (4.39) properly, as in Chapter 3 we split it into two parts:

$$\sum_{I \cap \Omega^c \neq \emptyset} + \sum_{I \cap \Omega^c = \emptyset} := I + II. \tag{4.41}$$

***Estimates for term I in* (4.41)**

As before, we observe first that, since $I \cap \Omega^c \neq \emptyset$, one requires that

$$\frac{|I \cap \Omega_0|}{|I|} \leq \frac{1}{100}$$

or, in other words,

$$|I \cap \Omega_0^c| > \frac{99}{100}|I|.$$

Again our approach will be to use three independent *stopping-time arguments* for the functions f, g, h and then to combine them, as we did in Chapter 3.

Start by defining

$$\Omega_1 = \left\{ x \,\middle|\, M^{\mathfrak{n}_1}(f)(x) > \frac{C \log <\mathfrak{n}_1>}{2^1} \right\}$$

and then set

$$\mathbb{I}_1 = \left\{ I \in \mathbb{I} \,\middle|\, |I \cap \Omega_1| > \frac{1}{100}|I| \right\}.$$

Now define

$$\Omega_2 = \left\{ x \,\middle|\, M^{\mathfrak{n}_1}(f)(x) > \frac{C \log <\mathfrak{n}_1>}{2^2} \right\}$$

and set

$$\mathbb{I}_2 = \left\{ I \in \mathbb{I} \setminus \mathbb{I}_1 \,\middle|\, |I \cap \Omega_2| > \frac{1}{100}|I| \right\},$$

and so on. The constant $C > 0$ is precisely that used in the definition of the set E' given before. Since $\mathbb{I}$ is finite this process stops after a while, producing the sets $(\Omega_n)_n$ and $(\mathbb{I}_n)_n$.

Independently of all the above, define

$$\Omega_1' = \left\{x \,\middle|\, S^{\mathfrak{n}_2}(g)(x) > \frac{C \log <\mathfrak{n}_2>}{2^1}\right\},$$

set

$$\mathbb{I}_1' = \left\{I \in \mathbb{I} \,\middle|\, |I \cap \Omega_1'| > \frac{1}{100}|I|\right\};$$

now define

$$\Omega_2' = \left\{x \,\middle|\, S^{\mathfrak{n}_2}(g)(x) > \frac{C \log <\mathfrak{n}_2>}{2^2}\right\}$$

and set

$$\mathbb{I}_2' = \left\{I \in \mathbb{I} \setminus \mathbb{I}_1' \,\middle|\, |I \cap \Omega_2'| > \frac{1}{100}|I|\right\},$$

and so on, thus constructing finitely many sets $(\{\Omega_n'\})_n$ and $(\{\mathbb{I}_n'\})_n$. We want to have a similar decomposition for the function h and, as before, we need first to construct the analogue of the set Ω_0 for h. Choose a large enough integer $N > 0$ such that, for every $I \in \mathbb{I}$, one has

$$|I \cap \Omega''^{c}_{-N}| > \frac{99}{100}|I|,$$

where we define analogously

$$\Omega''_{-N} = \{x \,|S(h)(x) > C2^N\}.$$

This is clearly possible since there are only finitely many intervals in our collection $\mathbb{I}$. Then, precisely as before, we define

$$\Omega''_{-N+1} = \left\{x \,\middle|\, S(h)(x) > \frac{C2^N}{2^1}\right\}$$

and set

$$\mathbb{I}''_{-N+1} = \left\{I \in \mathbb{I} \,\middle|\, |I \cap \Omega''_{-N+1}| > \frac{1}{100}|I|\right\}.$$

Again we define

$$\Omega''_{-N+2} = \left\{x \,\middle|\, S(h)(x) > \frac{C2^N}{2^2}\right\}$$

and set

$$\mathbb{I}''_{-N+2} = \left\{ I \in \mathbb{I} \setminus \mathbb{I}''_{-N+1} \;\middle|\; |I \cap \Omega''_{-N+2}| > \frac{1}{100}|I| \right\},$$

and so on, generating finitely many sets $(\Omega''_n)_n$ and $(\mathbb{I}''_n)_n$.

Combining these decompositions, we can split the term I in (4.41) as

$$\sum_{\ell_1,\ell_2>0,\ell_3>-N} \sum_{I\in\mathbb{I}_{\ell_1,\ell_2,\ell_3}} \frac{1}{|I|^{3/2}} |\langle f, \Phi^1_{I_{\mathfrak{n}_1}}\rangle||\langle g, \Phi^2_{I_{\mathfrak{n}_2}}\rangle||\langle h, \Phi^3_I\rangle||I|, \tag{4.42}$$

where

$$\mathbb{I}_{\ell_1,\ell_2,\ell_3} := \mathbb{I}_{\ell_1} \cap \mathbb{I}'_{\ell_2} \cap \mathbb{I}''_{\ell_3}.$$

Observe that since I belongs to $\mathbb{I}_{\ell_1,\ell_2,\ell_3}$ this implies that it has not been selected at any of the previous $\ell_1 - 1$, $\ell_2 - 1$, and $\ell_3 - 1$ steps respectively, and this means that $|I \cap \Omega_{\ell_1-1}|$, $|I \cap \Omega'_{\ell_2-1}|$, and $|I \cap \Omega''_{\ell_3-1}|$ are in consequence all less than or equal to $\frac{1}{100}|I|$. Equivalently, one requires that

$$|I \cap \Omega^c_{\ell_1-1}| > \frac{99}{100}|I|,$$

$$|I \cap \Omega'^c_{\ell_2-1}| > \frac{99}{100}|I|$$

and

$$|I \cap \Omega''^c_{\ell_3-1}| > \frac{99}{100}|I|$$

and, in particular, that

$$|I \cap \Omega^c_{\ell_1-1} \cap \Omega'^c_{\ell_2-1} \cap \Omega''^c_{\ell_3-1}| > \frac{97}{100}|I|. \tag{4.43}$$

Using this in (4.42) one can estimate that expression by

$$\begin{aligned}
&\sum_{\ell_1,\ell_2>0,\ell_3>-N} \sum_{I\in\mathbb{I}_{\ell_1,\ell_2,\ell_3}} \frac{1}{|I|^{3/2}} |\langle f, \Phi^1_{I_{\mathfrak{n}_1}}\rangle|\, |\langle g, \Phi^2_{I_{\mathfrak{n}_2}}\rangle|\, |\langle h, \Phi^3_I\rangle| \\
&\qquad\qquad \times |I \cap \Omega^c_{\ell_1-1} \cap \Omega'^c_{\ell_2-1} \cap \Omega''^c_{\ell_3-1}| \\
&= \sum_{\ell_1,\ell_2>0,\ell_3>-N} \int_{\Omega^c_{\ell_1-1}\cap\Omega'^c_{\ell_2-1}\cap\Omega''^c_{\ell_3-1}} \sum_{I\in\mathbb{I}_{\ell_1,\ell_2,\ell_3}} \frac{|\langle f, \Phi^1_{I_{\mathfrak{n}_1}}\rangle|}{|I|^{1/2}} \frac{|\langle g, \Phi^2_{I_{\mathfrak{n}_2}}\rangle|}{|I|^{1/2}} \frac{|\langle h, \Phi^3_I\rangle|}{|I|^{1/2}} \chi_I(x)\,dx \\
&\lesssim \sum_{\ell_1,\ell_2>0,\ell_3>-N} \int_{\Omega^c_{\ell_1-1}\cap\Omega'^c_{\ell_2-1}\cap\Omega''^c_{\ell_3-1}\cap\Omega_{\mathbb{I}_{\ell_1,\ell_2,\ell_3}}} M^{\mathfrak{n}_1}(f)(x) S^{\mathfrak{n}_2}(g)(x) S(h)(x)\,dx \\
&\lesssim \sum_{\ell_1,\ell_2>0,\ell_3>-N} \log \langle\mathfrak{n}_1\rangle \log \langle\mathfrak{n}_2\rangle\, 2^{-\ell_1}2^{-\ell_2}2^{-\ell_3} |\Omega_{\mathbb{I}_{\ell_1,\ell_2,\ell_3}}|,
\end{aligned} \tag{4.44}$$

where

$$\Omega_{\mathbb{I}_{\ell_1,\ell_2,\ell_3}} := \bigcup_{I \in \mathbb{I}_{\ell_1,\ell_2,\ell_3}} I.$$

Observe that we also have

$$\begin{aligned} |\Omega_{\mathbb{I}_{\ell_1,\ell_2,\ell_3}}| \le |\Omega_{\mathbb{I}_{\ell_1}}| &\le \left| \left\{ x \,\middle|\, M(\chi_{\Omega_{\ell_1}})(x) > \frac{1}{100} \right\} \right| \\ &\lesssim |\Omega_{\ell_1}| = \left| \left\{ x \,\middle|\, M^{\mathfrak{n}_1}(f)(x) > \frac{C \log <\mathfrak{n}_1>}{2^{\ell_1}} \right\} \right| \lesssim 2^{\ell_1 p}. \end{aligned}$$

Similarly, we may also deduce that

$$|\Omega_{\mathbb{I}_{\ell_1,\ell_2,\ell_3}}| \lesssim 2^{\ell_2 q}$$

and thus

$$|\Omega_{\mathbb{I}_{\ell_1,\ell_2,\ell_3}}| \lesssim 2^{\ell_3 \alpha}$$

for every $\alpha > 1$. Here we have used the boundedness of the operators $M^{\mathfrak{n}_1}$, $S^{\mathfrak{n}_2}$, and S on L^s for any $1 < s < \infty$ and, in addition, that $|E_3'| \simeq 1$. Together these imply that

$$|\Omega_{\mathbb{I}_{\ell_1,\ell_2,\ell_3}}| \lesssim 2^{\ell_1 p \theta_1} 2^{\ell_2 q \theta_2} 2^{\ell_3 \alpha \theta_3} \tag{4.45}$$

for any parameters $0 \le \theta_1, \theta_2, \theta_3 < 1$ with the property that $\theta_1 + \theta_2 + \theta_3 = 1$.

Now split (4.44) as follows:

$$\begin{aligned} &\sum_{\ell_1,\ell_2>0,\ell_3\ge 0} \log <\mathfrak{n}_1> \log <\mathfrak{n}_2> 2^{-\ell_1} 2^{-\ell_2} 2^{-\ell_3} |\Omega_{\mathbb{I}_{\ell_1,\ell_2,\ell_3}}| \\ &\quad + \sum_{\ell_1,\ell_2>0,0>\ell_3>-N} \log <\mathfrak{n}_1> \log <\mathfrak{n}_2> 2^{-\ell_1} 2^{-\ell_2} 2^{-\ell_3} |\Omega_{\mathbb{I}_{\ell_1,\ell_2,\ell_3}}|. \end{aligned} \tag{4.46}$$

We estimate the first term in (4.46) using the inequality (4.45) for $\theta_1, \theta_2, \theta_3$ such that $p\theta_1 < 1$, $q\theta_2 < 1$, and $\alpha\theta_3 < 1$. To estimate the other term we use (4.45) for $\theta_1, \theta_2, \theta_3$ such that $p\theta_1 < 1$, $q\theta_2 < 1$, and $\alpha\theta_3 > 1$. These indeed make the sum in (4.46) $O(\log <\mathfrak{n}_1> \log <\mathfrak{n}_2>)$ as desired. This completes the discussion of term I in (4.41).

Estimates for term II in (4.41)

The second term in (4.41) is simpler to understand. Observe at the start that the intervals which appear in our sum are those in Ω. Split the collection of them as $\bigcup_{d\geq 0} \mathbb{I}_d$, where

$$\mathbb{I}_d := \left\{ I \in \mathbb{I} \,\middle|\, I \subseteq \Omega \text{ and } 2^d \leq \frac{\text{dist}(I, \Omega^c)}{|I|} < 2^{d+1} \right\}.$$

It is important to realize that, since we are on the real line, for any $d \geq 0$ we have

$$\sum_{I \in \mathbb{I}_d} |I| \lesssim |\Omega| \lesssim 1.$$

In particular, for any $I \in \mathbb{I}_d$ we know that $2^d I \cap \Omega^c = \emptyset$. Therefore there must exist another interval $\widetilde{I}$ that is dyadic, has the same length, and lies ℓ steps of length $|I|$ away from I (with $2^d \leq |\ell| < 2^{d+1}$) and also has the property that $\widetilde{I} \cap \Omega^c \neq \emptyset$. This means that $I_{\mathfrak{n}_1}$ and $I_{\mathfrak{n}_2}$ are $\mathfrak{n}_1 - \ell$ and $\mathfrak{n}_2 - \ell$ steps of the same length $|I|$ away from $\widetilde{I}$. Putting all this together, we finally estimate term II by

$$\begin{aligned}
&\sum_{d\geq 0} \sum_{I \in \mathbb{I}_d} \frac{|\langle f, \Phi^1_{I_{\mathfrak{n}_1}} \rangle|}{|I|^{1/2}} \frac{|\langle g, \Phi^2_{I_{\mathfrak{n}_2}} \rangle|}{|I|^{1/2}} \frac{|\langle h, \Phi^3_I \rangle|}{|I|^{1/2}} |I| \\
&\lesssim \sum_{d\geq 0} \sum_{2^d \leq |\ell| < 2^{d+1}} \sum_{I \in \mathbb{I}_d} (\log <\mathfrak{n}_1 - \ell>) 2^{5d} (\log <\mathfrak{n}_2 - \ell>) 2^{5d} 2^{-Md} |I| \\
&\lesssim \log <\mathfrak{n}_1> \log <\mathfrak{n}_2>,
\end{aligned} \tag{4.47}$$

which completes our proof of Theorem 4.4.

4.2.4. Logarithmic estimates for the shifted maximal operator

We are left with proving the logarithmic bounds for the *shifted* maximal and square operators used before. We start this subsection with a theorem regarding the shifted maximal operator.

Theorem 4.5 *For any integer $\mathfrak{n} \in \mathbb{Z}$ the shifted maximal operator $M^{\mathfrak{n}}$ is bounded on every L^p space for $1 < p < \infty$, with an upper bound of the type $O(\log <\mathfrak{n}>)$.*

Proof Let us first observe that it is enough to prove the theorem for the corresponding *sharp* maximal operator $\widetilde{M}^{\mathfrak{n}}$ defined by

$$\widetilde{M}^{\mathfrak{n}} f(x) := \sup_{x \in I} \frac{1}{|I_{\mathfrak{n}}|} \int_{I_{\mathfrak{n}}} |f(y)|\, dy, \tag{4.48}$$

where, as before, the supremum is taken only over dyadic intervals. Now consider a point x and an interval I such that $x \in I$. One has

$$\frac{1}{|I_{\mathrm{n}}|}\int_{\mathbb{R}} |f(y)|\widetilde{\chi}_{I_{\mathrm{n}}}\,dy \lesssim \sum_{\ell\in\mathbb{Z}}\left(\frac{1}{|I_{\mathrm{n}}^{\ell}|}\int_{I_{\mathrm{n}}^{\ell}} |f(y)|\,dy\right)\frac{1}{<\ell>^{100}},$$

where we denote by I_{n}^{ℓ} the dyadic interval having the same length as I_{n} but lying ℓ steps of length $|I_{\mathrm{n}}|$ away from it. Assuming that the theorem holds for $\widetilde{M}^{\mathrm{n}}$, one has

$$\begin{aligned}\|M^{\mathrm{n}}f\|_p &\lesssim \sum_{\ell\in\mathbb{Z}}\frac{1}{<\ell>^{100}}\left\|\widetilde{M}^{\mathrm{n}+\ell}f\right\|_p \lesssim \sum_{\ell\in\mathbb{Z}}\frac{1}{<\ell>^{100}}(\log <\mathrm{n}+\ell>)\|f\|_p\\ &\lesssim \sum_{\ell\in\mathbb{Z}}\frac{1}{<\ell>^{100}}(\log(<\mathrm{n}><\ell>))\|f\|_p \lesssim \log <\mathrm{n}>\ \|f\|_p,\end{aligned}$$

as desired. We therefore need to prove the theorem for $\widetilde{M}^{\mathrm{n}}$.

Fix $\lambda > 0$. We claim that the following inequality holds:

$$\left|\left\{x\,|\widetilde{M}^{\mathrm{n}}f(x) > \lambda\right\}\right| \lesssim (\log <\mathrm{n}>)|\{x\,|Mf(x) > \lambda\}|, \tag{4.49}$$

where M is now the maximal Hardy–Littlewood operator. Assuming (4.49), the theorem for $\widetilde{M}^{\mathrm{n}}$ follows immediately from the Hardy–Littlewood theorem, by interpolation with an obvious L^{∞} estimate.

To prove (4.49) we denote by $\mathbb{I}_{\mathrm{n}}^{\lambda}$ the collection of all dyadic intervals I_{n} that are maximal with respect to inclusion and for which

$$\frac{1}{|I_{\mathrm{n}}|}\int_{I_{\mathrm{n}}} |f(y)|\,dy > \lambda.$$

In particular, all these intervals are disjoint, and

$$\bigcup_{I_{\mathrm{n}}\in\mathbb{I}_{\mathrm{n}}^{\lambda}} I_{\mathrm{n}} = \{x\,|Mf(x) > \lambda\}.$$

For every selected maximal dyadic interval I_{n} consider its dyadic subintervals, which could be of length $|I_{\mathrm{n}}|$, $|I_{\mathrm{n}}|/2$, $|I_{\mathrm{n}}|/2^2$, and so on. It is important to realize at this point that there exist only $(\log <\mathrm{n}>) + 1$ disjoint dyadic intervals $I_{\mathrm{n}}^1, I_{\mathrm{n}}^2, \ldots, I_{\mathrm{n}}^{(\log<\mathrm{n}>)+1}$ having the same length as $|I_{\mathrm{n}}|$ and such that, on translation by $-\mathrm{n}$ corresponding units of every smaller dyadic subinterval of I_{n}, they become subintervals in $I_{\mathrm{n}}^1, I_{\mathrm{n}}^2, \ldots, I_{\mathrm{n}}^{(\log<\mathrm{n}>)+1}$. We now claim that

$$\left\{x\,|\widetilde{M}^{\mathrm{n}}f(x) > \lambda\right\} \subseteq \bigcup_{I_{\mathrm{n}}\in\mathbb{I}_{\mathrm{n}}^{\lambda}}\left(I_{\mathrm{n}}\cup I_{\mathrm{n}}^1\cup\cdots\cup I_{\mathrm{n}}^{(\log<\mathrm{n}>)+1}\right).$$

In order to check this, consider a point z such that $\widetilde{M}^{\mathfrak{n}} f(z) > \lambda$. In particular, there exists a dyadic interval J containing z, so that

$$\frac{1}{|J_{\mathfrak{n}}|} \int_{J_{\mathfrak{n}}} |f(y)|\, dy > \lambda.$$

By the previous construction one can clearly find $I_{\mathfrak{n}} \in \mathbb{I}^{\lambda}_{\mathfrak{n}}$ with the property that $J_{\mathfrak{n}} \subseteq I_{\mathfrak{n}}$. This implies that the interval J itself will be a subset of one of the previous $I^{j}_{\mathfrak{n}}$ for $1 \leq j \leq (\log <\mathfrak{n}>) + 1$. This proves our claim.

Now using the disjointness of the maximal intervals $I_{\mathfrak{n}}$ we obtain (4.49), as desired. □

4.2.5. Logarithmic estimates for the shifted square function operator

As we know, we are left with the proof of the following theorem.

Theorem 4.6 *For any integer $\mathfrak{n} \in \mathbb{Z}$, the shifted square function operator $S^{\mathfrak{n}}$ is bounded on every L^p space for $1 < p < \infty$, with an upper bound of the type $O(\log <\mathfrak{n}>)$.*

Proof Clearly the proof can be based on a classical Calderón–Zygmund decomposition similar to that used in Theorem 2.5. However, as in the case of the shifted maximal functions, one has to be much more careful now since we need upper bounds that grow at most logarithmically.

Let us start by observing that $S^{\mathfrak{n}}$ is bounded on L^2, with a bound that is independent of $\mathfrak{n}$. The reason is that

$$\|S^{\mathfrak{n}} f\|_2 = \left(\sum_I \langle f, \Phi_{I_{\mathfrak{n}}} \rangle^2 \right)^{1/2}$$

and this expression is comparable with the L^2 norm of the standard discretized Littlewood–Paley square function operator, which is bounded on L^2.

The main part of the proof is devoted to showing that

$$\|S^{\mathfrak{n}} f\|_{1,\infty} \lesssim (\log <\mathfrak{n}>) \|f\|_1 \tag{4.50}$$

or, more explicitly, that

$$|\{x \in \mathbb{R} | S^{\mathfrak{n}} f(x) > \lambda\}| \lesssim \log <\mathfrak{n}> \frac{1}{\lambda} \|f\|_1. \tag{4.51}$$

Fix $\lambda > 0$ and perform a Calderón–Zygmund decomposition for the given function f at level λ. Pick, one at a time, maximal dyadic intervals J with the property

$$\frac{1}{|J|} \int_J |f(y)|\, dy > \lambda.$$

All these intervals are disjoint; we denote their union Ω. In particular,

$$|\Omega| = \sum_J |J| < \frac{1}{\lambda} \sum_J \int_J |f(y)|\, dy \le \frac{1}{\lambda} \|f\|_1. \tag{4.52}$$

Now we decompose the function f as a sum of a good and a bad function,

$$f = g + b,$$

where

$$g := f\chi_{\Omega^c} + \sum_J \left(\frac{1}{|J|} \int_J f(y)\, dy \right) \chi_J,$$
$$b := f - g = \sum_J b_J,$$

and

$$b_J := \left(f - \frac{1}{|J|} \int_J f(y)\, dy \right) \chi_J.$$

Notice that $\operatorname{supp} b_J \subseteq J$. We also have

$$|f(x)| \le \lambda$$

for every $x \in \Omega^c$, and so

$$\|g\|_\infty \lesssim \lambda,$$

since in addition we observe that

$$\left| \frac{1}{|J|} \int_J f(y)\, dy \right| \le \frac{1}{|J|} \int_J |f(y)|\, dy \le \frac{2}{|\widetilde{J}|} \int_{\widetilde{J}} |f(y)|\, dy \le 2\lambda,$$

where $\widetilde{J}$ is dyadic, contains J, and is twice as long as J. We observe also that

$$\int_{\mathbb{R}} b_J(y)\, dy = 0,$$

using the definition of b_J, and that

$$\|b_J\|_1 = \int_J |b_J(y)|\, dy \le \int_J |f(y)|\, dy + \left(\frac{1}{|J|} \int_J |f(y)|\, dy \right) |J|$$
$$\lesssim \int_J |f(y)|\, dy \lesssim \lambda |J|.$$

These properties imply that

$$|\{x \in \mathbb{R} | S^{\mathrm{n}} f(x) > \lambda\}|$$
$$\le \left| \left\{ x \in \mathbb{R} \,\middle|\, S^{\mathrm{n}} g(x) > \frac{\lambda}{2} \right\} \right| + \left| \left\{ x \in \mathbb{R} \,\middle|\, S^{\mathrm{n}} b(x) > \frac{\lambda}{2} \right\} \right|. \tag{4.53}$$

To estimate the first term in (4.53) we have to use the L^2-boundedness of S^n and write

$$\begin{aligned}\left|\left\{x \in \mathbb{R} \,\middle|\, S^n g(x) > \frac{\lambda}{2}\right\}\right| &\lesssim \frac{1}{\lambda^2}\|S^n g\|_2^2 \\ &\lesssim \frac{1}{\lambda^2}\|g\|_2^2 = \frac{1}{\lambda^2}\int_{\mathbb{R}} |g(x)|^2\,dx \lesssim \frac{1}{\lambda^2}\lambda\int_{\mathbb{R}} |g(x)|\,dx \\ &= \frac{1}{\lambda}\|g\|_1 \lesssim \frac{1}{\lambda}\left(\int_{\Omega^c} |f(x)|\,dx + \sum_J \int_J |f(x)|\,dx\right) = \frac{1}{\lambda}\|f\|_1,\end{aligned}$$

as desired.

For the second term in (4.53) we proceed as follows. For any interval J, consider the *associated* intervals $J^1, J^2, \ldots, J^{(\log<n>)+1}$, as defined in the previous section, and the set Ω_J given by

$$\Omega_J := 5J \cup 5J^1 \cup 5J^2 \cup \cdots \cup 5J^{(\log<n>)+1}.$$

One has

$$\begin{aligned}\left|\left\{x \in \mathbb{R} \,\middle|\, S^n b(x) > \frac{\lambda}{2}\right\}\right| \leq &\left|\left\{x \in \bigcup_J \Omega_J \,\middle|\, S^n b(x) > \frac{\lambda}{2}\right\}\right| \\ &+ \left|\left\{x \in \left(\bigcup_J \Omega_J\right)^c \,\middle|\, S^n b(x) > \frac{\lambda}{2}\right\}\right|. \qquad (4.54)\end{aligned}$$

We estimate the first term by

$$\begin{aligned}\left|\left\{x \in \bigcup_J \Omega_J \,\middle|\, S^n b(x) > \frac{\lambda}{2}\right\}\right| \leq \left|\bigcup_J \Omega_J\right| &\lesssim (\log<n>)\sum_J |J| \\ &\lesssim (\log<n>)\frac{1}{\lambda}\|f\|_1.\end{aligned}$$

The second term can be estimated by

$$\frac{1}{\lambda}\int_{(\bigcup_J \Omega_J)^c} S^n b(x)\,dx \leq \frac{1}{\lambda}\sum_J \int_{(\bigcup_J \Omega_J)^c} S^n b_J(x)\,dx \leq \frac{1}{\lambda}\sum_J \int_{(\Omega_J)^c} S^n b_J(x)\,dx.$$

We claim now that, for every interval J,

$$\int_{(\Omega_J)^c} S^n b_J(x)\,dx \lesssim \lambda|J|. \qquad (4.55)$$

Assuming (4.55), one can improve on the previous inequality, so that we obtain

$$\frac{1}{\lambda}\lambda\sum_J |J| \lesssim |\Omega| \lesssim \frac{1}{\lambda}\|f\|_1,$$

as desired.

In consequence we are left with proving the claim (4.55). Since the left-hand side of (4.55) is smaller than

$$\int_{(\Omega_J)^c} \sum_I \frac{|\langle b_J, \Phi_{I_{\mathfrak{n}}}\rangle|}{|I|^{1/2}} 1_I(x)\,dx = \sum_I \int_{(\Omega_J)^c} \frac{|\langle b_J, \Phi_{I_{\mathfrak{n}}}\rangle|}{|I|^{1/2}} 1_I(x)\,dx$$

$$= \sum_{|I|\le|J|} \int_{(\Omega_J)^c} \frac{|\langle b_J, \Phi_{I_{\mathfrak{n}}}\rangle|}{|I|^{1/2}} 1_I(x)\,dx$$

$$+ \sum_{|I|>|J|} \int_{(\Omega_J)^c} \frac{|\langle b_J, \Phi_{I_{\mathfrak{n}}}\rangle|}{|I|^{1/2}} 1_I(x)\,dx := A + B,$$

we need to understand both these terms.

Estimates for A. The crucial observation here is that since $|I| \le |J|$ and $I \cap (\Omega_J)^c \ne \emptyset$ one must have $I_{\mathfrak{n}} \cap 3J = \emptyset$. Using this, one can estimate A by

$$\sum_{|I_{\mathfrak{n}}|\le|J|} \left(1 + \frac{\mathrm{dist}(I_{\mathfrak{n}}, J)}{|I_{\mathfrak{n}}|}\right)^{-10} \int_{\mathbb{R}} |b_J(y)|\,dy \lesssim \lambda|J| \sum_{|I_{\mathfrak{n}}|\le|J|} \left(1 + \frac{\mathrm{dist}(I_{\mathfrak{n}}, J)}{|I_{\mathfrak{n}}|}\right)^{-10}$$

$$\lesssim \lambda|J|,$$

as required by (4.55).

Estimates for B. The fact that

$$\int_{\mathbb{R}} b_J(y)\,dy = 0 \tag{4.56}$$

will play an important role. First one estimates B by

$$\sum_{|I_{\mathfrak{n}}|>|J|} |\langle b_J, \widetilde{\Phi}_{I_{\mathfrak{n}}}\rangle|,$$

where $\widetilde{\Phi}_{I_{\mathfrak{n}}} := |I_{\mathfrak{n}}|^{1/2}\Phi_{I_{\mathfrak{n}}}$ is this time L^∞-normalized. The dependence on $\mathfrak{n}$ is irrelevant now so we can rewrite the above expression as

$$\sum_{|L|>|J|} |\langle b_J, \widetilde{\Phi}_L\rangle|,$$

where the sum runs over generic dyadic intervals L.

Let L satisfy $|L| > |J|$ and notice that

$$|\langle b_J, \widetilde{\Phi}_L\rangle| = \left|\int_{\mathbb{R}} b_J(z)\overline{\widetilde{\Phi}_L}(z)\,dz\right| = \left|\int_J b_J(z)\left(\overline{\widetilde{\Phi}_L}(z) - \overline{\widetilde{\Phi}_L}(c_J)\right)dz\right|,$$

where c_J is the center of the interval J.

Observe also that for $z \in J$ one has

$$\left|\overline{\widetilde{\Phi}_L(z)} - \overline{\widetilde{\Phi}_L(c_J)}\right| \lesssim |J| \frac{1}{|L|} \left(1 + \frac{\operatorname{dist}(L, J)}{|L|}\right)^{-10}$$

and so, in particular, the term $|\langle b_J, \widetilde{\Phi}_L \rangle|$ is smaller than

$$|J| \frac{1}{|L|} \left(1 + \frac{\operatorname{dist}(L, J)}{|L|}\right)^{-10} \int_J |b_J(y)|\, dy \lesssim |J| \frac{1}{|L|} \left(1 + \frac{\operatorname{dist}(L, J)}{|L|}\right)^{-10} \lambda |J|.$$

Finally, the claim (4.55) follows from the observation that

$$\sum_{|L| > |J|} \frac{|J|}{|L|} \left(1 + \frac{\operatorname{dist}(L, J)}{|L|}\right)^{-10} \lesssim 1.$$

By interpolating now between weak-L^1 and L^2 we automatically obtain the theorem for every $1 < p \leq 2$. To prove the rest of the estimates we argue, as usual, by duality. Consider $2 < p < \infty$. Using Khinchine's inequality one finds that

$$\begin{aligned}
\|S^{\mathrm{n}} f\|_p^p &= \int_{\mathbb{R}} \left(\sum_I \frac{|\langle f, \Phi_{I_{\mathrm{n}}} \rangle|^2}{|I|} \chi_I(x) \right)^{p/2} dx \\
&\lesssim \int_{\mathbb{R}} \int_0^1 \left| \sum_I r_I(t) \langle f, \Phi_{I_{\mathrm{n}}} \rangle h_I(x) \right|^p dx dt \\
&= \int_0^1 \left\| \sum_I r_I(t) \langle f, \Phi_{I_{\mathrm{n}}} \rangle h_I \right\|_p^p dt, \qquad (4.57)
\end{aligned}$$

where the $(r_I)_I$ are as usual the Rademacher functions and the $(h_I)_I$ are the L^2-normalized Haar functions. Their definitions and properties may be recalled from Section I.8.4. Now fix $t \in [0, 1]$ and consider the operator

$$f \mapsto \sum_I r_I(t) \langle f, \Phi_{I_{\mathrm{n}}} \rangle h_I.$$

Given that S^{n} and the Littlewood–Paley square function associated with $(h_I)_I$ are both bounded below L^2, an argument similar to that in Theorem 4.4 proves that this operator is also bounded below L^2 and, by duality, above L^2 as well, with bounds that are uniform in t and which grow at most logarithmically in <n>. Using this fact in (4.57) completes our proof. □

4.3. Generalizations

First let us remark that the method presented above proves more, namely the following theorem.

Theorem 4.7 *The first Calderón commutator C_1 extends naturally as a bilinear operator bounded from $L^p \times L^q$ into L^r for every $1 < p, q \leq \infty$ with $1/p + 1/q = 1/r$ and $1/2 < r < \infty$.*

Indeed, this is a straightforward consequence of the fact that the series

$$\sum_{\mathfrak{n}_1, \mathfrak{n}_2 \in \mathbb{Z}} |C(\mathfrak{n}_1, \mathfrak{n}_2)|^r \log <\mathfrak{n}_1> \log <\mathfrak{n}_2> \tag{4.58}$$

is convergent for every $r > 1/2$, as long as the sequence $C(\mathfrak{n}_1, \mathfrak{n}_2)$ decays at least quadratically in $\mathfrak{n}_1$ and $\mathfrak{n}_2$.

Exercise 4.7 Review the proof of Calderón's theorem, Theorem 4.1, and convince yourself of the validity of its general form.

Now observe that a simple change of variables allows one to rewrite C_1 as

$$C_1(f)(x) = p.v. \int_{\mathbb{R}} \left(\frac{\Delta_t}{t} A(x) \right) f(x+t) \frac{dt}{t}, \tag{4.59}$$

where Δ_t is the *finite difference* operator at scale t, defined by

$$\Delta_t g(x) := g(x+t) - g(x).$$

Given also that C_1 coincides with the commutator $[|D|, A]$ it is natural to ask what can be said about the double commutator $[|D|, [|D|, A]]$.

A direct calculation shows that modulo a universal constant the expression $[|D|, [|D|, A]](f)(x)$ is equal to

$$p.v. \int_{\mathbb{R}^2} \left(\frac{\Delta_t}{t} \circ \frac{\Delta_s}{s} A(x) \right) f(x+t+s) \frac{dt}{t} \frac{ds}{s}. \tag{4.60}$$

The above formula can be viewed as a bilinear operator, this time depending on f and A''. Its symbol can be calculated easily and is given by

$$\left(\int_0^1 \operatorname{sgn}(\xi + \alpha \xi_1)\, d\alpha \right)^2$$

which is precisely the square of the symbol of the first commutator C_1 (see (4.59)).

Theorem 4.8 *Let $a \neq 0$ and $b \neq 0$ and consider the expression*

$$p.v. \int_{\mathbb{R}^2} \left(\frac{\Delta_{at}}{t} \circ \frac{\Delta_{bs}}{s} A(x) \right) f(x+t+s) \frac{dt}{t} \frac{ds}{s}.$$

Viewed as a bilinear operator in A'' and f, it extends naturally as a bounded operator from $L^p \times L^q$ into L^r for every $1 < p, q \leq \infty$ with $1/p + 1/q = 1/r$ and $1/2 < r < \infty$.

To prove this theorem one can use the method described earlier for the first commutator. First, an easy calculation shows that the symbol of this operator is given by

$$\left(\int_0^1 \operatorname{sgn}(\xi + \alpha a \xi_1)\, d\alpha\right)\left(\int_0^1 \operatorname{sgn}(\xi + \beta b \xi_1)\, d\beta\right). \tag{4.61}$$

Then, one needs to realize that each factor in (4.61) satisfies the quadratic estimates as before. In other words, now one has to decompose each factor separately as a double Fourier series and after that to continue the previous argument. The details are left to the reader.

Another generalization comes from the Leibnitz identity for functions A and B:

$$A'B' = (AB)'' - (BA')' - (AB')' + A'B'. \tag{4.62}$$

In particular, the right-hand side of (4.62) satisfies Hölder estimates of the type

$$\|(AB)'' - (BA')' - (AB')' + A'B'\|_r \lesssim \|A'\|_p \|B'\|_q$$

for p, q, r as in the theorem. A natural question is whether this inequality continues to hold if one replaces every derivative D by its modulus $|D|$. A direct calculation shows that the resulting expression,

$$|D|^2(AB) - |D|\big(B\,|D|\,A\big) - |D|\big(A\,|D|\,B\big) + (|D|A)(|D|B),$$

can be also written as

$$p.v. \int_{\mathbb{R}^2} \left(\frac{\Delta_t}{t} A(x+s)\right)\left(\frac{\Delta_s}{s} B(x+t)\right) \frac{dt}{t}\frac{ds}{s}. \tag{4.63}$$

As before, we can view this formula as a bilinear operator in A' and B'. Its symbol can be calculated immediately; it is given by

$$\left(\int_0^1 \operatorname{sgn}(\xi_1 + \alpha \xi_2)\, d\alpha\right)\left(\int_0^1 \operatorname{sgn}(\xi_2 + \beta \xi_1)\, d\beta\right), \tag{4.64}$$

which this time is a symmetric function in the variables ξ_1 and ξ_2. We call such expressions *circular commutators.*

Theorem 4.9 *Let $a \neq 0$ and $b \neq 0$ and consider the expression*

$$p.v. \int_{\mathbb{R}^2} \left(\frac{\Delta_{at}}{t} A(x+s)\right)\left(\frac{\Delta_{bs}}{s} B(x+t)\right) \frac{dt}{t}\frac{ds}{s}.$$

Viewed as a bilinear operator in A' and B', it extends naturally as a bounded operator from $L^p \times L^q$ into L^r for every $1 < p, q \le \infty$ with $1/p + 1/q = 1/r$ and $1/2 < r < \infty$.

The proof uses the same technique as before, since each factor of the symbol satisfies the same *quadratic estimates.*

4.4. The Cauchy integral on Lipschitz curves

Let us start by recalling the definition of this operator. Consider a Lipschitz function A on the real line $\mathbb{R}$ that defines a Lipschitz curve Γ in the complex plane by the parametrization $x \mapsto x + iA(x)$. The singular integral operator

$$C_\Gamma f(x) := p.v. \int_{\mathbb{R}} \frac{f(y)}{(x-y) + i(A(x) - A(y))}\, dy \tag{4.65}$$

is called the Cauchy integral associated to the Lipschitz curve C_Γ.

The goal of this section is to prove the following theorem of Coifman, McIntosh, and Meyer.

Theorem 4.10 *The operator C_Γ extends naturally as a bounded linear operator from L^p into L^p for any $1 < p < \infty$.*

As we pointed out before, one can reduce this problem to that of proving polynomial bounds for the corresponding Calderón commutators, defined by

$$C_d f(x) := p.v. \int_{\mathbb{R}} \frac{(A(x) - A(y))^d}{(x-y)^{d+1}} f(y)\, dy. \tag{4.66}$$

More specifically, it is enough to show that

$$\|C_d f\|_p \le C(d)C(p)\|f\|_p \|A'\|_\infty^d \tag{4.67}$$

for every $f \in L^p$, where the constant $C(d)$ grows at most polynomially with respect to d. Indeed, this is a simple consequence of the decomposition (4.15) and of the fact that $\|A'\|_\infty$ can always be assumed to be strictly smaller than 1, as we have seen already.

Standard calculations, similar to those performed in the case of the first commutator, C_1, show that if both $a := A'$ and f are Schwartz functions then (4.66) exists and can be rewritten as

$$\int_{\mathbb{R}^{d+1}} \left(\int_{[0,1]^d} \operatorname{sgn}(\xi + \alpha_1\xi_1 + \cdots + \alpha_d\xi_d)\, d\alpha_1 \cdots d\alpha_d \right)$$
$$\times\, \widehat{f}(\xi)\widehat{a}(\xi_1) \cdots \widehat{a}(\xi_d) e^{2\pi i x(\xi+\xi_1+\cdots+\xi_d)}\, d\xi d\xi_1 \cdots d\xi_d. \tag{4.68}$$

Exercise 4.8 Check the formula (4.68).

In particular, the operator C_d can be interpreted as a $(d+1)$-linear operator. However, as in the case of C_1 its symbol

$$m_d(\xi, \xi_1, \ldots, \xi_d) := \int_{[0,1]^d} \operatorname{sgn}(\xi + \alpha_1\xi_1 + \cdots + \alpha_d\xi_d)\, d\alpha_1 \cdots d\alpha_d \quad (4.69)$$

is not a classical Marcinkiewicz–Mikhlin–Hörmander symbol, and because of this there are no estimates for C_d that can be reduced to the multilinear paraproduct theorem of Coifman and Meyer from Chapter 2. Proving (4.67), even without polynomial bounds, is a more challenging problem.

Remember that the proof of the boundedness of the first commutator was based on the observation that, even though $m_1(\xi, \xi_1)$ is not a classical symbol, when we smoothly restrict it to Whitney squares the Fourier coefficients of the corresponding functions decay at least quadratically. This important property, together with the logarithmical bounds that we derived for the shifted Hardy–Littlewood maximal functions and Littlewood–Paley square functions, helped us to achieve our goal.

However, it should not be difficult to realize that these ideas alone would not be enough to prove the desired polynomial bounds in (4.67). Indeed, even if one could prove quadratic estimates for the Fourier coefficients of $m_d(\xi, \xi_1, \ldots, \xi_d)$, the mere fact that one would eventually have to sum $O(d)$ power series would produce an exponential upper bound of the type C^d.

To avoid this problem, the main new idea is to realize that instead of treating m_d as a multiplier depending on each of its $d+1$ variables one should always see it as being a multiple average of various m_1-type multipliers. This may sound a little vague and also surprising, but it turns out to be the correct point of view. In fact, as will be pointed out later on, the symbols of C_d for $d \geq 2$ do not seem to satisfy the same quadratic estimates that are available for m_1.

Coming back to (4.67), we will prove the following theorem.

Theorem 4.11 *Let $1 < p_1, \ldots, p_{d+1} \leq \infty$ and $1 \leq p < \infty$ be such that $1/p_1 + \cdots + 1/p_{d+1} = 1/p$. Denote also by ℓ the number of indices j for which $p_j \neq \infty$. Then C_d extends naturally as a $(d+1)$-linear operator bounded from $L^{p_1} \times \cdots \times L^{p_{d+1}}$ into L^p with an operator bound of the type*

$$C(d)C(\ell)C(p_1)\cdots C(p_{d+1}) \quad (4.70)$$

where $C(d)$ grows at most polynomially in d and $C(p_j) = 1$ if $p_j = \infty$ for $1 \leq j \leq d+1$.

Let us observe that (4.67) follows easily from Theorem 4.11 in the particular case $p_1 = p$ and $p_2 = \cdots = p_{d+1} = \infty$.

Recall now that since C_d is a $(d+1)$-linear operator, it has $d+1$ natural *adjoints*. Consider first the associated $(d+2)$-linear form Λ_d defined by

$$\int_{\mathbb{R}} C_d(f_1, \ldots, f_{d+1})(x) f_{d+2}(x)\, dx = \Lambda_d(f_1, \ldots, f_{d+2}), \tag{4.71}$$

for Schwartz functions $f_1, \ldots, f_{d+2}$.

Then, for every $1 \le j \le d+1$ one defines the adjoint C_d^{*j} by

$$\int_{\mathbb{R}} C_d^{*j}(f_1, \ldots, f_{j-1}, f_{j+1}, \ldots, f_{d+2})(x) f_j(x)\, dx = \Lambda_d(f_1, \ldots, f_{d+2}) \tag{4.72}$$

again for Schwartz functions $f_1, \ldots, f_{d+2}$. For notational symmetry, we set $C_d := C_d^{*d+2}$.

In order to prove Theorem 4.11 we will prove that for every $1 \le j \le d+2$ and Schwartz functions $\phi_1, \ldots, \phi_{d+1}$, one has

$$\left\| C_d^{*j}(\phi_1, \ldots, \phi_{d+1}) \right\|_p \le C(d)C(\ell)C(p_1)\cdots C(p_{d+1}) \|\phi_1\|_{p_1} \cdots \|\phi_{d+1}\|_{p_{d+1}}, \tag{4.73}$$

where p_j, $1 \le j \le d+1$, and p are exactly as before. From these inequalities one can extend C_d^{*j} by the density of the class of Schwartz functions in the space L^{p_j}, whenever $1 < p_j < \infty$, and also in $\mathbf{S}^\infty$ (the closure of the set of Schwartz functions in L^∞) if it should happen that $p_j = \infty$.

In the next subsection we will use duality arguments to show how one can define C_d^{*j} on arbitrary products of L^{p_j} and L^∞ spaces. These duality arguments will also explain the appearance of the larger range of estimates in Theorem 4.11 and (4.73) for C_d and all its adjoints.

4.4.1. Extension by duality

For every $\mathfrak{l} = 0, 1, \ldots, d$ let us denote by $S(\mathfrak{l})$ the *statement* that the estimates (4.73) for C_d and all its adjoints can be extended naturally to the case when at most $\mathfrak{l}$ of the L^{p_j} spaces are equal to L^∞ and the rest are either $\mathbf{S}^\infty$ or correspond to an index j for which $1 < p_j < \infty$. Since above we promised to prove the $S(0)$ case later (this is equivalent to (4.73)), proceeding by induction it is enough to prove that $S(d)$ holds true. To demonstrate that $S(\mathfrak{l})$ implies $S(\mathfrak{l}+1)$, fix indices $1 < p_1, \ldots, p_{d+1} \le \infty$ as in the assumption of Theorem 4.11. For symmetry reasons (in particular, that all the adjoints can be analyzed in the same way) we can assume that we may extend (4.73) for C_d when the first $\mathfrak{l}+1$ functions $f_1, \ldots, f_{\mathfrak{l}+1}$ belong to L^∞ while all the other $\phi_{\mathfrak{l}+2}, \ldots, \phi_{d+1}$ are Schwartz functions.

Case I: $p > 1$. This case is simpler since $(L^p)^* = L^{p'}$ for $1/p + 1/p' = 1$ and, as a consequence, $p' > 1$ as well. Using duality we define $C_d(f_1, \ldots, f_{\mathfrak{l}+1}, \phi_{\mathfrak{l}+2}, \ldots, \phi_{d+1})$ to be the unique L^p function having the property that

$$\int_{\mathbb{R}} C_d(f_1, \ldots, f_{\mathfrak{l}+1}, \phi_{\mathfrak{l}+2}, \ldots, \phi_{d+1})(x)\, \phi_{d+2}(x)\, dx$$
$$= \int_{\mathbb{R}} C_d^{*1}(f_2, \ldots, f_{\mathfrak{l}+1}, \phi_{\mathfrak{l}+2}, \ldots, \phi_{d+1}, \phi_{d+2})(x) f_1(x)\, dx$$

for any $L^{p'}$ normalized Schwartz function ϕ_{d+2}. Notice that the last expression is clearly well defined since we know $S(\mathfrak{l})$ for C_d^{*1}.

Case II: $p = 1$. This case is more difficult since the dual of L^1 is L^∞ and the Schwartz functions are not dense in it. Let us first remark that, since $p = 1$, there exist at least two indices i_1 and i_2 with $1 < p_{i_1}, p_{i_2} < \infty$. By the symmetry of our discussion, we can assume without loss of generality that these indices are $\mathfrak{l} + 2$ and $\mathfrak{l} + 3$. To define $C_d(f_1, \ldots, f_{\mathfrak{l}+1}, \phi_{\mathfrak{l}+2}, \ldots, \phi_{d+1})$ as a function of L^1, we first need to observe that one can define it as a function of, for instance, L^2, using the fact that the functions ϕ_j are Schwartz functions and therefore belong to all the L^s spaces, for $1 < s < \infty$. One can for instance treat $\phi_{\mathfrak{l}+2}$ and $\phi_{\mathfrak{l}+3}$ as being functions in L^4 and the rest of the ϕ_j as being in the space $\mathbf{S}^\infty$. Then, as before, define $C_d(f_1, \ldots, f_{\mathfrak{l}+1}, \phi_{\mathfrak{l}+2}, \ldots, \phi_{d+1})$ to be the unique function in L^2 having the property that

$$\int_{\mathbb{R}} C_d(f_1, \ldots, f_{\mathfrak{l}+1}, \phi_{\mathfrak{l}+2}, \ldots, \phi_{d+1})(x)\, \phi_{d+2}(x)\, dx$$
$$= \int_{\mathbb{R}} C_d^{*1}(f_2, \ldots, f_{\mathfrak{l}+1}, \phi_{\mathfrak{l}+2}, \ldots, \phi_{d+1}, \phi_{d+2})(x) f_1(x)\, dx$$

for any L^2-normalized Schwartz function ϕ_{d+2}, since we can use again $S(\mathfrak{l})$ for the case C_d^{*1}.

So now we know that $C_d(f_1, \ldots, f_{\mathfrak{l}+1}, \phi_{\mathfrak{l}+2}, \ldots, \phi_{d+1})$ is a well-defined L^2 function, and we would like to prove that it is in fact in L^1. We can write, for any $\mu > 0$,

$$\int_{-\mu}^{\mu} |C_d(f_1, \ldots, f_{\mathfrak{l}+1}, \phi_{\mathfrak{l}+2}, \ldots, \phi_{d+1})(x)|\, dx \tag{4.74}$$
$$= \int_{\mathbb{R}} C_d(f_1, \ldots, f_{\mathfrak{l}+1}, \phi_{\mathfrak{l}+2}, \ldots, \phi_{d+1})(x) \widetilde{\chi}_{[-\mu,\mu]}(x)\, dx$$

where clearly $|\widetilde{\chi}_{[-\mu,\mu]}(x)| = \chi_{[-\mu,\mu]}(x)$ almost everywhere.

Consider now a sequence of smooth and compactly supported functions $(g_{d+2}^n)_n$ with the property that $g_{d+2}^n \to \widetilde{\chi}_{[-\mu,\mu]}$ weakly and also such that $\|g_{d+2}^n\|_\infty \leq 1$ (by convolving $\widetilde{\chi}_{[-\mu,\mu]}$ with a smooth approximation of identity, for instance, one can construct such a sequence). As a consequence, one can estimate (4.74) by

$$\lim_n \left| \int_{\mathbb{R}} C_d(f_1, \ldots, f_{l+1}, \phi_{l+2}, \ldots, \phi_{d+1})(x)\, g_{d+2}^n(x)\, dx \right|$$

$$\leq \sup_n \left| \int_{\mathbb{R}} C_d^{*1}(f_2, \ldots, f_{l+1}, \phi_{l+2}, \ldots, \phi_{d+1}, g_{d+2}^n)(x) f_1(x)\, dx \right|$$

and, given that $g_{d+2}^n \in \mathbf{S}^\infty$ and $\|g_{d+2}^n\|_\infty \leq 1$, one can use as before the induction hypothesis to complete the argument.

A careful look at the previous argument shows that if we assume that in (4.73) $C(d)$ grows polynomially then this property will be preserved if all the $\mathbf{S}^\infty$ spaces are replaced by the corresponding L^∞ spaces.

We therefore need to prove (4.73) for C_d and all its adjoints. This inequality is very convenient since the operators C_d^{*i} behave well on Schwartz functions. In particular they are given by well-defined formulae similar to (4.68). Later, we will be able to decompose and discretize them even further (as we did with paraproducts) and reduce (4.73) to similar estimates but for simpler *finite model operators*.

4.4.2. A few remarks on the symbols of C_d for $d \geq 2$

Now we will make a few observations about the symbols of the commutators C_d for $d \geq 2$, which are not particularly encouraging but which it is important to notice from the start.

Consider the symbol corresponding to the second commutator C_2. We want to understand whether its Fourier coefficients satisfy the same quadratic estimates as the coefficients of the symbol for C_1. To check this, choose Schwartz functions $\widehat{\phi}(\xi)$, $\widehat{\phi}(\xi_1)$, and $\widehat{\phi}(\xi_2)$ whose supports are the intervals $[-2, -1]$, $[1, 2]$, and $[-1/2, 1/2]$. Since the function

$$(\xi, \xi_1, \xi_2) \mapsto \widehat{\phi}(\xi)\widehat{\phi}(\xi_1)\widehat{\phi}(\xi_2)$$

is supported on a Whitney cube with respect to the origin in $\mathbb{R}^3$, the corresponding Fourier coefficients to be analyzed are given by (see the start of subsection 4.2.1)

$$\int_{\mathbb{R}^3} \left(\int_{[0,1]^2} 1_{\mathbb{R}_+}(\xi + \alpha\xi_1 + \beta\xi_2)\, d\alpha d\beta \right) \widehat{\varphi}(\xi)\widehat{\varphi}(\xi_1)\widehat{\phi}(\xi_2)$$
$$\times e^{-2\pi i n\xi} e^{-2\pi i n_1\xi_1} e^{-2\pi i n_2\xi_2}\, d\xi d\xi_1 d\xi_2 \tag{4.75}$$

for arbitrary integers n, n_1, n_2. Given that ξ_1 cannot be zero, we can rewrite the symbol in (4.75) as

$$\int_0^1 \frac{1}{\xi_1} \int_0^{\xi_1} 1_{\mathbb{R}_+}(\xi + \alpha + \beta\xi_2)\, d\alpha d\beta. \tag{4.76}$$

Observe that if one differentiates (4.76) with respect to ξ_1 then the inner integral becomes

$$\int_0^1 1_{\mathbb{R}_+}(\xi + \xi_1 + \beta\xi_2)\, d\beta$$

which is precisely $\frac{1}{2}m_1(\xi + \xi_1, \xi_2) + \frac{1}{2}$. Since $\xi + \xi_1$ lies now within the interval $[-1, 1]$ and ξ_2 within $[-1/2, 1/2]$, and they both contain the origin, this expression cannot be differentiated any further. As a consequence, the Fourier coefficients in (4.75) seem to decay only *linearly*, not *quadratically* as before. These simple remarks show also that the extension from C_1 to C_d for $d \geq 2$ will require new ideas.

4.4.3. Some heuristical arguments

Before starting the actual proof of (4.73), we will describe an easier situation, to give the reader some hints about the procedure. Suppose that instead of proving (4.73) we consider the simpler and more particular $L^p \times L^\infty \times \cdots \times L^\infty \to L^p$ estimate for a generic $(d+1)$-linear paraproduct Π_{d+1} whose $(d+2)$-linear form is given by

$$\int_{\mathbb{R}} \sum_k (f_1 * \Phi_k^1)(x) \cdots (f_{d+2} * \Phi_k^{d+2})(x)\, dx, \tag{4.77}$$

where $f_1 \in L^p$ and $f_j \in L^\infty$ for $2 \leq j \leq d+1$ while $f_{d+2} \in L^{p'}$ with $1/p + 1/p' = 1$. We would like these estimates to grow at most polynomially with respect to d and even to be independent of d if possible.

As always, the functions $(\Phi_k^j)_k$ above are smooth L^1-normalized bump functions, adapted to intervals of the type $[-2^{-k}, 2^{-k}]$ for $k \in \mathbb{Z}$ and such that for at least two indices i_1, i_2 one has $\int_{\mathbb{R}} \Phi_k^{i_1}(x)\, dx = \int_{\mathbb{R}} \Phi_k^{i_2}(x)\, dx = 0$. The standard terminology that we will use, is to say that the families corresponding to i_1 and

i_2 are Ψ *type* while the families corresponding to the remaining indices are Φ *type*. We emphasize that it is not necessary to make the stronger assumption (used throughout Chapter 2) that the Fourier transforms of the functions in the Ψ families are supported on the corresponding lacunary intervals $[2^{k-1}, 2^{k+1}]$ for $k \in \mathbb{Z}$; the mere fact that their integrals are zero is enough to guarantee that the *continuous* and *discretized* square functions associated naturally with them still satisfy the usual L^s estimates for every $1 < s < \infty$. The proof of these two statements constitutes Problem 4.1 at the end of the chapter.

Because of these facts, later, when we discretize expressions similar to (4.77), the *adapted* families that the Ψ functions generate will still be called lacunary while those generated by the Φ functions will be called nonlacunary as before.

Several cases can occur, and it is instructive to analyze them one by one.

Case A: $i_1 = 1$ ***and*** $i_2 = d + 2$. We start by making the further essential assumption that the L^1 norms of the functions from the Φ families are bounded by 1. As a consequence of this, for any $2 \le j \le d+1$ in our particular case, one can write

$$\left| f_j * \Phi_k^j(x) \right| \le \left\| \Phi_k^j \right\|_1 \|f_j\|_\infty \le \|f_j\|_\infty. \tag{4.78}$$

We then estimate (4.77) by

$$\begin{aligned}
&\prod_{j=2}^{d+1} \|f_j\|_\infty \int_{\mathbb{R}} \sum_k \left| f_1 * \Phi_k^1(x) \right| \left| f_{d+2} * \Phi_k^{d+2}(x) \right| dx \\
&\le \prod_{j=2}^{d+1} \|f_j\|_\infty \int_{\mathbb{R}} \left(\sum_k \left| f_1 * \Phi_k^1(x) \right|^2 \right)^{1/2} \left(\sum_k \left| f_{d+2} * \Phi_k^{d+2}(x) \right|^2 \right)^{1/2} dx \\
&= \prod_{j=2}^{d+1} \|f_j\|_\infty \int_{\mathbb{R}} S(f_1)(x) S(f_{d+2})(x)\, dx \\
&\le \prod_{j=2}^{d+1} \|f_j\|_\infty \|S(f_1)\|_p \|S(f_{d+2})\|_{p'} \lesssim \prod_{j=2}^{d+1} \|f_j\|_\infty \|f_1\|_p \|f_{d+2}\|_{p'},
\end{aligned}$$

as desired, using the usual boundedness properties of the square functions.

Case B: $i_1 = 1$ ***and*** $i_2 = 2$. This case is a little more complicated. The most natural way to estimate (4.77) this time would appear to be by

$$\prod_{j=3}^{d+1} \|f_j\|_\infty \int_{\mathbb{R}} \sum_k \left|f_1 * \Phi_k^1(x)\right| \left|f_2 * \Phi_k^2(x)\right| \left|f_{d+2} * \Phi_k^{d+2}(x)\right| dx$$

$$\leq \prod_{j=3}^{d+1} \|f_j\|_\infty \int_{\mathbb{R}} \left(\sum_k \left|f_1 * \Phi_k^1(x)\right|^2\right)^{1/2} \left(\sum_k \left|f_2 * \Phi_k^2(x)\right|^2\right)^{1/2} \times \left(\sup_k \left|f_{d+2} * \Phi_k^{d+2}(x)\right|\right) dx$$

$$= \prod_{j=3}^{d+1} \|f_j\|_\infty \int_{\mathbb{R}} S(f_1)(x)\, S(f_2)(x)\, M(f_{d+2})(x)\, dx$$

$$\leq \prod_{j=3}^{d+1} \|f_j\|_\infty \|S(f_1)\|_{r_1} \|S(f_2)\|_{r_2} \|M(f_{d+2})\|_{r_3'}$$

$$\lesssim \prod_{j=3}^{d+1} \|f_j\|_\infty \|f_1\|_{r_1} \|f_2\|_{r_2} \|f_{d+2}\|_{r_3'},$$

which holds for $1 < r_1, r_2, r_3 < \infty$ with $1/r_1 + 1/r_2 = 1/r_3$. The estimate for which we are aiming corresponds to the particular case $r_1 = r_3 = p$ and $r_2 = \infty$ but unfortunately it cannot be obtained as above since the square function operator S is known to be unbounded on L^∞.

To get around this problem we freeze the functions $f_3, \ldots, f_{d+1}$ and regard the expression (4.77) as a three-linear form depending only on f_1, f_2, and f_{d+2}. In particular, the above estimate proves that its associated bilinear operator $\Pi_2(f_1, f_2)$ is bounded from $L^{r_1} \times L^{r_2}$ into L^{r_3}. Because of the symmetry of this operator, the same estimates remain true for both of its adjoints, Π_2^{*1} and Π_2^{*2}. Now the estimate that we are seeking becomes equivalent to the boundedness of

$$\Pi_2 : L^p \times L^\infty \to L^p; \tag{4.79}$$

to be able to deduce it we would need, besides the previous Banach estimates, to prove some quasi-Banach estimates as well, of the type $\Pi_2^{*2} : L^{r_1} \times L^{r_2} \to L^{r_3}$ for any $1 < r_1, r_2 < \infty, 0 < r_3 < \infty$ with $1/r_1 + 1/r_2 = 1/r_3$. Such estimates can be proved using methods from Chapters 2 or 3. At the end one just uses multilinear interpolation between the Banach and quasi-Banach estimates to deduce the intermediate step (4.79). The multilinear interpolation argument to which we have referred is explained in detail in the appendix at the end of the book.

In the present case one can use two Banach and one quasi-Banach estimates, having implicit boundedness constants C_B^1, C_B^2, C_{q-B}, so that if $C_B := \max\{C_B^1, C_B^2, C_{q-B}\}$ is the maximum of the three then this constant is certainly an upper bound for the boundedness constant of (4.79), as desired.

Case C: $i_1 = 2$ ***and*** $i_2 = 3$. This last case is similar to case B but it is again a step harder. We will describe it in full detail. Again, the natural way to majorize (4.77) would appear to be by

$$
\begin{aligned}
&\prod_{j=4}^{d+1} \|f_j\|_\infty \int_{\mathbb{R}} \sum_k |f_1 * \Phi_k^1(x)|\,|f_2 * \Phi_k^2(x)|\,|f_3 * \Phi_k^3(x)|\,|f_{d+2} * \Phi_k^{d+2}(x)|\,dx \\
&\leq \prod_{j=4}^{d+1} \|f_j\|_\infty \int_{\mathbb{R}} \left(\sum_k |f_2 * \Phi_k^2(x)|^2\right)^{1/2} \left(\sum_k |f_3 * \Phi_k^3(x)|^2\right)^{1/2} \\
&\qquad\qquad \times \left(\sup_k |f_1 * \Phi_k^1(x)|\right)\left(\sup_k |f_{d+2} * \Phi_k^{d+2}(x)|\right) dx \\
&= \prod_{j=4}^{d+1} \|f_j\|_\infty \int_{\mathbb{R}} S(f_2)(x)\, S(f_3)(x)\, M(f_1)(x)\, M(f_{d+2})(x)\, dx \\
&\leq \prod_{j=4}^{d+1} \|f_j\|_\infty \|M(f_1)\|_{r_1} \|S(f_2)\|_{r_2} \|S(f_3)\|_{r_3} \|M(f_{d+2})\|_{r_4'} \\
&\lesssim \prod_{j=4}^{d+1} \|f_j\|_\infty \|f_1\|_{r_1} \|f_2\|_{r_2} \|f_3\|_{r_3} \|f_{d+2}\|_{r_4'},
\end{aligned}
$$

which is valid for every $1 < r_1, r_2, r_3, r_4 < \infty$ such that $1/r_1 + 1/r_2 + 1/r_3 = 1/r_4$. However, our particular estimate corresponds to $r_1 = r_4 = p$ and $r_2 = r_3 = \infty$ and, as in case B, it cannot be deduced in this way.

What one does this time is to freeze the functions $f_4, \ldots, f_{d+1}$ and treat the expression (4.77) as a four-linear form depending on f_1, f_2, f_3, and f_{d+2}. The estimates derived earlier show that its associated three-linear operator $\Pi_3(f_1, f_2, f_3)$ is bounded from $L^{r_1} \times L^{r_2} \times L^{r_3}$ into L^{r_4}. Using symmetry, we can see that these bounds continue to hold for its adjoints, Π_3^{*1}, Π_3^{*2}, and Π_3^{*3}. The estimate that we are seeking then becomes equivalent to the boundedness of

$$\Pi_3 : L^p \times L^\infty \times L^\infty \to L^p. \tag{4.80}$$

As before, we would like to obtain it by interpolation from various Banach and quasi-Banach estimates. The quasi-Banach estimates that we need now are of the type $\Pi_3^{*2}, \Pi_3^{*3} : L^{r_1} \times L^{r_2} \times L^{r_3} \to L^{r_4}$, for every $1 < r_1, r_2, r_3 < \infty$ and $0 < r_4 < \infty$ with $1/r_1 + 1/r_2 + 1/r_3 = 1/r_4$. As mentioned above, such estimates can be obtained using the methods of Chapters 2 and 3.

Finally, one uses multilinear interpolation between two Banach and two quasi-Banach estimates to obtain the intermediate estimate (4.80). Moreover, as we pointed out before, this method guarantees that if C_B^1, C_B^2, C_{q-B}^1, and C_{q-B}^2 denote the corresponding boundedness constants then the maximum of these, denoted C_B, becomes an upper bound for the boundedness constant of (4.80).

This ends our discussion of the boundedness of the paraproduct Π_{d+1} from $L^p \times L^\infty \times \cdots \times L^\infty$ into L^p, since it is not difficult to see that all other possible cases can be reduced to one of the three that we have discussed.

It is also clear that an analogous argument works in the general case, $\Pi_{d+1} : L^{p_1} \times \cdots \times L^{p_{d+1}} \to L^p$. The only difference is that instead of the minimal bilinear or trilinear operators that appeared before one needs to consider ℓ-linear operators for some $1 \leq \ell \leq d+1$. However, the multilinear interpolation argument works in a completely similar way.

We have learned some important facts from this heuristical argument. First, we note that all the bounds that have been obtained are independent of d. This is a consequence of the essential assumption that the L^1 norms of the Φ families were taken to be less than 1, which implied the crucial relation (4.78).

Then, we recall also that after using (4.78) several times we were able to reduce our analysis to the study of several Banach and quasi-Banach estimates, for some *minimal* bilinear, trilinear, or, in the general case, ℓ-linear operators.

4.4.4. Discrete minimal models

From now on our goal is to make the above heuristical arguments work in the case of the Calderón commutator C_d. We claim that, in spite of all the weaknesses of its symbol, this operator can be treated in essentially the same way. First, the plan is to decompose it into many *paraproduct-like pieces* whose number grows at most polynomially and then to estimate each individual piece independently of d. To obtain the desired estimates we need as before to interpolate *multilinearly* between Banach and quasi-Banach estimates for certain minimal operators that will be proved directly. The Banach estimates are easy, as always, but the quasi-Banach estimates are hard. To obtain them we need to discretize these minimal operators carefully, as in Chapters 2 and 3.

The *minimal* and *discretized* ℓ-linear operators are defined as follows.

Consider a positive integer $1 \leq \ell \leq d+1$, arbitrary integers $\mathbf{n}_1, \ldots, \mathbf{n}_\ell$, and families $(\Phi^1_{I_{\mathbf{n}_1}})_I, (\Phi^2_{I_{\mathbf{n}_2}})_I, \ldots, (\Phi^\ell_{I_{\mathbf{n}_\ell}})_I$ of L^2-normalized bump functions adapted to dyadic intervals of the type $I_{\mathbf{n}_j}$ (recall that, given a dyadic interval I, we denote by $I_{\mathbf{n}_j}$ the dyadic interval having the same length as I but lying $\mathbf{n}_j$ units of length $|I|$ away from I) such that at least two of these families of intervals are lacunary.

We will be a little more precise here and say that a smooth function Φ is said to be adapted to an interval I if one has

$$|\partial^\alpha \Phi(x)| \lesssim \frac{1}{|I|^{|\alpha|}} \frac{1}{(1 + \text{dist}(x, I)/|I|)^M},$$

for any derivative α such that $|\alpha| \leq 5$. Recall that previously we have never needed the above inequality to hold for α greater than 2. For every fixed family $\mathbb{I}$ of dyadic intervals, define the *minimal* ℓ-linear discrete operator $T_{\mathbb{I}}$ by the formula

$$T_{\mathbb{I}}(f_1, \ldots, f_\ell) = \sum_{I \in \mathbb{I}} \frac{1}{|I|^{(\ell-1)/2}} \langle f_1, \Phi^1_{I_{\mathbf{n}_1}} \rangle \cdots \langle f_\ell, \Phi^\ell_{I_{\mathbf{n}_\ell}} \rangle \Phi^{\ell+1}_I. \tag{4.81}$$

The following theorem holds.

Theorem 4.12 *For any finite family $\mathbb{I}$, the ℓ-linear operator $T_{\mathbb{I}}$ maps $L^{p_1} \times \cdots \times L^{p_\ell}$ into L^p for any $1 < p_1, \ldots, p_\ell < \infty$ such that $1/p_1 + \cdots + 1/p_\ell = 1/p$, $0 < p < \infty$, with an upper bound of the type*

$$O(\log \langle \mathbf{n}_1 \rangle \times \cdots \times \log \langle \mathbf{n}_\ell \rangle).$$

The implicit constants are allowed to depend on ℓ but, as usual, are otherwise independent of the cardinality of $\mathbb{I}$.

Clearly, this theorem is the ℓ-linear generalization of the bilinear version, Theorem 4.4. Its proof does not require any new ideas and is therefore left to the reader.

It is also important to recall that, as before, Theorem 4.12 follows (by scale invariance and interpolation) from the more particular statement that, for every $f_j \in L^{p_j}$ such that $\|f_j\|_{p_j} = 1$ for $1 \leq j \leq \ell$ and every measurable set $E \subseteq \mathbb{R}$ with $|E| = 1$, there exists a subset $E' \subseteq E$ of comparable measure such that the inequality

$$\sum_{I\in\mathbb{I}} \frac{1}{|I|^{(\ell-1)/2}} |\langle f_1, \Phi^1_{I_{\mathfrak{n}_1}}\rangle| \cdots |\langle f_\ell, \Phi^\ell_{I_{\mathfrak{n}_\ell}}\rangle| |\langle f_{\ell+1}, \Phi^{\ell+1}_I\rangle| \tag{4.82}$$

$$\lesssim \log <\mathfrak{n}_1> \times \cdots \times \log <\mathfrak{n}_\ell>$$

holds for $f_{\ell+1} = \chi_{E'}$. The fact that the boundedness constants above grow only logarithmically will be very helpful, as before.

4.4.5. Reduction to the discrete minimal model

As we mentioned earlier, we are left with explaining how (4.73) can be reduced to the discretized estimate (4.82). Basically, this will occupy the rest of the chapter. Since the operator C_d and its adjoints have the same $(d+2)$-linear form, it will be clearly enough to treat the case of C_d. To be able to decompose the multilinear form, as promised, we will use Littlewood–Paley decompositions for each individual function. However, since we want to keep the essential inequalities (4.78) available, we will need to work with noncompact (in frequency) Littlewood–Paley projections most of the time. This fact alone will cause technical difficulties later, as we will see.

We define these noncompact Littlewood–Paley decompositions in detail in the next subsection.

4.4.6. Noncompact Littlewood–Paley projections

Let $\Phi(x)$ be a Schwartz function that is positive and even and has the property that $\int_{\mathbb{R}} \Phi(x)\, dx = 1$. Define $\Psi(x)$ by

$$\Psi(x) = \Phi(x) - \frac{1}{2}\Phi\left(\frac{x}{2}\right)$$

and notice that $\int_{\mathbb{R}} \Psi(x)\, dx = 0$. For every integer $k \in \mathbb{Z}$, consider as usual the functions $\Psi_k(x)$ and $\Phi_k(x)$ given by $2^k\Psi(2^k x)$ and $2^k\Phi(2^k x)$ respectively. It is important to observe from the beginning that all the L^1 norms of the Φ_k functions are equal to 1. Observe also that

$$\Psi_k(x) = \Phi_k(x) - \Phi_{k-1}(x)$$

and that, for every $k_0 \in \mathbb{Z}$,

$$\sum_{k\le k_0} \Psi_k = \Phi_{k_0}. \tag{4.83}$$

In particular

$$\sum_{k\in\mathbb{Z}} \Psi_k = \delta_0, \tag{4.84}$$

or, equivalently,

$$\sum_{k\in\mathbb{Z}} \widehat{\Psi_k}(\xi) = 1 \tag{4.85}$$

for every $\xi \in \mathbb{R} \setminus \{0\}$. Observe also that

$$\widehat{\Psi}(0) = \int_{\mathbb{R}} \Psi(x)dx = 1 - 1 = 0.$$

Then, given that $\widehat{\Psi}(\xi) = \int_{\mathbb{R}} \Psi(x)e^{-2\pi i x\xi}\, dx$, one has

$$\widehat{\Psi}'(\xi) = -2\pi i \int_{\mathbb{R}} x\Psi(x)e^{-2\pi i x\xi}\, dx$$

and, as a consequence,

$$\widehat{\Psi}'(0) = -2\pi i \int_{\mathbb{R}} x\Psi(x)\, dx = 0$$

using the fact that Ψ was chosen to be even. In particular, $\widehat{\Psi}(\xi)$ can be written as

$$\widehat{\Psi}(\xi) = \xi^2 \widehat{\phi}(\xi) \tag{4.86}$$

for another smooth and rapidly decaying function ϕ.

These are our *noncompact Littlewod–Paley decompositions*. Recall that the *compact* ones are constructed similarly, the only difference being that the Schwartz function Φ has the property that $\operatorname{supp} \widehat{\Phi} \subseteq [-1, 1]$ and $\widehat{\Phi} = 1$ on $[-1/2, 1/2]$.

4.4.7. The generic decomposition of C_d

Let us start by writing the $(d+2)$-linear form $\Lambda_d(f, f_1, \ldots, f_{d+1})$ associated with C_d as

$$\begin{aligned} &\int_{\xi+\xi_1+\cdots+\xi_{d+1}=0} \left(\int_{[0,1]^d} 1_{\mathbb{R}^+}(\xi + \alpha_1\xi_1 + \cdots + \alpha_d\xi_d)\, d\alpha_1 \cdots d\alpha_d \right) \\ &\quad \times \widehat{f}(\xi)\widehat{f_1}(\xi_1)\cdots \widehat{f_{d+1}}(\xi_{d+1})\, d\xi\, d\xi_1 \cdots d\xi_{d+1}, \end{aligned} \tag{4.87}$$

for Schwartz functions $f, f_1, \ldots, f_{d+1}$. This is a consequence of (4.68). Notice that we have replaced the sign function by $1_{\mathbb{R}^+}$, as before. Using $d+2$ Littlewood–Paley decompositions (4.85), one can decompose the identity as

$$1 = \sum_{k_0,k_1,\cdots,k_d,k_{d+1}\in\mathbb{Z}} \widehat{\Psi_{k_0}}(\xi)\widehat{\Psi_{k_1}}(\xi_1)\cdots \widehat{\Psi_{k_d}}(\xi_d)\widehat{\Psi_{k_{d+1}}}(\xi_{d+1}). \tag{4.88}$$

Now, given any $(d+2)$-tuple $(k_0, k_1, \ldots, k_d, k_{d+1}) \in \mathbb{Z}^{d+2}$, one must have $k_0 \geq k_1, \ldots, k_d, k_{d+1}$ or $k_1 \geq k_0, k_2, \ldots, k_{d+1} \cdots$ or $k_{d+1} \geq k_0, k_1, \ldots, k_d$. If one replaces some of the above inequalities with the corresponding strict ones, one can also guarantee that all the $(d+2)$ regions of $\mathbb{Z}^{d+2}$ are disjoint.

Then, in each of these $d+2$ situations, fix the largest index and sum over the rest. Use (4.83) multiple times and rewrite the constant 1 in (4.88) as

$$\sum_k \widehat{\Psi_k}(\xi)\widehat{\Phi_k}(\xi_1)\cdots\widehat{\Phi_k}(\xi_d)\widehat{\Phi_k}(\xi_{d+1}) \\ +\cdots+\sum_k \widehat{\Phi_k}(\xi)\widehat{\Phi_k}(\xi_1)\cdots\widehat{\Phi_k}(\xi_d)\widehat{\Psi_k}(\xi_{d+1}). \tag{4.89}$$

Clearly, to be completely rigorous, some of the indices k in (4.89) should in fact be equal to $k-1$, but for simplicity we will leave them as they are.

Notice that we have $d+2$ factors in each term in the decomposition (4.89), each containing a single function of the type Ψ. We also assume that for the ξ and ξ_{d+1} variables we use the *compact* Littlewood–Paley decomposition while for the rest of the variables we use the *noncompact* decomposition (4.85). This may seem artificial at this point but it will be used later.

Assume now that in addition we have $\xi+\xi_1+\cdots+\xi_{d+1}=0$ and consider, for example, the second sum in (4.89). We rewrite its $k=0$ term for simplicity as

$$\widehat{\Phi}(\xi)\widehat{\Psi}(\xi_1)\cdots\widehat{\Phi}(\xi_d)\widehat{\Phi}(\xi_{d+1}). \tag{4.90}$$

Recall that from (4.86) we know that $\widehat{\Psi}(\xi_1)=\xi_1^2\widehat{\phi}(\xi_1)$. Rewrite this as

$$\begin{aligned}\widehat{\Psi}(\xi_1) &= \xi_1\widehat{\phi}(\xi_1)(-\xi-\xi_2-\cdots-\xi_{d+1}) \\ &= -\xi_1\xi\widehat{\phi}(\xi_1)-\xi_1\xi_2\widehat{\phi}(\xi_1)-\cdots-\xi_1\xi_{d+1}\widehat{\phi}(\xi_1).\end{aligned} \tag{4.91}$$

Using this new decomposition in (4.90), one can write the latter as a sum of $O(d)$ terms, each containing two functions of Ψ type because, besides $\xi_1\widehat{\phi}(\xi_1)$, one sees in addition expressions either of the type $\xi_j\widehat{\Phi}(\xi_j)$ for $j=2,\ldots,d+1$ or of the type $\xi\widehat{\Phi}(\xi)$.

One can clearly do this for every scale $k\in\mathbb{Z}$ and every term in (4.89), obtaining in consequence a decomposition of $1_{\{\xi+\xi_1+\cdots+\xi_{d+1}=0\}}$ as a large sum of $O(d^2)$ expressions whose generic terms now contain precisely two functions of the Ψ type. More specifically, the Ψ functions (at scale 1) are of the form $\eta\widehat{\phi}(\eta)$. If one finally uses this new decomposition in the formula (4.87) one obtains $O(d^2)$ $(d+2)$-linear forms. These will be studied in detail in the rest of the chapter. This completes our *generic decomposition*.

It should be clear at this point that if one wants to go further then one needs to understand the symbol corresponding to C_d.

The indices corresponding to the positions of the Ψ functions will play an important role. We denote them by i_1, i_2 for $0 \le i_1, i_2 \le d+1$. There are three distinct cases, which will be analyzed one by one.

4.4.8. Case I: $i_1 = 0$ and $i_2 = 1$

We start by rewriting (for symmetry) the corresponding $(d+2)$-linear form as

$$\sum_k \int_{\xi+\xi_1+\ldots+\xi_{d+1}=0} \left(\int_{[0,1]^d} 1_{\mathbb{R}^+}(\xi + \alpha_1\xi_1 + \cdots + \alpha_d\xi_d)\, d\alpha_1 \cdots d\alpha_d \right)$$

$$\times \widehat{\Phi_k^0}(\xi)\widehat{\Phi_k^1}(\xi_1) \cdots \widehat{\Phi_k^d}(\xi_d)\widehat{\Phi_k^{d+1}}(\xi_{d+1})\widehat{f}(\xi)\widehat{f_1}(\xi_1) \cdots \widehat{f_{d+1}}(\xi_{d+1})$$

$$\times\, d\xi d\xi_1 \cdots d\xi_{d+1}. \tag{4.92}$$

Recall that both the families $(\Phi_k^0)_k$ and $(\Phi_k^1)_k$ are of the Ψ type. Since the variable ξ_1 corresponds to a Ψ function, it is in some sense special. The idea now is to view the symbol of C_d as a *multiple average* of symbols of the C_1 type that depend on ξ_1 and on a new variable $\widetilde{\xi}$, which we define by $\widetilde{\xi} = \xi + \alpha_2\xi_2 + \cdots + \alpha_d\xi_d$. Ideally, one would then like to decompose the new symbol $m_1(\widetilde{\xi}, \xi_1)$ as we did when we studied the first Calderón commutator. However, given that the functions $\widehat{\Phi_k^j}(\xi_j)$ for $1 \le j \le d$ are not compactly supported, one has to insert two other *compact* Littlewood–Paley decompositions into (4.92) for such an argument to be possible. Write[1]

$$1 = \sum_{k_0,k_1} \widehat{\Psi_{k_0}}(\widetilde{\xi})\widehat{\Psi_{k_1}}(\xi_1) = \sum_{k_0 \ll k_1} \cdots + \sum_{k_0 \simeq k_1} \cdots + \sum_{k_0 \gg k_1} \cdots \tag{4.93}$$

which modulo the usual minor, harmless, errors can be rewritten as

$$\sum_r \widehat{\Phi_r}(\widetilde{\xi})\widehat{\Psi_r}(\xi_1) + \sum_r \widehat{\Psi_r}(\widetilde{\xi})\widehat{\Psi_r}(\xi_1) + \sum_r \widehat{\Psi_r}(\widetilde{\xi})\widehat{\Phi_r}(\xi_1).$$

By using this expression in (4.92), the latter splits into a sum of three distinct terms. We will denote them by I_a, I_b, and I_c and will analyze them separately. The term I_a is given in the following equation. Terms I_b and I_c are similar.

$$\sum_r \sum_k \int_{\xi+\xi_1+\cdots+\xi_{d+1}=0} \left(\int_{[0,1]^d} 1_{\mathbb{R}^+}(\xi + \alpha_1\xi_1 + \cdots + \alpha_d\xi_d)\, d\alpha_1 \cdots d\alpha_d \right)$$

$$\times \widehat{\Phi_r}(\tilde{\xi})\widehat{\Psi_r}(\xi_1)\widehat{\Phi_k^0}(\xi)\widehat{\Phi_k^1}(\xi_1) \cdots \widehat{\Phi_k^d}(\xi_d)\widehat{\Phi_k^{d+1}}(\xi_{d+1})\widehat{f}(\xi)\widehat{f_1}(\xi_1) \cdots \widehat{f_{d+1}}(\xi_{d+1})$$

$$\times\, d\xi d\xi_1 \cdots d\xi_{d+1} = I_a. \tag{4.94}$$

[1] As earlier, $k_a \ll k_b$ means $k_a < k_b - 100$ and $k_a \simeq k_b$ means $k_a - 100 \le k_b \le k_a + 100$.

Term I_a

Term I_a of (4.92) is given by (4.94). To understand the effect of the decomposition over r, consider for simplicity the particular term corresponding to $k = 0$. All our arguments will be *scale invariant* in any case. Let us first ignore the symbol $\int_{[0,1]^d} 1_{\mathbb{R}^+}(\xi + \alpha_1\xi_1 + \cdots + \alpha_d\xi_d)d\alpha_1 \cdots d\alpha_d$ and rewrite the rest of the expression as

$$\sum_r \left(\widehat{\Phi_r}(\tilde{\xi})\widehat{\Psi_r}(\xi_1)\right) \widehat{\Phi_0^0}(\xi)\widehat{\Phi_0^1}(\xi_1)\cdots \widehat{\Phi_0^d}(\xi_d)\widehat{\Phi_0^{d+1}}(\xi_{d+1})$$
$$= \sum_{r\le 0}\cdots + \sum_{r>0}\cdots = I_a' + I_a''. \tag{4.95}$$

***Term I_a' of* (4.95)** Given that $\widehat{\Phi_0^1}(\xi_1)$ is of the Ψ type (recall that it is of the form $\xi_1\widehat{\phi}(\xi_1)$) and, using also the fact that $\widehat{\Psi_r}(\xi_1)$ is compactly supported, one can rewrite the term I_a' as

$$\sum_{r\le 0} 2^r \widehat{\Phi_r}(\tilde{\xi})\widehat{\Phi_0^0}(\xi)\widehat{\Psi_r^1}(\xi_1)\cdots \widehat{\Phi_0^d}(\xi_d)\widehat{\Phi_0^{d+1}}(\xi_{d+1})$$
$$= \sum_{r\le 0} 2^r \left(\widehat{\widetilde{\Phi}_r}(\tilde{\xi})\widehat{\widetilde{\Psi}_r^1}(\xi_1)\right) \widehat{\Phi_0^0}(\xi)\widehat{\Psi_r^1}(\xi_1)\cdots \widehat{\Phi_0^d}(\xi_d)\widehat{\Phi_0^{d+1}}(\xi_{d+1})\widehat{\Phi_r}(\tilde{\xi})$$

for compactly supported functions $\widehat{\widetilde{\Phi}_r}(\tilde{\xi})$, $\widehat{\Psi_r^1}(\xi_1)$, and $\widehat{\widetilde{\Psi}_r^1}(\xi_1)$ (clearly, the first is of the Φ type while the other two are of the Ψ type).

In consequence, one can split the symbol

$$\left(\int_0^1 1_{\mathbb{R}^+}(\tilde{\xi} + \alpha_1\xi_1)\,d\alpha_1\right)\widehat{\widetilde{\Phi}_r}(\tilde{\xi})\widehat{\widetilde{\Psi}_r^1}(\xi_1)$$

as a *double Fourier series* (exactly as in the case of the first Caldeórn commutator, studied earlier)

$$\sum_{n,n_1} C_{n,n_1}^r e^{2\pi i n\tilde{\xi}/2^r} e^{2\pi i n_1\xi_1/2^r}, \tag{4.96}$$

where

$$C_{n,n_1}^r = \frac{1}{2^r}\frac{1}{2^r}\int_{\mathbb{R}^2}\left(\int_0^1 1_{\mathbb{R}^+}(\tilde{\xi} + \alpha_1\xi_1)\,d\alpha_1\right)\widehat{\widetilde{\Phi}_r}(\tilde{\xi})\widehat{\widetilde{\Psi}_r^1}(\xi_1)e^{-2\pi i n\tilde{\xi}/2^r}e^{-2\pi i n_1\xi_1/2^r}\,d\tilde{\xi}\,d\xi_1$$
$$= \int_{\mathbb{R}^2}\left(\int_0^1 1_{\mathbb{R}^+}(\tilde{\xi} + \alpha_1\xi_1)\,d\alpha_1\right)\widehat{\widetilde{\Phi}_0}(\tilde{\xi})\widehat{\widetilde{\Psi}_0^1}(\xi_1)e^{-2\pi i n\tilde{\xi}}e^{-2\pi i n_1\xi_1}\,d\tilde{\xi}\,d\xi_1.$$

Note that the above expression is independent of r.

Remember that these Fourier coefficients satisfy the crucial *quadratic estimates*

$$|C^r_{n,n_1}| =: |C_{n,n_1}| \lesssim \frac{1}{<n>^2} \frac{1}{<n_1>^{100}}, \tag{4.97}$$

where instead of 100 one can take an arbitrarily large integer.

Using all of the above, the contribution of I'_a in (4.92) is equal to

$$\begin{aligned}
&\int_{[0,1]^{d-1}} \sum_{r\leq 0} 2^r \sum_{n,n_1} C^r_{n,n_1} \int_{\xi+\xi_1+\cdots+\xi_d+\xi_{d+1}=0} \left(\widehat{\Phi^0_0}(\xi) e^{2\pi i n\xi/2^r}\right) \\
&\times \left(\widehat{\Psi^1_r}(\xi_1) e^{2\pi i n_1 \xi_1/2^r}\right) \left(\widehat{\Phi^2_0}(\xi_2) e^{2\pi i n\alpha_2\xi_2/2^r}\right) \cdots \left(\widehat{\Phi^d_0}(\xi_d) e^{2\pi i n\alpha_d\xi_d/2^r}\right) \\
&\times \widehat{\Phi^{d+1}_0}(\xi_{d+1}) \widehat{\Phi_r}(\widetilde{\xi}) \widehat{f}(\xi) \widehat{f_1}(\xi_1) \cdots \widehat{f_d}(\xi_d) \\
&\times \widehat{f_{d+1}}(\xi_{d+1})\, d\xi\, d\xi_1 \cdots d\xi_{d+1} d\alpha_2 \cdots d\alpha_d.
\end{aligned} \tag{4.98}$$

Fix all the parameters $\alpha_2, \ldots, \alpha_d \in [0, 1]$, r, n, and n_1 and consider only the inner expression

$$\begin{aligned}
&\int_{\xi+\xi_1+\cdots+\xi_d+\xi_{d+1}=0} (\widehat{f}(\xi)\widehat{\Phi^0_0}(\xi) e^{2\pi i n\xi/2^r})(\widehat{f_1}(\xi_1)\widehat{\Psi^1_r}(\xi_1) e^{2\pi i n_1\xi_1/2^r}) \\
&\times (\widehat{f_2}(\xi_2)\widehat{\Phi^2_0}(\xi_2) e^{2\pi i n\alpha_2\xi_2/2^r}) \cdots (\widehat{f_d}(\xi_d)\widehat{\Phi^d_0}(\xi_d) e^{2\pi i n\alpha_d\xi_d/2^r}) \\
&\times (\widehat{f_{d+1}}(\xi_{d+1})\widehat{\Phi^{d+1}_0}(\xi_{d+1}))\widehat{\Phi_r}(\xi + \alpha_2\xi_2 + \cdots + \alpha_d\xi_d)\, d\xi\, d\xi_1 \cdots d\xi_{d+1}.
\end{aligned} \tag{4.99}$$

Let us pause for a moment and prove the following lemma, which will be very helpful in handling expression (4.99).

Lemma 4.13 *Consider Schwartz functions $F, F_1, \ldots, F_{d+1}$ and Φ. Then one has the identity*

$$\begin{aligned}
&\int_{\xi+\xi_1+\cdots+\xi_d+\xi_{d+1}=0} \widehat{F}(\xi)\widehat{F_1}(\xi_1)\cdots \widehat{F_{d+1}}(\xi_{d+1}) \\
&\qquad \times \widehat{\Phi}(a\xi + a_1\xi_1 + \cdots + a_{d+1}\xi_{d+1})\, d\xi\, d\xi_1 \cdots d\xi_{d+1} \\
&\quad = \int_{\mathbb{R}^2} F(x - at)F_1(x - a_1 t)\cdots F_{d+1}(x - a_{d+1}t)\Phi(t)\, dt dx,
\end{aligned} \tag{4.100}$$

valid for arbitrary real numbers $a, a_1, \ldots, a_{d+1}$.

Proof The proof is a straightforward calculation in Fourier analysis. Let Γ be an arbitrary vector subspace in $\mathbb{R}^{d+2}$ and let δ_Γ denote the *Dirac distribution* associated with Γ; δ_Γ is given by

$$\delta_\Gamma(\phi) = \int_\Gamma \phi(\gamma)\,d\gamma$$

for every Schwartz function ϕ. Then one has $\widehat{\delta_\Gamma} = \delta_{\Gamma^\perp}$ in the sense of distributions. In our particular situation,

$$\Gamma = \{(\xi, \xi_1, \ldots, \xi_{d+1}) \in \mathbb{R}^{d+2} \,|\, \xi + \xi_1 + \cdots + \xi_d + \xi_{d+1} = 0\} \tag{4.101}$$

and so $\Gamma^\perp$ must be the one-dimensional line along the vector $(1, \ldots, 1) \in \mathbb{R}^{d+2}$. Using all these facts and Plancherel's theorem, the left-hand side of our identity (4.100) can be written as

$$\int_{\mathbb{R}^{d+3}} \widehat{F}(\xi)\widehat{F_1}(\xi_1)\cdots\widehat{F_{d+1}}(\xi_{d+1}) \\ \times \widehat{\Phi}(a\xi + a_1\xi_1 + \cdots + a_{d+1}\xi_{d+1})e^{2\pi i x(\xi+\xi_1+\cdots+\xi_{d+1})}\,d\xi d\xi_1\cdots d\xi_{d+1}dx.$$

If one also recalls that

$$\widehat{\Phi}(a\xi + a_1\xi_1 + \cdots + a_{d+1}\xi_{d+1}) = \int_{\mathbb{R}} \Phi(t)e^{-2\pi i t(a\xi+a_1\xi_1+\cdots+a_{d+1}\xi_{d+1})}\,dt$$

then one obtains (4.100) easily, using the Fourier inversion formula multiple times. □

Exercise 4.9 Prove the equality $\widehat{\delta_\Gamma} = \delta_{\Gamma^\perp}$ used in the above proof.

It is also important to record the following generalization of (4.100), which will be useful later.

Lemma 4.14 *The following identity, similar to* (4.100), *holds:*

$$\int_{\xi+\xi_1+\ldots+\xi_d+\xi_{d+1}=0} \widehat{F}(\xi)\widehat{F_1}(\xi_1)\cdots\widehat{F_{d+1}}(\xi_{d+1}) \\ \times \widehat{\Phi_1}(a\xi + a_1\xi_1 + \cdots + a_{d+1}\xi_{d+1})\widehat{\Phi_2}(b\xi + b_1\xi_1 + \cdots + b_{d+1}\xi_{d+1}) \\ \times d\xi d\xi_1\cdots d\xi_{d+1} = \int_{\mathbb{R}^3} F(x - at - bs)F_1(x - a_1t - b_1s)\ldots F_{d+1} \\ \times (x - a_{d+1}t - b_{d+1}s)\Phi_1(t)\Phi_2(s)\,dtdsdx. \tag{4.102}$$

Its proof is similar to that of Lemma 4.13 and is left to the reader.

Exercise 4.10 Prove the above formula and show that it can be extended to an arbitrary number of Φ factors.

Let us denote generically by H^a the function given by

$$\widehat{H^a}(\xi) = \widehat{H}(\xi)e^{2\pi i a\xi}. \tag{4.103}$$

Notice that one can write $H^a(x) = H(x+a)$ alternatively. We now use this notation and the formula (4.100) to rewrite (4.99) as

$$\int_{\mathbb{R}^2} (f * \Phi_0^{0,n/2^r})(x-t)(f_1 * \Psi_r^{1,n_1/2^r})(x)$$

$$\times \prod_{j=2}^{d} (f_j * \Phi_0^{j,n\alpha_j/2^r})(x-\alpha_j t)(f_{d+1} * \Phi_0^{d+1})(x)\, \Phi_r(t)\, dt dx$$

$$= \int_{\mathbb{R}^2} (f * \Phi_0^{0,n/2^r})(x-2^{-r}t)\,(f_1 * \Psi_r^{1,n_1/2^r})(x)$$

$$\times \prod_{j=2}^{d} (f_j * \Phi_0^{j,n\alpha_j/2^r})(x-2^{-r}\alpha_j t)\big(f_{d+1} * \Phi_0^{d+1}\big)(x)\, \Phi_0(t)\, dt dx$$

$$= \int_{\mathbb{R}^2} (f * \Phi_0^{0,(n-t)/2^r})(x)\,(f_1 * \Psi_r^{1,n_1/2^r})(x)$$

$$\times \prod_{j=2}^{d} (f_j * \Phi_0^{j,(n-t)\alpha_j/2^r})(x)(f_{d+1} * \Phi_0^{d+1})(x)\, \Phi_0(t)\, dt dx. \tag{4.104}$$

Clearly, one can perform a similar calculation at any scale $k \neq 0$. The formula analogous to (4.104) is

$$\int_{\mathbb{R}^2} (f * \Phi_k^{0,(n-t)/2^{r+k}})(x)\,(f_1 * \Psi_{r+k}^{1,n_1/2^{r+k}})(x)$$

$$\times \prod_{j=2}^{d} (f_j * \Phi_k^{j,(n-t)\alpha_j/2^{r+k}})(x)(f_{d+1} * \Phi_k^{d+1})(x)\, \Phi_0(t)\, dt dx. \tag{4.105}$$

Exercise 4.11 Check carefully the generic formula (4.105).

To conclude, if we write $\vec{\alpha} = (\alpha_2, \ldots, \alpha_d)$ then it is not difficult to see that the part of C_d that corresponds to the term I'_a becomes

$$\int_{[0,1]^{d-1}} \int_{\mathbb{R}} \left(\sum_{r \leq 0} 2^r \sum_{n,n_1} C_{n,n_1}^r\, C_d^{r,n,n_1,\vec{\alpha},t} \right) \Phi_0(t)\, dt d\vec{\alpha}, \tag{4.106}$$

where $C_d^{r,n,n_1,\vec{\alpha},t}$ is the $(d+1)$-linear operator whose $(d+2)$-linear form is given by the sum over $k \in \mathbb{Z}$ of the corresponding integrands in (4.105).

Looking again at formula (4.106), we can see that we have finally been able to unfold the structure of the symbol of the commutator C_d. However, the proof does not end here since some intricate points still need to be addressed.

First, if one wants to prove (4.73) for the operator in (4.106), it is clearly necessary to do this for the operator $C_d^{r,n,n_1,\vec{\alpha},t}$ with some upper bounds that are summable with respect to r, n, n_1 and integrable in t and $\vec{\alpha}$. We plan to treat this operator using our previous heuristical method. At the outset one needs to reduce it to its minimal variant. Recall that we are aiming to apply (4.78) to all the indices $2 \leq j \leq d$ for which $p_j = \infty$. Let us denote by T the set of indices $2 \leq j \leq d$ with the property that $p_j \neq \infty$. Then define $\ell := |T| + 2$ and freeze, as before, the L^∞-normalized Schwartz functions corresponding to the indices in the complement $\{2, \ldots, d\} \setminus T$. The resulting operator is the minimal ℓ-linear operator naturally associated with the original $C_d^{r,n,n_1,\vec{\alpha},t}$, which we will denote by $C_d^{\ell,r,n,n_1,\vec{\alpha},t}$. To complete our strategy we need to prove Banach estimates for this operator and then quasi-Banach estimates, and, in the end to interpolate carefully between all these estimates.

Banach estimates for the minimal $C_d^{\ell,r,n,n_1,\vec{\alpha},t}$. Consider indices $1 < r_1, \ldots, r_{\ell+1} < \infty$ satisfying $1/r_1 + \ldots + 1/r_\ell = 1/r_{\ell+1}$. As we have learned already, the boundedness constants of

$$C_d^{\ell,r,n,n_1,\vec{\alpha},t} : L^{r_1} \times \ldots \times L^{r_\ell} \to L^{r_{\ell+1}} \tag{4.107}$$

depend directly on the corresponding boundedness constants of the square functions

$$\left(\sum_k \left|f * \Phi_k^{0,(n-t)/2^{r+k}}(x)\right|^2\right)^{1/2} \quad \text{and} \quad \left(\sum_k \left|f_1 * \Psi_{r+k}^{1,n_1/2^{r+k}}(x)\right|^2\right)^{1/2}$$

and those of various maximal functions of the type

$$\sup_k \left|f_j * \Phi_k^{j,(n-t)\alpha_j/2^{r+k}}(x)\right|$$

for $j \in T$.

The square functions are the continuous analogue of the shifted discrete square functions $S^{[(n-t)/2^r]}$ and S^{n_1} that we studied earlier, and therefore they are bounded on every L^q space for $1 < q < \infty$, with upper bounds of the type $O(\log <[(n-t)/2^r]>)$ and $O(\log <n_1>)$ respectively (here $[x]$ refers to the integer part of the real number x).

Similarly, the maximal functions can be pointwise estimated by the shifted maximal functions $M^{[(n-t)\alpha_j/2^r]}$ and so they also are bounded on every L^q space for $1 < q < \infty$, with upper bounds of the type $O(\log <[(n-t)\alpha_j/2^r]>)$.

From all this we deduce that the boundedness constants of (4.107) are no greater than

$$C <r>^\ell (\log <n>)^\ell (\log <n_1>)^\ell (\log <[t]>)^\ell. \tag{4.108}$$

Exercise 4.12 Show that the *continuous* square functions that appeared earlier satisfy the same logarithmic bounds as their *discrete* analogues $S^{[(n-t)/2^r]}$ and S^{n_1}.

Quasi-Banach estimates for the minimal $C_d^{\ell,r,n,n_1,\vec{\alpha},t}$. Consider indices $1 < r_1, \ldots, r_\ell < \infty$ and $0 < r_{\ell+1} < \infty$ with the property that $1/r_\ell + \cdots + 1/r_\ell = 1/r_{\ell+1}$. Remember that $r_{\ell+1}$ can be subunitary now.

The goal is to estimate the boundedness constants of

$$C_d^{\ell,r,n,n_1,\vec{\alpha},t} : L^{r_1} \times \cdots \times L^{r_\ell} \to L^{r_{\ell+1}} \tag{4.109}$$

and of its adjoint operators. As usual, these quasi-Banach estimates are tricky and we need to discretize (4.105) with respect to the x variable to be able to treat them in the same way as before. This allows us to rewrite the operator $C_d^{\ell,r,n,n_1,\vec{\alpha},t}$ in a form similar to (4.81), for which one can then apply the corresponding version of (4.82). We notice that now we face the following extra difficulty. The bump Schwartz functions corresponding to the index 1 in (4.105) are in fact adapted to scales that are 2^{-r} times larger than the scales of the bump functions corresponding to the other indices. Recall that r is a negative integer. This fact does not matter in the Banach case, as we have seen but here, in the quasi-Banach case, one has to be more careful.

The natural idea is of course to discretize using a larger scale. In order to do this correctly we need to observe the following general fact. If a generic function Φ happens to be a smooth bump function adapted to the dyadic interval that is K, and if $K \subseteq \widetilde{K}$ is the unique dyadic interval that is 2^{-r} times longer than K then $2^{5r}\Phi$ is a bump function adapted to $\widetilde{K}$ (notice that "5" corresponds to the number of derivatives in our *new* definition of adaptedness).

All these facts, together with the usual standard averaging and approximation arguments reduce our problem to that of estimating expressions of the type

$$\frac{1}{2^{6rl}} \sum_I \frac{1}{|I|^{(\ell-1)/2}} |\langle f, \Phi^0_{I_{[n-t]}} \rangle| \, |\langle f_1, \Phi^1_{I_{n_1}} \rangle| \prod_{j \in T} |\langle f_j, \Phi^j_{I_{[(n-t)\alpha_j]}} \rangle| \, |\langle f_{d+1}, \Phi_I^{d+1} \rangle|, \tag{4.110}$$

where the functions f, $(f_j)_j$ are as in (4.82) and the indices, say $(p_j)_j$, that appear there are the same as the indices, say $(r_j)_j$, in (4.110). Also, the power 6 in the earlier expression should be thought of as being equal to $5 + 1$, where the number 5 comes from the adaptedness condition and the number 1 is a consequence of the scaling (remember that all our smooth bump functions are L^2-normalized now). From (4.82) and interpolation, we can conclude that the boundedness constants in (4.109) are no greater than

$$2^{-6\ell r}(\log <n>)^{\ell}(\log <n_1>)^{\ell}(\log <[t]>)^{\ell}. \tag{4.111}$$

By symmetry, the same conclusion can be drawn for the adjoints of the operator $C_d^{\ell,r,n,n_1,\vec{\alpha},t}$.

The final interpolation. It is important to focus our discussion on the dependence on r of all these estimates, since the other parameters contribute at most logarithmically. Notice that the Banach bound of $<r>^{\ell}$ in (4.108) is excellent since in (4.106) there is also a factor of the type 2^r, but the quasi-Banach bound of $2^{-6\ell r}$ in (4.111) is clearly too large. However, let us not forget that the desired estimates in (4.73) are on the borderline of the Banach estimates, and this suggests that they should grow only a little faster than the latter.

Now let us fix indices p, $(p_j)_j$ as in (4.73). One can first use standard arguments based on convexity (by freezing all but one function) to interpolate linearly between the perfect Banach estimates and the quasi-Banach estimates, which are too weak; as a result one obtains many quasi-Banach estimates whose bounds do not grow too much with respect to r (say, at a rate of at most $2^{-\epsilon r}$ for some small ϵ).

Finally, one can use multilinear interpolation theory and interpolate between the improved quasi-Banach estimates and the Banach estimates in (4.108) to finally deduce that (4.73) for the operator $C_d^{r,n,n_1,\vec{\alpha},t}$ and the indices p, $(p_j)_j$ comes with a bound which is acceptable in (4.106). This completes the description of term I_a'.

The remaining cases follow a similar strategy. Before discussing them we note that, besides the quadratic logarithmic arguments that have played an important role, we have also taken advantage of the smallness of the factor 2^r in (4.106). The only difference in the other cases will come from the way in which a similarly small factor appears. We will therefore describe the other cases one by one and point out explicitly the changes that one has to make sometimes in order to produce this decay factor. For the sake of brevity we will do this only for the scale-1 terms (which correspond to $k = 0$) since, as always, our argument is scale invariant.

Term I_a'' of (4.95) This term corresponds to $r > 0$ and it is actually easier than term I_a'. The interaction between the two functions $\widehat{\Psi_r}(\xi_1)$ and $\widehat{\Phi_0^1}(\xi_1)$ gives, this time,

$$\widehat{\Psi_r}(\xi_1)\widehat{\Phi_0^1}(\xi_1) = \frac{1}{2^{rM}}\widehat{\widetilde{\Psi}_r}(\xi_1)$$

for some large constant $M > 0$, where $\widehat{\widetilde{\Psi_r}}(\xi_1)$ is another Ψ function adapted to the same scale as the original $\widehat{\Psi_r}(\xi_1)$. The fact that now we have a strong decay factor, together with the previous method, solves this situation in precisely the same way as before.

Term I_b

Term I_b is obtained from (4.94) by replacing $\widehat{\Phi_r}$ by $\widehat{\Psi_r}$. Thus it is almost identical to term I_a. The only difference is that the corresponding Fourier coefficients in (4.97) can be now estimated by

$$|C_{n,n_1}| \lesssim \frac{1}{<n>^2}\frac{1}{<n-n_1>^{\mu}} + \frac{1}{<n>^{\mu}}\frac{1}{<n_1>^{\mu}},$$

as it was shown for the first Calderón commutator and this is still enough to give a contribution summable over the parameters n, n_1.

Term I_c

Term I_c is obtained from (4.94) by interchanging $\widehat{\Phi_r}$ by $\widehat{\Psi_r}$. Now, our first observation is that on the support of the function $\widehat{\Psi_r}(\widetilde{\xi})\widehat{\Phi_r}(\xi_1)$ the symbol $\int_0^1 1_{\mathbb{R}^+}(\widetilde{\xi} + \alpha_1\xi_1)\,d\alpha_1$ is a classical Marcinkiewicz–Mikhlin–Hörmander symbol, and so for its Fourier coefficients one has arbitrary decay of the type

$$\frac{1}{<n>^{\mu}}\frac{1}{<n_1>^{\mu}}.$$

There are two subterms, which we will denote by I'_c and I''_c; as before, they correspond to $r < 0$ and $r \geq 0$ respectively. We will discuss them separately.

Term I'_c: $r < 0$ This case is the simpler of the two: one just has to observe that

$$\widehat{\Phi_r}(\xi_1)\widehat{\Phi_0^1}(\xi_1) = \widehat{\Phi_r}(\xi_1)\xi_1\widehat{\phi_0^1}(\xi_1) = \widehat{\widetilde{\Phi_r}}(\xi_1)\xi_1 = 2^r\widehat{\widetilde{\Phi_r}}(\xi_1)\frac{\xi_1}{2^r} = 2^r\widehat{\widetilde{\Psi_r}}(\xi_1),$$

where the new function $\widehat{\widetilde{\Psi_r}}(\xi_1)$ is also Ψ type. The presence of the factor 2^r makes this case similar to case I'_a.

Term I''_c: $r \geq 0$ This situation is trickier, and requires a lengthier discussion. Observe that when the two functions $\widehat{\Phi_r}(\xi_1)$ and $\widehat{\Phi_0^1}(\xi_1)$ are multiplied, no decay factor comes out of this since we have only

$$\widehat{\Phi_r}(\xi_1)\widehat{\Phi_0^1}(\xi_1) = \widehat{\widetilde{\Phi_0^1}}(\xi_1),$$

where, as before, $\widehat{\widetilde{\Phi_0^1}}(\xi_1)$ is another Ψ-type function. This is a consequence of the fact that $\widehat{\Phi_0^1}(\xi_1)$ is adapted to an interval contained within that corresponding to $\widehat{\Phi_r}(\xi_1)$ (remember that $r \geq 0$ in the present case). To extract a similar decay factor, one would have to argue in a different way.

For simplicity, let us assume as before that $k = 0$, since the argument that we will describe is scale invariant. Consider again the term

$$\widehat{\Phi_0^0}(\xi)\widehat{\Phi_0^1}(\xi_1)\cdots\widehat{\Phi_0^{d+1}}(\xi_{d+1}). \tag{4.112}$$

We now recall that we used *compact* Littlewood–Paley decompositions for the 0 and $d+1$ positions and, because of this, the function $\widehat{\Phi_0^0}(\xi)$ is not only Ψ type but also has compact support. Choose a Φ-type function $\widehat{\widetilde{\Phi_0^0}}(\xi)$, supported on a slightly wider interval and identically equal to 1 on the support of the original $\widehat{\Phi_0^0}(\xi)$. Split (4.112) as follows:

$$\widehat{\Phi_0^0}(\xi)\widehat{\Phi_0^1}(\xi_1)\cdots\widehat{\Phi_0^{d+1}}(\xi_{d+1})\widehat{\widetilde{\Phi_0^0}}(\widetilde{\xi}) \tag{4.113}$$

$$+\,\widehat{\Phi_0^0}(\xi)\widehat{\Phi_0^1}(\xi_1)\cdots\widehat{\Phi_0^{d+1}}(\xi_{d+1})\left(1-\widehat{\widetilde{\Phi_0^0}}(\widetilde{\xi})\right). \tag{4.114}$$

If one considers now the $(d+2)$-linear form generated by (4.113) (and its analogues for each scale $k \neq 0$) one sees that it is very similar to the form in the I_a'' case. In fact it is even simpler, since one may observe that the product $\widehat{\Psi_r}(\widetilde{\xi})\widehat{\widetilde{\Phi_0^0}}(\widetilde{\xi})$ is zero except possibly when $r = 1, 2, 3$.

However, the form that (4.114) determines is somewhat more complicated. First, let us rewrite (4.114):

$$\widehat{\Phi_0^0}(\xi)\widehat{\Phi_0^1}(\xi_1)\cdots\widehat{\Phi_0^{d+1}}(\xi_{d+1})\left(\widehat{\widetilde{\Phi_0^0}}(\xi)-\widehat{\widetilde{\Phi_0^0}}(\widetilde{\xi})\right)$$

$$= -\widehat{\Phi_0^0}(\xi)\widehat{\Phi_0^1}(\xi_1)\cdots\widehat{\Phi_0^{d+1}}(\xi_{d+1})\left(\int_0^1 \widehat{\widetilde{\Phi_0^0}}'\left((1-s)\xi+s\widetilde{\xi}\right)ds\right)(\widetilde{\xi}-\xi)$$

$$= -\widehat{\Phi_0^0}(\xi)\widehat{\Phi_0^1}(\xi_1)\cdots\widehat{\Phi_0^{d+1}}(\xi_{d+1})\left(\int_0^1 \widehat{\widetilde{\Phi_0^0}}'(\xi+s(\alpha_2\xi_2+\cdots+\alpha_d\xi_d))\,ds\right)$$

$$\times(\alpha_2\xi_2+\cdots+\alpha_d\xi_d). \tag{4.115}$$

Observe that because of the extra parenthesis $(\alpha_2\xi_2+\cdots+\alpha_d\xi_d)$, we will lose another factor d. Then note that each new $O(d)$ expression is of the form

$$\widehat{\Phi_0^0}(\xi)\widehat{\Phi_0^1}(\xi_1)\cdots\widehat{\Phi_0^{d+1}}(\xi_{d+1})\left(\int_0^1 \widehat{\widetilde{\Phi_0^0}}'(\xi+s(\alpha_2\xi_2+\cdots+\alpha_d\xi_d))\,ds\right) \tag{4.116}$$

where, in addition, for indices $2 \leq j \leq d$ one has an extra factor of the type $\alpha_j\xi_j$ besides the previous $\widehat{\Phi_0^j}(\xi_j)$. This adds another, harmless, Ψ function to that particular term, so it suffices to analiyze (4.116). The important observation this time is to realize that, when one multiplies the expression (4.116) by a factor of the type $\widehat{\Psi_r}(\xi)\widehat{\widetilde{\Phi}_r}(\xi_1)$, one must have the constraint $0 \leq s \leq C/2^r$ in order for that corresponding term to be different from zero. In particular, one can simply replace the integral

$$\int_0^1 \widehat{\widetilde{\Phi_0^0}}'(\xi + s(\alpha_2\xi_2 + \cdots + \alpha_d\xi_d))\,ds$$

in (4.116) by

$$\int_0^{C/2^r} \widehat{\widetilde{\Phi_0^0}}'(\xi + s(\alpha_2\xi_2 + \ldots + \alpha_d\xi_d))\,ds.$$

Fix $r \geq 0$ and $0 \leq s \leq C/2^r$ and notice that the corresponding $(d+2)$-linear form can be handled, exactly as before, in a way that is independent of the parameter s. The extra decay factor that we were looking for will come out naturally after integration over the interval $[0, C/2^r]$. Notice also that because of the identity (4.102) the extra factor $\widehat{\widetilde{\Phi_0^0}}'(\xi + s(\alpha_2\xi_2 + \cdots + \alpha_d\xi_d))$ just adds another harmless average to the original generic form.

This completes the discussion of case I_c'' and also of case I.

4.4.9. Case *II*: $j_1 = 0$ and $j_2 = d+1$

Observe that now both Ψ functions have *extremal* positions. Our goal is to describe a procedure that (modulo some simpler terms) will allow us to reduce case II to the case I, where at least one Ψ function was in an *intermediate* position.

As usual by now, we consider a generic $k = 0$ term, such as that (4.112). The argument will be *scale invariant* as before.

Split (4.112) as follows:

$$\widehat{\Phi_0^0}(\xi)\widehat{\phi_0^1}(\xi_1)\cdots\widehat{\Phi_0^{d+1}}(\xi_{d+1}) + \widehat{\Phi_0^0}(\xi)\widehat{\psi_0^1}(\xi_1)\cdots\widehat{\Phi_0^{d+1}}(\xi_{d+1}) = A + B,$$

where clearly $\widehat{\phi_0^1}(\xi_1)$ is Φ type and compactly supported at scale 1, while $\widehat{\psi_0^1}(\xi_1)$ is Ψ type and also adapted at scale 1. Observe that the B term (together with all the others corresponding to $k \neq 0$) generates $(d+2)$-linear forms very similar to those discussed in case I, so it is enough to understand the A term only. Here it is important to realize from the start that, by construction, at least one of the two Ψ functions $\widehat{\Phi_0^0}(\xi)$ and $\widehat{\Phi_0^{d+1}}(\xi_{d+1})$ has its support away from zero. Moreover, we claim that without loss of generality we can assume in addition that $\widehat{\Phi_0^{d+1}}(\xi_{d+1})$ is that function. To see this, one just has to notice that the roles of the two variables ξ and ξ_{d+1} are completely symmetric. Indeed, given that

$\xi + \xi_1 + \cdots + \xi_{d+1} = 0$, a simple change of variables gives

$$\int_{[0,1]^d} 1_{\mathbb{R}_+}(\xi + \alpha_1\xi_1 + \cdots + \alpha_d\xi_d)\, d\alpha_1 \cdots d\alpha_d$$
$$= (-1)^d \int_{[0,1]^d} 1_{\mathbb{R}_-}(\xi_{d+1} + \beta_1\xi_1 + \cdots + \beta_d\xi_d)\, d\beta_1 \cdots d\beta_d,$$

which is clearly a similar symbol.

Using this fact, we can then rewrite term A as

$$\widehat{\Phi_0^0}(\xi)\widehat{\phi_0^1}(\xi_1)\cdots\widehat{\Phi_0^{d+1}}(\xi_{d+1})\widehat{\widetilde{\Psi_0^{d+1}}}(\xi + \xi_1 + \cdots + \xi_d) \tag{4.117}$$

for another compactly supported Ψ function, $\widehat{\widetilde{\Psi_0^{d+1}}}$. After that we decompose (4.117) further as

$$\begin{aligned}
&\widehat{\Phi_0^0}(\xi)\widehat{\phi_0^1}(\xi_1)\cdots\widehat{\Phi_0^{d+1}}(\xi_{d+1})\widehat{\widetilde{\Psi_0^{d+1}}}(\xi + \alpha_2\xi_2 + \cdots + \alpha_d\xi_d)\\
&\quad + \widehat{\Phi_0^0}(\xi)\widehat{\phi_0^1}(\xi_1)\cdots\widehat{\Phi_0^{d+1}}(\xi_{d+1})\\
&\quad \times \left(\widehat{\widetilde{\Psi_0^{d+1}}}(\xi + \xi_1 + \cdots + \xi_d) - \widehat{\widetilde{\Psi_0^{d+1}}}(\xi + \alpha_2\xi_2 + \cdots + \alpha_d\xi_d)\right).
\end{aligned} \tag{4.118}$$

Using a previous notation the first term can be also written as

$$\widehat{\Phi_0^0}(\xi)\widehat{\phi_0^1}(\xi_1)\cdots\widehat{\Phi_0^{d+1}}(\xi_{d+1})\widehat{\widetilde{\Psi_0^{d+1}}}(\widetilde{\xi}).$$

Then, one simply remarks that

$$\int_0^1 1_{\mathbb{R}^+}(\widetilde{\xi} + \alpha_1\xi_1)\, d\alpha_1$$

is a classical Marcinkiewicz–Mikhlin–Hörmander symbol on the support of the above term, and therefore the whole analysis becomes simpler in this case. In particular, it should be also clear that one no longer needs to consider a decomposition over r to study this term.

We are as a consequence left with the second term in (4.118). As before, one rewrites this term as

$$\begin{aligned}
&\widehat{\Phi_0^0}(\xi)\widehat{\phi_0^1}(\xi_1)\cdots\widehat{\Phi_0^{d+1}}(\xi_{d+1})\\
&\times \left(\int_0^1 \widehat{\widetilde{\Psi_0^{d+1}}}'((1-s)(\xi + \xi_1 + \cdots + \xi_d) + s(\xi + \alpha_2\xi_2 + \cdots + \alpha_d\xi_d))\, ds\right)\\
&\times (\xi_1 + (1-\alpha_2)\xi_2 + \cdots + (1-\alpha_d)\xi_d).
\end{aligned}$$

The pleasant fact is that this expression decomposes as a sum of $O(d)$ terms, each containing an extra factor of the type ξ_j for some index $1 \le j \le d$. This decomposition transforms that jth function into an intermediate Ψ function exactly as in case I. Also as before, the extra factor

$$\int_0^1 \widehat{\Psi_0^{d+1}}'((1-s)(\xi+\xi_1+\cdots+\xi_d)+s(\xi+\alpha_2\xi_2+\cdots+\alpha_d\xi_d))\,ds$$

is harmless because of (4.102).

4.4.10. Case *III*: $j_1 = 2$ and $j_2 = 3$

Let us just mention here that case *III* can be addressed in the same way as case I, as now we have two Ψ-type functions in intermediate positions.

The proof of how one can reduce (4.73) to the discretized estimate (4.92) is now complete since, by symmetry, any other possibility can be reduced to one of the three cases described above.

4.5. Generalizations

Let us first of all observe that the *Cauchy integral on Lipschitz curves* operator can be rewritten in a suggestive way as

$$C_\Gamma f(x) = p.v. \int_{\mathbb{R}} F\left(\frac{\Delta_t}{t}A(x)\right) f(x+t)\frac{dt}{t}, \tag{4.119}$$

where F is given by $F(z) = (1+iz)^{-1}$ for $|z| < 1$. If one takes $F(z) = z^k$ in (4.119) then one obtains Calderón's commutators and, as a consequence, the Taylor series expansion of $z \to (1+iz)^{-1}$ corresponds to the earlier decomposition of C_Γ as a series of the type $\sum_k (-i)^k C_k$. Suppose now that F is a generic analytic function of one complex variable defined on a certain disk centered at the origin. Assume also that A is a Lipschitz function such that $\|A'\|_\infty$ is strictly smaller than the radius of convergence of F. The previous polynomial bounds for the commutators allow one to obtain in similar way the following result, due to Coifman, McIntosh, and Meyer.

Theorem 4.15 *Under the above assumptions, the linear operator*

$$f \mapsto p.v. \int_{\mathbb{R}} F\left(\frac{\Delta_t}{t}A(x)\right) f(x+t)\frac{dt}{t}$$

is bounded on every L^p *for* $1 < p < \infty$.

Assume now that A has the properties that $A'' \in L^\infty$ and that $\|A''\|_\infty$ is, as mentioned above, strictly smaller than the radius of convergence of F. The natural extension of Theorem 4.8 is the following result.

Theorem 4.16 *Let $a \neq 0$ and $b \neq 0$ be two fixed real numbers. Under the above assumptions, the linear operator*

$$f \mapsto p.v. \int_{\mathbb{R}^2} F\left(\frac{\Delta_{at}}{t} \circ \frac{\Delta_{bs}}{s} A(x)\right) f(x+t+s)\, \frac{dt}{t}\frac{ds}{s} \tag{4.120}$$

is bounded on every L^p for $1 < p < \infty$.

To see why this theorem holds true, suppose for simplicity that $a = b = 1$. Then, if $F(z) = z^k$, a simple calculation shows that the corresponding expression for f can be viewed as a $(k+1)$-linear operator whose symbol is precisely the square of the symbol of C_k. In particular, one can prove polynomial bounds for such operators as well, using the same method.

Finally, we consider the following generalization of the result in Theorem 4.9 on *circular commutators*.

Theorem 4.17 *Let $a \neq 0$, $b \neq 0$, and $c \neq 0$ be three fixed real numbers. Consider also three Lipschitz functions A, B, and C. Then, the following expression*

$$p.v. \int_{\mathbb{R}^3} \left(\frac{\Delta_{at_1}}{t_1} A(x+t_2)\right) \left(\frac{\Delta_{bt_2}}{t_2} B(x+t_3)\right) \left(\frac{\Delta_{ct_3}}{t_3} C(x+t_1)\right) \frac{dt_1}{t_1}\frac{dt_2}{t_2}\frac{dt_3}{t_3},$$

viewed as a trilinear operator in A', B', and C', maps $L^{p_1} \times L^{p_2} \times L^{p_3}$ into L^p boundedly, for every $1 < p_1, p_2, p_3 \leq \infty$ with $1/p_1 + 1/p_2 + 1/p_3 = 1/p$ and $1/2 < p < \infty$.

The proof of this theorem uses the same method as before. Assume for simplicity that $a = b = c = 1$. Then the symbol of the corresponding trilinear operator is given by the circular product

$$m_2(\xi_1, \xi_2, \xi_3) m_2(\xi_2, \xi_3, \xi_1) m_2(\xi_3, \xi_1, \xi_2)$$

and such product symbols can be studied in a similar fashion.

Notes

The original proofs of the theorems on Calderón commutators and the Cauchy integral on Lipschitz curves are due to Calderón [10, 11], Coifman and Meyer [24], and Coifman, McIntosh, and Meyer [29]. Different proofs have been found by David and Journé [33], Christ and Journé [17], Murai [80], Mitrea [79], Coifman, Jones, and Semmes [30], and Verdera [118]. A standard text, in which many related topics are described is the book by Coifman and Meyer [26]. For more about the connection of these results to elliptic PDEs see Kenig [63], Jerison and Kenig [58], Dahlberg [32], Verchota [117] or the more recent work on the Kato's problem by Auscher, Hofmann, Lacey *et al.* [2].

The history of these topics given in Section 4.1 in part follows Calderón [11] and Verdera [118].

The proofs presented in this chapter come from Muscalu [86, 87] and also contain the new results in the generalization sections. For bidisk and polydisk extensions see Journè [59, 61] and Muscalu [88]. For an extension of Calderón's calculations to functions of arbitrary polynomial growth, see Muscalu [87, 88]. It is also of interest that in the case of the new operators (such as those in (4.120)) the standard arguments based on $T(1)$ and $T(b)$ theorems become ineffective, as observed in [87].

The logarithmic theorem on the shifted maximal function can be found in Stein [105]. Other works on commutators include those of Rochberg and Weiss [103] and of Coifman and Rochberg [27]. For connections with the water-wave equation, see the recent articles of Wu [120, 121].

Problems

Problem 4.1 Prove that the following statements hold true.

(a) Let $(\Phi_k)_k$ be a family of L^1-normalized smooth bump functions adapted to the intervals $[-2^{-k}, 2^{-k}]$ and having the property that $\int_{\mathbb{R}} \Phi_k(x)\,dx = 0$ for every $k \in \mathbb{Z}$. Prove that the *continuous* square function

$$f \mapsto \left(\sum_{k\in\mathbb{Z}} (f * \Phi_k)^2 \right)^{1/2}$$

is bounded on every L^p space for $1 < p < \infty$.

(b) Consider similarly $(\Phi_I)_I$, a family of L^2-normalized smooth bump functions adapted to dyadic intervals I and having the property that $\int_{\mathbb{R}} \Phi_I(x)\,dx = 0$ for every I. Prove that the *discretized* square function

$$f \mapsto \left(\sum_I \frac{|\langle f, \Phi_I \rangle|^2}{|I|} \chi_I \right)^{1/2}$$

is also bounded on every L^p space for $1 < p < \infty$.

Problem 4.2 Use the $T(1)$ theorem described in Sections I.9.4 and I.9.5 to prove (inductively) that Calderón commutators are bounded on L^p for every $1 < p < \infty$.

Problem 4.3 Let F, G be analytic functions and let A, B be such that $\|A'\|_\infty$, $\|B''\|_\infty$ are strictly smaller than the radii of convergence of F and G respectively. Then, the linear operator

$$f \mapsto p.v. \int_{\mathbb{R}} F\left(\frac{\Delta_t}{t} A(x)\right) G\left(\frac{\Delta_t}{t} \circ \frac{\Delta_t}{t} B(x)\right) f(x+t)\, \frac{dt}{t}$$

is bounded on every L^p space for $1 < p < \infty$.

Problem 4.4 Let $d \geq 1$ and let F be as in Problem 4.3. Assume that A is a function such that $\|A^{(d)}\|_\infty$ is strictly smaller than the radius of convergence of F. Then the linear operator

$$f \mapsto p.v. \int_{\mathbb{R}} F\left(\frac{A(x) - T_y^{d-1} A(x)}{(x-y)^d}\right) \frac{f(y)}{x-y}\, dy$$

is bounded on every L^p space for $1 < p < \infty$, where $T_y^{d-1} A(x)$ is the Taylor polynomial of order $d-1$ of the function A about the point y.

(The case $F(x) = x$ is discussed in Coifman and Meyer [26]; see also related work by Cohen and Gosselin [22]. The general case can be found in Muscalu [87].)

Problem 4.5 Suppose that F is as before while A is a complex-valued function in $\mathbb{R}^n$, so that we have

$$\frac{\partial^n A}{\partial x_1 \cdots \partial x_n} \in L^\infty(\mathbb{R}^n)$$

with an L^∞ norm strictly smaller than the radius of convergence of F. Define the linear operator

$$f \mapsto p.v. \int_{\mathbb{R}^n} f(x+t) F\left(\frac{\Delta_{t_1}^{(1)}}{t_1} \circ \cdots \circ \frac{\Delta_{t_n}^{(n)}}{t_n} A(x)\right) \frac{dt_1}{t_1} \cdots \frac{dt_n}{t_n}$$

for functions $f(x)$ of n variables for which the principal-value integral exists; here $\Delta_s^{(i)}$ denotes the finite difference operator at scale s in the direction of e_i and is given by

$$\Delta_s^{(i)} B(x) := B(x + s e_i) - B(x);$$

$e_1, \ldots, e_n$ is the standard basis in $\mathbb{R}^n$.

Prove that this linear operator is bounded on every L^p space for every $1 < p < \infty$. (The case $n = 2$ is due to Journé [61]; the general case appeared in Muscalu [88].)

5

Iterated Fourier series and physical reality

On the one hand, we learned in the first volume of the book that Carleson's maximal operator appears quite naturally when one tries to study the problem of the almost everywhere convergence of Fourier series. On the other hand, we saw in Chapter 4 that the bilinear Hilbert transform was discovered by Calderón in his quest to find a general approach to his commutators and, eventually, the Cauchy integral on Lipschitz curves. It turns out that these two operators are closely related, and Chapters 6 and 7 will be devoted to their study.

The main aim of the present chapter is to describe a completely distinct problem, where both these operators appear together in a natural way.

5.1. Iterated Fourier series

Let f be a 2π-periodic function on the real line that belongs to $L^p([0, 2\pi])$ for a certain $1 < p < \infty$. Then, it is well known that

$$\sum_{-N \leq n \leq N} \widehat{f}(n) e^{inx} \xrightarrow[N \to \infty]{} f(x) \tag{5.1}$$

either in the L^p topology or almost everywhere. Since both statements are trivial for smooth functions, the first is equivalent to the L^p-boundedness of the linear operator

$$f \mapsto \sum_{-N \leq n \leq N} \widehat{f}(n) e^{inx}, \tag{5.2}$$

independently of N, while the second is equivalent to the L^p-boundedness of the maximal operator

$$f \mapsto \sup_N \left| \sum_{-N \leq n \leq N} \widehat{f}(n) e^{inx} \right|. \tag{5.3}$$

Both (5.2) and (5.3) are classical results in analysis and are due to M. Riesz and to Carleson and Hunt respectively. The first statement was proved in two different ways in the first volume of the book, while the second will be proved in Chapter 7 of this second volume.

For the rest of the discussion let us assume for simplicity that $p = 2$.

If one squares (5.1), one obtains

$$\sum_{-N \leq n_1, n_2 \leq N} \widehat{f}(n_1) \widehat{f}(n_2) e^{in_1 x} e^{in_2 x} \xrightarrow[N \to \infty]{} f^2(x) \tag{5.4}$$

either in the topology of L^1 or almost everywhere, as a consequence of the Hölder inequality. Now split the left hand side of (5.4) as

$$\begin{aligned} & \sum_{-N \leq n_1 < n_2 \leq N} \widehat{f}(n_1) \widehat{f}(n_2) e^{in_1 x} e^{in_2 x} \\ & + \sum_{-N \leq n_2 < n_1 \leq N} \widehat{f}(n_1) \widehat{f}(n_2) e^{in_1 x} e^{in_2 x} \\ & + \sum_{-N \leq n_1 = n_2 \leq N} \widehat{f}(n_1) \widehat{f}(n_2) e^{in_1 x} e^{in_2 x}. \end{aligned} \tag{5.5}$$

Observe that the last term of (5.5) equals

$$\sum_{-N \leq n \leq N} \widehat{f * f}(n) e^{2inx}$$

and, in particular, since the convolution $f * f$ is still in $L^2([0, 2\pi])$ it converges to $f * f(2x)$ either in the topology of L^2 or almost everywhere. Since the topology of L^2 is stronger than the topology of L^1, the convergence holds in L^1 as well.

However, the first two terms of (5.5) are identical and this implies that the series

$$\sum_{-N \leq n_1 < n_2 \leq N} \widehat{f}(n_1) \widehat{f}(n_2) e^{in_1 x} e^{in_2 x}$$

converges in the topology of L^1 and also almost everywhere.

If instead of squaring (5.1) one multiplies two similar statements corresponding to possibly distinct functions $f, g \in L^2([0, 2\pi])$ and then performs a similar decomposition to that in (5.5), one immediately realizes that the problem of the convergence of *iterated Fourier series* of the type

$$\sum_{-N \leq n_1 < n_2 \leq N} \widehat{f}(n_1)\widehat{g}(n_2)e^{in_1x}e^{in_2x} \tag{5.6}$$

no longer has an easy answer. As before, the L^1 convergence is essentially equivalent to the $L^2 \times L^2 \to L^1$ boundedness of the bilinear operator

$$(f, g) \mapsto \sum_{-N \leq n_1 < n_2 \leq N} \widehat{f}(n_1)\widehat{g}(n_2)e^{in_1x}e^{in_2x}, \tag{5.7}$$

independently of N, while the almost everywhere convergence reduces to the $L^2 \times L^2 \to L^1$ boundedness of the maximal operator

$$(f, g) \mapsto \sup_N \left| \sum_{-N \leq n_1 < n_2 \leq N} \widehat{f}(n_1)\widehat{g}(n_2)e^{in_1x}e^{in_2x} \right|. \tag{5.8}$$

Exercise 5.1 Prove that the boundedness of the operators in (5.7) and (5.8) does indeed imply the convergence in L^1, and also almost everywhere, of the iterated Fourier series (5.6).

The operator in (5.7) is essentially the periodic version of the bilinear Hilbert transform mentioned in Chapter 4 (if one simply sends N to infinity). Recall that its symbol is given by $\mathrm{sgn}(\xi_1 + \alpha\xi_2)$, where α can be any number different from 0 and 1. If one takes α to be -1 (as is conventionally the case) and recalls also that the product of two functions can be seen as a bilinear operator with symbol 1, one can see that the bilinear operator with symbol $\chi_{\xi_1 < \xi_2}$ is of a very similar nature; this explains the analogy with (5.7).

The boundedness of the bilinear operator in (5.7) was proved by Lacey and Thiele and that of the maximal operator in (5.8) by Muscalu, Tao, and Thiele. More accurately, these two theorems were proved in the continuous setting of iterated Fourier integrals but it is a standard fact that the proofs hold equally well in this periodic setting. More recently, Honzík and Grafakos showed that the almost everywhere convergence can be transferred to the corresponding continuous setting.

More generally, one can ask whether generic iterated Fourier series of the type

$$\sum_{-N \leq n_1 < \cdots < n_d \leq N} \widehat{f_1}(n_1) \cdots \widehat{f_d}(n_d)e^{in_1x} \cdots e^{in_dx} \tag{5.9}$$

converge either in $L^{2/d}$ or almost everywhere and, as before, one is led to the study of the $L^2 \times \cdots \times L^2 \to L^{2/d}$ boundedness of the operator

$$(f_1, \ldots, f_d) \mapsto \sum_{-N \le n_1 < \cdots < n_d \le N} \widehat{f_1}(n_1) \cdots \widehat{f_d}(n_d) e^{in_1 x} \cdots e^{in_d x}, \tag{5.10}$$

independently of N, and that of its maximal analogue

$$(f_1, \ldots, f_d) \mapsto \sup_N \left| \sum_{-N \le n_1 < \cdots < n_d \le N} \widehat{f_1}(n_1) \cdots \widehat{f_d}(n_d) e^{in_1 x} \cdots e^{in_d x} \right|. \tag{5.11}$$

These boundedness properties were proved in a series of papers by Muscalu, Tao, and Thiele.

5.2. Physical reality

The goal of this section is to show that the previous iterated Fourier series, or rather their natural continuous analogues, are closely related to physical reality.

Let us start by assuming that n bodies or particles are moving freely on circular orbits in a fixed plane with distinct speeds $d_1 \neq \cdots \neq d_n$ around a fixed point called the origin. Clearly, their motions can be parametrized by the curves $t \mapsto C_j e^{id_j t}$ for $j = 1, \ldots, n$. Alternatively, if $u_j(t)$ denotes the position of the jth body at time t, it satisfies the differential equation

$$u_j'(t) = i d_j u_j(t) \tag{5.12}$$

for $j = 1, \ldots, n$.

Now let us assume that these bodies start to interact and that the velocity $u_j'(t)$ of the jth body is influenced by the positions $u_k(t)$, $k \neq j$, of the other bodies, as follows:

$$u_j'(t) = i d_j u_j(t) + \sum_{k \neq j} a_{jk}(t) u_k(t) \tag{5.13}$$

for $j = 1, \ldots, n$, where $a_{jk}(t)$ are all complex-valued measurable functions. The system of differential equations (5.13) can be rewritten more compactly as

$$u' = iDu + Au, \tag{5.14}$$

where $u = [u_1 \ldots u_n]^t$ is a vector-valued function, D is a diagonal $n \times n$ constant matrix with distinct real entries $d_1 \neq \cdots \neq d_n$, and $A = (a_{jk})_{j,k=1}^n$ is a matrix-valued potential having the property that $a_{jj} \equiv 0$ for $j = 1, \ldots, n$. Of course, it may be the case that the entries a_{jk} depend in a nonlinear manner on the positions $u_j(t)$ and $u_k(t)$. Because of this nonlinearity in the potential, systems such as (5.14) can be very difficult to solve. However, assuming that a solution exists, a very natural and fundamental question is the following.

What type of conditions should the interaction matrix A satisfy in order that none of the bodies moves off its orbit, i.e. $\|u_j\|_\infty$ remains finite for every $j = 1, \ldots, n$?

As we will see, for this question to be meaningful from an analytical point of view one has to embed the system (5.14) into the continuous family of systems

$$u' = i\lambda Du + Au, \tag{5.15}$$

which depend on a real *spectral parameter* λ, and ask instead whether for almost every $\lambda \in \mathbb{R}$ the solutions u_j^λ of (5.15) remain bounded for $j = 1, \ldots, n$. We need to specify that now the potential matrix A in (5.15) is precisely the one in (5.14) with the conditions given in the paragraph following (5.14).

Of particular importance is the case when the entries of the matrix A are functions that are bounded and decay as slowly as possible at infinity (think of the *solar system*, for example).

The family (5.15) is called an $AKNS$ (Ablowitz, Kaup, Newell, and Segur) *system*. Such systems play an important role in mathematical physics and scattering theory. The particular case $n = 2$ is closely connected with the spectral theory of classical time-independent Schrödinger operators. It is also a remarkable fact that there are deep connections between these systems and many fundamental PDEs such as the Korteweg–de Vries equation (KdV), the nonlinear Schrödinger equation, and the sine Gordon equation. We will not describe these very interesting issues in detail here; the reader is invited to consult any standard text in scattering theory.

When $A \not\equiv 0$ one can use a variation-of-constants type of argument and write

$$u_k(x) := e^{i\lambda d_k x} v_k(x)$$

for $k = 1, \ldots, n$. In particular, the-column vector $v = [v_1 \cdots v_n]^t$ becomes the solution of a new system,

$$v' = Wv \tag{5.16}$$

where the entries of the matrix-valued function W are now given by $w_{lm}(x) = a_{lm}(x)e^{i\lambda(d_m - d_l)x}$.

It is therefore enough to understand when the solutions of (5.16) are bounded.

Case 1. Let us first consider the easiest possible case, that of 2×2 upper triangular matrices, in other words, $n = 2$ and $a_{11} = a_{22} = a_{21} \equiv 0$ while

$a_{12}(x) := f(x)$ is an arbitrary measurable function. Then, the system (5.16) becomes

$$\begin{bmatrix} v_1' \\ v_2' \end{bmatrix} = \begin{bmatrix} 0 & f(x)e^{i\lambda(d_1-d_2)x} \\ 0 & 0 \end{bmatrix} \begin{bmatrix} v_1 \\ v_2 \end{bmatrix},$$

which implies that

$$v_1' = v_2(x) f(x) e^{i\lambda(d_1-d_2)x},$$
$$v_2' = 0.$$

Clearly, v_2 is bounded since it equals a constant (which we call C_λ) while $v_1 =: v_1^\lambda$ can be written as

$$v_1^\lambda(x) = C_\lambda \int_{-\infty}^{x} f(y)e^{i\lambda(d_1-d_2)y}\, dy + \widetilde{C}_\lambda$$

for some other constant $\widetilde{C}_\lambda$. In particular, we have

$$\|v_1^\lambda\|_\infty \leq |C_\lambda| \sup_x \left| \int_{-\infty}^{x} f(y)e^{i\lambda(d_1-d_2)y}\, dy \right| + |\widetilde{C}_\lambda| \tag{5.17}$$

and so the boundedness question reduces to the problem of analyzing the maximal integral expression (5.17). If the function f were in L^1 then clearly v_1^λ would be bounded, and this would be the most that one could say if d_1 were equal to d_2. However, we know that they are distinct and because of this we will see that the class of *good potential functions* f can be enlarged dramatically. Assume for simplicity that $d_1 - d_2 = 1$.

Let us denote now by C the Carleson maximal operator, defined by

$$Cg(x) = \sup_N \left| \int_{\xi<N} \widehat{g}(\xi)e^{ix\xi}\, d\xi \right|$$

which is clearly the continuous analogue of the maximal operator in (5.3). The similarity between C and (5.17) is of course striking. The theorem of Carleson and Hunt, mentioned before, guarantees that C is bounded on every L^r space as long as $1 < r < \infty$. Suppose now that g is a function such that $\widehat{g} \in L^p$ for some $1 \leq p < 2$ and let p' be the dual index of p. Using the Carleson–Hunt theorem followed by the Hausdorff–Young inequality, we can write

$$\|Cg\|_{p'} \lesssim \|g\|_{p'} \lesssim \|\widehat{g}\|_p, \tag{5.18}$$

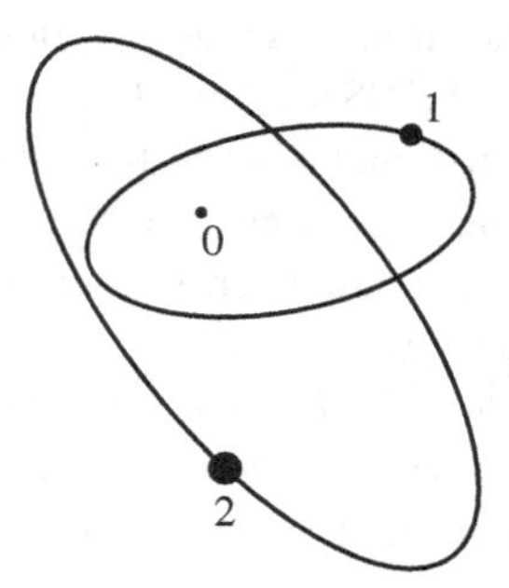

Figure 5.1. The Carleson case – first mass much smaller than the second mass.

and this proves in particular that $Cg(x) < \infty$ for almost every $x \in \mathbb{R}$. Then, from (5.17) we can see that $\|v_1^\lambda\|_\infty$ is also finite for almost every real λ, as long as $f \in L^p$ for $1 \leq p < 2$.

Let us have another look at inequality (5.18). It says that the maximal operator

$$f \mapsto \sup_N \left| \int_{x<N} f(x) e^{i\xi x} \, dx \right| \tag{5.19}$$

is bounded from L^p into $L^{p'}$ for every $1 \leq p < 2$, a fact that is known as the Menshov–Zygmund theorem. It can be considered as a maximal variant of the Hausdorff–Young inequality and one does not need to use the Carleson–Hunt theorem to prove it. Menshov and Zygmund proved this a long time before the almost everywhere convergence of Fourier series was understood. We will come back to this theorem and present the original, simpler, proof in the next section.

Suppose now that the potential function f is of the form $f = \widehat{g}$ for some $g \in L^q$ with $1 < q \leq 2$. The Carleson–Hunt theorem implies that the corresponding $\|v_1^\lambda\|_\infty$ is again finite for almost every λ, and this basically ends the discussion about the simplest 2×2 upper triangular case. To conclude, the Menshov–Zygmund theorem and the Carleson–Hunt theorem have allowed us to understand many interesting aspects of this case.

From the physical reality point of view, the best situation would be to have a potential function f of the form $f = \widehat{g}$ with $g \in L^{1+\varepsilon}$ for some small $\varepsilon > 0$, since then f would decay very slowly at infinity. Recall that case 1 above corresponds to the situation when we have two bodies, the first influenced by the second one but not vice versa; for instance the second body could have a much bigger mass; see Figure 5.1.

It is natural to ask whether it is reasonable to hope that the same scenario occurs in general, for *generic* AKNS systems. The Menshov–Zygmund case has indeed been extended, by Christ and Kiselev, to cover all those AKNS systems defined by matrices whose entries belong to some L^p space for $1 \leq p < 2$. We will come back to this particular case later. We are therefore left with the study of the general Carleson–Hunt case.

Is it true then, that the solutions of (5.15) *are bounded for almost every real number* λ *provided that the entries of the matrix* A *are of the form* $\widehat{g}$ *for some* $g \in L^q$ *with* $1 < q \leq 2$*? In particular, is this true when the entries of* A *are* L^2 *functions?*

It is of interest that if a matrix A has entries that are in L^r for $r > 2$ then solutions can become unbounded. However, not every function in L^r for $r > 2$ is of the form $\widehat{g}$ for some $g \in L^q$ with $1 < q \leq 2$.

The only case where a completely positive answer to the above question is known is case 1, described above. There are, however, quite a few cases where at least the L^2 situation (and a little beyond) has been understood, and these include the cases of upper (or lower) triangular matrices.

Case 2. Let us now consider similarly the case of 3×3 upper triangular systems. So, this time $n = 3$ and $a_{12}(x) := f_1(x)$, $a_{13}(x) := f_2(x)$, $a_{23}(x) := f_3(x)$, all the other entries being identically equal to zero. Our system (5.16) now becomes

$$\begin{bmatrix} v_1' \\ v_2' \\ v_3' \end{bmatrix} = \begin{bmatrix} 0 & f_1(x)e^{i\lambda(d_1-d_2)x} & f_2(x)e^{i\lambda(d_1-d_3)x} \\ 0 & 0 & f_3(x)e^{i\lambda(d_2-d_3)x} \\ 0 & 0 & 0 \end{bmatrix} \begin{bmatrix} v_1 \\ v_2 \\ v_3 \end{bmatrix},$$

which implies that

$$\begin{aligned} v_1'(x) &= v_2(x) f_1(x) e^{i\lambda(d_1-d_2)x} + v_3(x) f_2(x) e^{i\lambda(d_1-d_3)x}, \\ v_2'(x) &= v_3(x) f_3(x) e^{i\lambda(d_2-d_3)x}, \\ v_3'(x) &= 0. \end{aligned}$$

Clearly, v_3 is bounded since it is a constant C_λ and $v_2 =: v_2^\lambda$ can be understood exactly as before. A new challenge will appear, however, in the study of v_1. Since

$$v_2(x) = C_\lambda \int_{-\infty}^{x} f_3(y) e^{i\lambda(d_2-d_3)y}\, dy + \widetilde{C}_\lambda$$

it follows that

$$v_1'(x) = C_\lambda \left(\int_{-\infty}^{x} f_3(y) e^{i\lambda(d_2-d_3)y}\, dy \right) f_1(x) e^{i\lambda(d_1-d_2)x}$$
$$+ \widetilde{C}_\lambda f_1(x) e^{i\lambda(d_1-d_2)x} + C_\lambda f_2(x) e^{i\lambda(d_1-d_3)x}.$$

Then, on integrating, we obtain for $v_1 =: v_1^\lambda$

$$v_1^\lambda(x) = C_\lambda \int_{-\infty}^{x} f_1(y) e^{i\lambda(d_1-d_2)y} \left(\int_{-\infty}^{y} f_3(z) e^{i\lambda(d_2-d_3)z}\, dz \right) dy$$
$$+ \widetilde{C}_\lambda \int_{-\infty}^{x} f_1(y) e^{i\lambda(d_1-d_2)y}\, dy + C_\lambda \int_{-\infty}^{x} f_2(y) e^{i\lambda(d_1-d_3)y}\, dy + \approx{C}_\lambda$$
$$=: I + II + III + \approx{C}_\lambda.$$

The terms II and III are of a famliar type while term I can be rewritten as

$$C_\lambda \int_{z<y<x} f_3(z) f_1(y) e^{i\lambda((d_2-d_3)z+(d_1-d_2)y)}\, dz dy \tag{5.20}$$

and can be seen to be of a more complex nature. One would like to know whether its supremum over x is finite for almost every λ, as before. Let us now consider the bi-Carleson operator C_2^α defined as follows:

$$C_2^\alpha(f,g)(x) := \sup_N \left| \int_{\xi_1<\xi_2<N} \widehat{f}(\xi_1)\widehat{g}(\xi_2) e^{ix(\alpha_1\xi_1+\alpha_2\xi_2)}\, d\xi_1 d\xi_2 \right|,$$

where $\alpha_1 \neq 0$ and $\alpha_2 \neq 0$. Clearly, this is the continuous analogue of the maximal operator (5.8). A theorem of Muscalu, Tao, and Thiele says that

Theorem 5.1 *The operator C_2^α maps $L^p(\mathbb{R}) \times L^q(\mathbb{R})$ into $L^r(\mathbb{R})$ provided that $\alpha_1 + \alpha_2 \neq 0$ and $1 < p, q \leq \infty$ with $1/p + 1/q = 1/r$ and $2/3 < r < \infty$.*

As before, the similarity between C_2^α and the integral expression (5.20) is evident. One can also check that $(d_2 - d_3) + (d_1 - d_2) = d_1 - d_3 \neq 0$. It is not known whether the above theorem holds for every $1/2 < r < \infty$. If that were true then one would have a completely positive answer to the generic AKNS question in the present case of 3×3 upper (and lower) triangular systems. We will come back to this particular point in Chapter 6, which is devoted to the bilinear Hilbert transform.

It is also of interest to note that if $\alpha_1 + \alpha_2 = 0$ then C_2^α does not satisfy any $L^p(\mathbb{R})$ estimate. Assume now that $\widehat{f} \in L^{p_1}$ and $\widehat{g} \in L^{p_2}$ for some $1 \leq p_1, p_2 < 2$ and denote by r the index defined by $1/r = 1/p_1' + 1/p_2'$, where p_1' and p_2' are the dual indexes of p_1 and p_2 respectively. Using the above theorem and the Hausdorff–Young inequality one can write

$$\|C_2^\alpha(f, g)\|_r \lesssim \|f\|_{p_1'} \|g\|_{p_2'} \lesssim \|\widehat{f}\|_{p_1} \|\widehat{g}\|_{p_2}, \tag{5.21}$$

and this shows that in particular $C_2^\alpha(f, g)(x) < \infty$ for almost every $x \in \mathbb{R}$. Further more, it also shows that, for $f_1 \in L^{p_1}$, $f_2 \in L^{p_2}$, and $f_3 \in L^{p_3}$ with $1 \leq p_1, p_2, p_3 < 2$, the corresponding expressions $\|v_1^\lambda\|_\infty$, $\|v_2^\lambda\|_\infty$, and $\|v_3^\lambda\|_\infty$ are all finite for almost every real number λ, as desired. As in case 1, let us have another look at the inequality (5.21). It says that the maximal operator

$$(f, g) \mapsto \sup_N \left| \int_{x_1 < x_2 < N} f(x_1) g(x_2) e^{i\xi(\alpha_1 x_1 + \alpha_2 x_2)}\, dx_1 dx_2 \right| \tag{5.22}$$

is bounded from $L^{p_1} \times L^{p_2}$ into L^r, a result proved directly by Christ and Kiselev without relying on Theorem 5.1. This result can be considered as a bilinear maximal variant of the Menshov–Zygmund theorem mentioned previously. We will return to it in the next section of the chapter. Let us also mention that it holds true even when $\alpha_1 + \alpha_2 = 0$, when it is known that the corresponding C_2^α is unbounded on any pair of L^p spaces.

Using Theorem 5.1 itself, however, we see as before that provided that f_1, f_2, f_3 are all L^2 functions, the solutions of the 3×3 upper triangular system are all bounded, again for almost every λ. In fact, the same conclusion can be drawn if for instance $f_3 \in L^2$ while $f_1 = \widehat{g_1}$ and $f_2 = \widehat{g_2}$, with $g_1 \in L^{1+\varepsilon_1}$ and $g_2 \in L^{1+\varepsilon_2}$ for all arbitrarily small numbers $\varepsilon_1, \varepsilon_2 > 0$.

Recall that the present case corresponds to the situation when there are three bodies and the first body is influenced by the other two bodies but the second body is influenced only by the third, while the third is not influenced by either of the other two. For instance, one can imagine that the mass of the first is much smaller than the mass of the second, which is much smaller still than the mass of the third body (see Figure 5.2).

Case 3. The $n \times n$ upper or lower triangular case for $n \geq 2$ is similar and can be reduced to proving estimates for maximal operators of the form given by

$$C_n^\alpha(f_1, \ldots, f_n)(x) = \sup_N \left| \int_{\xi_1 < \ldots < \xi_n < N} \widehat{f_1}(\xi_1) \cdots \widehat{f_n}(\xi_n) e^{ix(\alpha_1 \xi_1 + \cdots + \alpha_n \xi_n)}\, d\xi \right|,$$

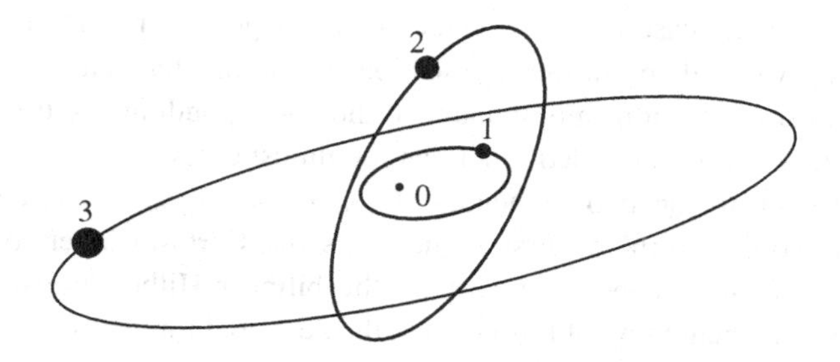

Figure 5.2. The bi-Carleson case – first mass much smaller than the second mass, which in turn is much smaller than the third mass.

where now $\alpha = (\alpha_1, \ldots, \alpha_m)$ satisfies the nondegeneracy condition

$$\alpha_j + \alpha_{j+1} \neq 0 \tag{5.23}$$

for every $1 \leq j \leq n$. If we denote by T_n^α the n-linear operator given by the same formula above but with $\sup_N$ dropped, the analogy between the operators C_n^α and T_n^α and those in (5.10) and (5.11) becomes trivial. To be exact, T_n^α now depends on the parameter N as well but for simplicity we do not write this dependence explicitly.

The operators analogous to (5.19) and (5.22) are clearly given by

$$(f_1, \ldots, f_n) \mapsto \sup_N \left| \int_{x_1 < \cdots < x_n < N} f_1(x_1) \cdots f_n(x_n) e^{i\xi(\alpha_1 x_1 + \cdots + \alpha_n x_n)} \, dx \right|. \tag{5.24}$$

Let us denote by $\widetilde{C}_n^\alpha$ the maximal operator (5.24) and by $\widetilde{T}_n^\alpha$ the n-linear operator obtained after dropping $\sup_N$ in (5.24).

As before, the natural estimates for $\widetilde{C}_n^\alpha$ solve the boundedness problem in the L^p case for $1 \leq p < 2$, while the natural estimates for C_n^α are enough to solve it in the L^2 situation.

The general case, when any two bodies influence each other and one has to consider generic $n \times n$ matrices, is even harder and not completely understood. Of course, one can first run a standard Picard iteration argument and write the generic solution of (5.16) as an infinite series of expressions of the above type. If one treats these expressions as $\widetilde{C}_n^\alpha$ operators, one can prove the desired L^p estimates for each of them and conclude the boundedness of the solutions in the case $1 \leq p < 2$.

The important difference now is that the above-mentioned expressions, when viewed as C_n^α operators, can be unbounded on $L^2(\mathbb{R})$ since for some (or sometimes for all) of them the nondegeneracy condition (5.23) can be violated.

This is why the L^2 case is not yet understood for generic potential matrices A. However, we will see in the last section of the chapter that even though the multilinear expressions are unbounded the corresponding solutions of the AKNS system are still bounded functions for almost every λ.

Coming back to the two sequences $(C_n^\alpha)_{n\geq 1}$ and $(T_n^\alpha)_{n\geq 2}$ it is clear that the simplest operator in the first sequence is the Carleson operator, while the simplest in the second is essentially the bilinear Hilbert transform. As mentioned at the beginning of this chapter, the next goal for us is to understand these two operators in detail.

Before doing that, however, as promised we will present more detail for some facts used during the above presentation.

5.3. Generic L^p AKNS systems for $1 \leq p < 2$

Let us consider now the following particular case of (5.16):

$$\begin{bmatrix} v_1'(x) \\ v_2'(x) \end{bmatrix} = \begin{bmatrix} 0 & f(x)e^{-i\lambda x} \\ \overline{f(x)}e^{i\lambda x} & 0 \end{bmatrix} \begin{bmatrix} v_1(x) \\ v_2(x) \end{bmatrix}. \tag{5.25}$$

We claim that this system is sufficiently complex to capture the essence of the general case. Notice that the above W matrix is self-adjoint. It is not difficult to see that if one assumes that initially $v_1(-\infty) = 1$ and $v_2(-\infty) = 0$ then a *standard Picard iteration argument* allows us to write both v_1 and v_2 as infinite series of expressions of type $\widetilde{T}_n^\alpha(f, \overline{f}, \ldots, f)$, where $\alpha_j = (-1)^j$ for $j = 1, \ldots, n$. More precisely, one has

$$v_1 =: v_1^\lambda = 1 + \sum_{n\geq 2:\ n \text{ even}} \widetilde{T}_n^\alpha(f, \overline{f}, \ldots, \overline{f}) \tag{5.26}$$

while

$$v_2 =: v_2^\lambda = \sum_{n\geq 1:\ n \text{ odd}} \widetilde{T}_n^\alpha(f, \overline{f}, \ldots, f). \tag{5.27}$$

Exercise 5.2 Prove that the expressions (5.26) and (5.27) are indeed the unique solutions of the system (5.25) satisfying the above initial conditions.

Now suppose that $f \in L^1$. It is easy to observe that one has the uniform bound

$$\left|\widetilde{C}_n^\alpha(f, \overline{f}, \ldots, f)(\lambda)\right| \leq \frac{1}{n!}\|f\|_1^n, \tag{5.28}$$

which implies in particular that both $\|v_1^\lambda\|_\infty$ and $\|v_2^\lambda\|_\infty$ are smaller than $\exp(\|f\|_1)$ for every real number λ. Hence, both solutions are bounded as desired.

The case $1 < p < 2$ is much trickier and was considered by Christ and Kiselev, as mentioned earlier. The previous discussions show that one needs to prove $L^{p'/n}$ estimates for the expressions $\widetilde{T}_n^\alpha(f, \overline{f}, \ldots, f)$ and $\widetilde{C}_n^\alpha(f, \overline{f}, \ldots, f)$ that are tight enough to be summable over n (as before, in (5.28)), in order to prove the boundedness of the solutions. This is indeed possible, but it is not our intention to describe all these interesting estimates here. Instead we will just prove the necessary bounds for the simplest terms $\widetilde{C}_1^\alpha(f)$ and $\widetilde{T}_2^\alpha(f, \overline{f})$, since these two expressions are closely related to the Carleson operator and the bilinear Hilbert transform, to be studied later. The first is the *maximal Menshov–Zygmund* operator.

Theorem 5.2 *Let $\alpha \neq 0$. The maximal operator $\widetilde{C}_1^\alpha(f)$ defined by*

$$\widetilde{C}_1^\alpha(f)(\xi) := \sup_N \left| \int_{x<N} f(x) e^{i\xi\alpha x}\, dx \right|$$

is bounded from L^p into $L^{p'}$ for every $1 \le p < 2$.

Proof First, one can assume that $\alpha = 1$. One can also assume without loss of generality that N belongs to some fixed arbitrary finite set, by using some standard approximation arguments. Let f be such that $\|f\|_p = 1$. Define the map $\varphi : \mathbb{R} \to [0, 1]$ by

$$\varphi(t) := \int_{-\infty}^{t} |f(s)|^p\, ds.$$

Observe that $\varphi(-\infty) = 0$ and $\varphi(+\infty) = 1$ and that φ is increasing. In particular, for every interval $I \subseteq [0, 1]$, $\varphi^{-1}(I)$ is also an interval and $\int_{\varphi^{-1}(I)} |f(s)|^p\, ds = |I|$ or, equivalently,

$$\|f \chi_{\varphi^{-1}(I)}\|_p = |I|^{1/p}. \tag{5.29}$$

For every $\xi \in \mathbb{R}$ let us denote by $N(\xi)$ the integer for which the supremum is attained in the formula defining $\widetilde{C}_1^\alpha(f)$. It is clearly enough to study the linear operator

$$f \mapsto \int_{x<N(\xi)} f(x) e^{i\xi x}\, dx. \tag{5.30}$$

Fix x and ξ. Then $x < N(\xi)$ if and only if $\varphi(x) < \varphi(N(\xi))$. Since both these numbers are within the interval $[0, 1]$, there exists a unique dyadic interval $\omega \subseteq [0, 1]$ that contains them both and such that the length of ω is minimal. In particular, it follows that $\varphi(x) \in \omega_\ell$ and $\varphi(N(\xi)) \in \omega_r$, where ω_ℓ and ω_r are the left-hand and right-hand halves of ω. Using this observation, one can decompose (5.30) as

$$\sum_\omega \left(\int_{x\,:\,\varphi(x) \in \omega_\ell} f(x) e^{i\xi x}\, dx \right) \chi_{\{\xi\,:\,\varphi(N(\xi)) \in \omega_r\}}, \tag{5.31}$$

where the sum is over all possible dyadic subintervals of $[0, 1]$. Thus the $L^{p'}$ norm of (5.31) is smaller than

$$\sum_{k\geq 0} \left\| \sum_{\omega : |\omega|=2^{-k}} \left(\int_{x : \varphi(x)\in\omega_\ell} f(x)e^{i\xi x}\, dx \right) \chi_{\{\xi \,:\, \varphi(N(\xi))\in\omega_r\}} \right\|_{p'} . \tag{5.32}$$

Fix $k \geq 0$ and consider the expression above. Observe that there are 2^k such dyadic intervals and they are all disjoint. In particular, the sets $\{\xi \mid \varphi(N(\xi)) \in \omega_r\}$ are disjoint. As a consequence, the $L^{p'}$ norm of the expression can be estimated by

$$\left(\sum_{\omega : |\omega|=2^{-k}} \left\| \widehat{f\chi_{\varphi^{-1}(\omega_\ell)}}(\xi) \right\|_{p'}^{p'} \right)^{1/p'} . \tag{5.33}$$

Using the Hausdorff–Young inequality with (5.29), one can continue and estimate the above expression further by

$$\begin{aligned}
\left(\sum_{\omega : |\omega|=2^{-k}} \left\| f\chi_{\varphi^{-1}(\omega_\ell)} \right\|_{p}^{p'} \right)^{1/p'} &= \left(\sum_{\omega : |\omega|=2^{-k}} |\omega_\ell|^{p'/p} \right)^{1/p'} \\
&= \left(\sum_{\omega : |\omega|=2^{-k}} 2^{-kp'/p} \right)^{1/p'} = 2^{-k(1/p-1/p')}
\end{aligned}$$

and, since $1 \leq p < 2$, this is clearly summable over $k \geq 0$. □

The following theorem can be proved in a similar way.

Theorem 5.3 *Let $\alpha_1 \neq 0$ and $\alpha_2 \neq 0$. The bilinear operator $\widetilde{T}_2^{\alpha}(f, g)$ defined by*

$$\widetilde{T}_2^{\alpha}(f, g)(\xi) = \int_{x_1<x_2} f(x_1)g(x_2)e^{i\xi(\alpha_1 x_1+\alpha_2 x_2)}\, dx_1 dx_2$$

maps $L^p \times L^p$ into $L^{p'/2}$ for every $1 \leq p < 2$.

Proof The method is the same as in the proof of Theorem 5.2, concerning the Menshov–Zygmund operator. Assume that f is L^p-normalized and define the φ map as before. Then, given real numbers x_1 and x_2, we have $x_1 \leq x_2$ if and only if $\varphi(x_1) \leq \varphi(x_2)$. Since these two numbers now belong to the interval $[0, 1]$, there exists a unique dyadic interval $\omega \subseteq [0, 1]$ that contains them both and is of minimal length. As before, this observation allows us to decompose

$\widetilde{T}_2^\alpha(f, g)(\xi)$ as

$$\sum_{\omega} \int_{\varphi(x_1)\in\omega_\ell\,;\,\varphi(x_2)\in\omega_r} f(x_1)g(x_2)e^{i\xi(\alpha_1 x_1+\alpha_2 x_2)}\,dx_1dx_2, \tag{5.34}$$

where the sum runs over all dyadic subintervals of $[0, 1]$. In particular, using the Hölder and Hausdorff–Young inequalities and the fact that dyadic intervals of the same length are disjoint, one can estimate the $L^{p'/2}$ norm (5.34) by

$$\begin{aligned}
&\sum_{k\geq 0}\sum_{\omega:|\omega|=2^{-k}} \left\| \widehat{f\chi_{\varphi^{-1}(\omega_\ell)}}(\alpha_1\xi)\, \widehat{g\chi_{\varphi^{-1}(\omega_r)}}(\alpha_2\xi)\right\|_{p'/2} \\
&\lesssim \sum_{k\geq 0}\sum_{\omega:|\omega|=2^{-k}} \left\| \widehat{f\chi_{\varphi^{-1}(\omega_\ell)}}(\alpha_1\xi)\right\|_{p'} \left\|\widehat{g\chi_{\varphi^{-1}(\omega_r)}}(\alpha_2\xi)\right\|_{p'} \\
&\lesssim \sum_{k\geq 0}\sum_{\omega:|\omega|=2^{-k}} \left\| f\chi_{\varphi^{-1}(\omega_\ell)}\right\|_{p} \left\|g\chi_{\varphi^{-1}(\omega_r)}\right\|_{p} \\
&= \sum_{k\geq 0}\sum_{\omega:|\omega|=2^{-k}} |\omega_\ell|^{1/p} \left\|g\chi_{\varphi^{-1}(\omega_r)}\right\|_{p} \\
&\leq \sum_{k\geq 0} 2^{-k/p} \sum_{\omega:|\omega|=2^{-k}} \left\|g\chi_{\varphi^{-1}(\omega_r)}\right\|_{p} \\
&\leq \sum_{k\geq 0} 2^{-k/p}2^{k/p'} \left(\sum_{\omega:|\omega|=2^{-k}} \left\|g\chi_{\varphi^{-1}(\omega_r)}\right\|_{p}^{p}\right)^{1/p} = \sum_{k\geq 0} 2^{-k(1/p-1/p')}\|g\|_p,
\end{aligned}$$

which as desired is bounded by $\|g\|_p$, again thanks to the fact that $1 \leq p < 2$. □

Exercise 5.3 Let $1 \leq p_1, p_2 < 2$ and define r by $1/r = 1/p_1' + 1/p_2'$. Use the method of the above proof to show that $\widetilde{T}_2^\alpha$ is bounded from $L^{p_1} \times L^{p_2}$ into L^r.

5.4. Generic L^2 AKNS systems

Recall that $\widetilde{T}_3^\alpha$ (see the previous section) is given by the formula

$$\widetilde{T}_3^\alpha(f, \overline{f}, f)(x) = \int_{x_1<x_2<x_3} f(x_1)\overline{f(x_2)}f(x_3)e^{ix(x_1-x_2+x_3)}\,dx_1dx_2dx_3. \tag{5.35}$$

Using arguments similar to those used to prove Theorems 5.2 and 5.3, one can show that the above expression is bounded from $L^p \times L^p \times L^p$ into $L^{p'/3}$ for every $1 \leq p < 2$. This constitutes Problem 5.1 at the end of this chapter.

However, we will start this section by proving that $\widetilde{T}_3^{\alpha}$ does not map $L^2 \times L^2 \times L^2$ into $L^{2/3}$, as one might expect. Let $g \in L^2$ be such that $\widehat{g} = f$. One then can rewrite (5.35) in a way that clarifies its structure:

$$\int_{\xi_1<\xi_2<\xi_3} \widehat{g}(\xi_1)\,\overline{\widehat{g}(\xi_2)}\,\widehat{g}(\xi_3)\, e^{ix(\xi_1-\xi_2+\xi_3)}\, d\xi_1 d\xi_2 d\xi_3$$
$$= -\int_{\xi_1<-\xi_2<\xi_3} \widehat{g}(\xi_1)\,\overline{\widehat{g}(-\xi_2)}\,\widehat{g}(\xi_3)\, e^{ix(\xi_1+\xi_2+\xi_3)}\, d\xi_1 d\xi_2 d\xi_3$$
$$= -\int_{\xi_1<-\xi_2<\xi_3} \widehat{g}(\xi_1)\,\widehat{\overline{g}}(\xi_2)\,\widehat{g}(\xi_3)\, e^{ix(\xi_1+\xi_2+\xi_3)}\, d\xi_a \xi_2 d\xi_3. \tag{5.36}$$

Notice that the above expression is a trilinear operator with symbol $-\chi_{\{\xi_1<-\xi_2<\xi_3\}}$, which can also be written as $-\chi_{\mathbb{R}_-}(\xi_1+\xi_2)\,\chi_{\mathbb{R}_+}(\xi_2+\xi_3)$. For simplicity, we would like to replace this symbol with one that is similar, $\operatorname{sgn}(\xi_1+\xi_2)\operatorname{sgn}(\xi_2+\xi_3)$, and prove a counterexample for the operator corresponding to this new symbol instead. Later, the reader will realize that the same argument holds for the original expression. The advantage of the replacement is that (modulo a dilation with a factor 2π coming from the Fourier transform) this new trilinear operator has the following nice kernel representation:

$$T(g,\overline{g},g)(x) = p.v. \int_{\mathbb{R}^2} g(x+t)\,\overline{g(x+t+s)}\,g(x+s)\,\frac{dt}{t}\frac{ds}{s}. \tag{5.37}$$

Exercise 5.4 Prove the above kernel representation for the trilinear operator with symbol $\operatorname{sgn}(\xi_1+\xi_2)\operatorname{sgn}(\xi_2+\xi_3)$.

The following theorem holds.

Theorem 5.4 *The operator given by the formula* (5.37) *does not map* $L^2 \times L^2 \times L^2$ *into* $L^{2/3}$.

Proof We start with a simple observation, on which a counterexample can be based. Let $g(x) = e^{2\pi i x^2}$. Then, one has formally

$$T(e^{2\pi i x^2}, e^{-2\pi i x^2}, e^{2\pi i x^2})(x) = e^{2\pi i x^2} \int_{\mathbb{R}^2} \frac{e^{2\pi i t s}}{ts}\, dt ds$$
$$= C e^{2\pi i x^2} \int_{\mathbb{R}} \frac{\operatorname{sgn}(t)}{t}\, dt = 2C e^{2\pi i x^2} \int_0^{\infty} \frac{1}{t}\, dt.$$

In other words, for this particular bounded function g we have

$$T(g,\overline{g},g)(x) = Cg(x)\overline{g}(x)g(x) \int_0^{\infty} \frac{1}{t}\, dt,$$

which is an infinite quantity. The idea now is to localize this observation and produce a rigorous counterexample. For every large positive number N, define

$g_N(x) = e^{2\pi i x^2} \chi_{[-N,N]}$. Then, observe that as long as $x \in [-N/1000, N/1000]$ one has

$$|T(g_N, \overline{g_N}, g_N)(x)| \geq C \left| \int_{-N/100}^{N/100} \int_{-N/100}^{N/100} \frac{e^{2\pi i t s}}{ts} \, dt ds \right| + O(1)$$
$$\geq C \left| \int_0^{N/100} \int_0^{N/100} \frac{\sin(ts)}{ts} \, dt ds \right| + O(1), \qquad (5.38)$$

where the two constants C may be different. We now need the following lemma.

Lemma 5.5 *There exist two universal constants $C_1, C_2 > 0$ such that*

$$\left| \int_0^N \int_0^N \frac{\sin(ts)}{ts} \, dt ds \right| \geq C_1 \log N$$

for every $N > C_2$.

If we accept this lemma for a moment and combine it with (5.38), it follows easily that T cannot map $L^2 \times L^2 \times L^2$ into $L^{2/3}$. In fact, the argument shows that T cannot satisfy *any* L^p estimates of the Hölder type.

We are therefore left with the proof of the lemma. Using the fact that

$$\int_0^\infty \frac{\sin t}{t} \, dt = \frac{\pi}{2},$$

there must be a constant $C > 0$ such that

$$\int_0^x \frac{\sin t}{t} \, dt \in \left[\frac{\pi}{4}, \frac{3\pi}{4} \right] \qquad (5.39)$$

whenever $x > C$. Then one has

$$\int_0^N \int_0^N \frac{\sin(xy)}{xy} \, dx dy = \int_0^N \left(\int_0^N \frac{\sin(xy)}{y} \, dy \right) \frac{dx}{x} = \int_0^N \left(\int_0^{Nx} \frac{\sin t}{t} \, dt \right) \frac{dx}{x}$$
$$= \int_0^{C/N} \left(\int_0^{Nx} \frac{\sin t}{t} \, dt \right) \frac{dx}{x} + \int_{C/N}^N \left(\int_0^{Nx} \frac{\sin t}{t} \, dt \right) \frac{dx}{x}$$
$$= \int_0^C \left(\int_0^x \frac{\sin t}{t} \, dt \right) \frac{dx}{x} + \int_{C/N}^N \left(\int_0^{Nx} \frac{\sin t}{t} \, dt \right) \frac{dx}{x}. \qquad (5.40)$$

Given that the function

$$x \mapsto \frac{1}{x} \int_0^x \frac{\sin t}{t} \, dt$$

is continuous on the interval $[0, C]$, it follows that the first term in (5.40) is in fact $O(1)$. To estimate the second term in (5.40) we observe that $x > C/N$

implies $Nx > C$ and so, using (5.39), we can finally write

$$\int_{C/N}^{N} \left(\int_{0}^{Nx} \frac{\sin t}{t}\, dt \right) \frac{dx}{x} \geq \frac{\pi}{4} \int_{C/N}^{N} \frac{dx}{x} = \frac{\pi}{4}(2 \log N - \log C),$$

and this ends the proof of the lemma, if N is a large enough constant. □

Exercise 5.5 Prove in detail the estimate (5.38).

Exercise 5.6 Prove that Theorem 5.4 also holds true for the trilinear operator with symbol $\chi_{\mathbb{R}_-}(\xi_1 + \xi_2)\, \chi_{\mathbb{R}_+}(\xi_2 + \xi_3)$.

For every large positive real number N as before, now consider the function

$$f_N(x) = \sum_{j=2}^{N} \varphi_j(x) e^{ijx}, \tag{5.41}$$

where the functions $\varphi_j(x)$ are smooth, real-valued, adapted to the intervals $[j-1, j+1]$, and supported on the smaller intervals $[j - 1/3, j + 1/3]$, for every $j = 2, \ldots, N$. If one recalls the phase-space portrait of the function $e^{2\pi i x^2}$ given in the first chapter, it is not difficult to see that the f_N are in fact *discrete variants* of the g_N functions before (recall also that the Fourier transform of a complex Gaussian function is also a complex Gaussian function). We claim that the functions f_N provide a counterexample to the L^2-boundedness of the operator $\widetilde{T}_3^{\alpha}$ given by (5.35).

Exercise 5.7 Prove, this time directly (using the original expression in (5.35)), that the functions f_N defined above do indeed provide a counterexample to the L^2-boundedness of $\widetilde{T}_3^{\alpha}$.

However, in spite of this fact we will see that the corresponding solutions of the system (5.25) are bounded, for every real λ, uniformly with respect to N.

To prove this it will be more convenient to study the *matrix* analogue of our system (5.25). Consider the 2×2 matrices defined by

$$G(x) = \begin{bmatrix} v_1(x) & \overline{v_2(x)} \\ v_2(x) & \overline{v_1(x)} \end{bmatrix}$$

and

$$W(x) = \begin{bmatrix} 0 & f_N(x) e^{-i\lambda x} \\ \overline{f_N(x)} e^{i\lambda x} & 0 \end{bmatrix}.$$

Observe that $v_1(x)$ and $v_2(x)$ satisfy (5.25) for the particular function f_N if and only if $G(x)$ satisfies the equation

$$G'(x) = W(x)G(x) \tag{5.42}$$

with the initial condition $G(-\infty) = I_2$, where I_2 is the 2×2 identity matrix. Fix $\lambda, x \in \mathbb{R}$. The goal is to show that, for this particular parameter λ, the operatorial norm $\|G(x)\|$ is bounded by a universal constant that is independent of x and of N. We will see that the bound is uniform with respect to λ as well. Assume that $\lambda \in [1, N]$. If this is not the case, we will see that the argument allows us to obtain even better bounds. Let j_0 be the unique integer in the interval $[1, N]$ with the property that $|\lambda - j_0| \leq 1/2$, and let $j_1(= j_1(x))$ be the largest integer in the interval $[1, N]$ that is still smaller than x. Notice that if x lies in the region to the left of the interval $[1, N]$ then $G(x)$ is constant in that region since $f_N(x) = 0$ there.

Let us denote by $W_j(x)$ the matrix obtained by replacing the function f_N in the definition of $W(x)$ by $\varphi_j(x)e^{ijx}$ for every $j = 2, \ldots, N$. Then let $G_j(x)$ be the unique solution to the equation

$$G_j'(x) = W_j(x)G_j(x)$$

satisfying the initial condition $G_j(-\infty) = I_2$. Observe that $G(x)$ can be split as follows:

$$G(x) = G_{j_1}(x)G_{j_1-1}(+\infty)\cdots G_2(+\infty). \tag{5.43}$$

This relation is a consequence of the fact that the $W_j(x)$ have disjoint supports and that every time one is presented with a matrix solution $G(x)$ of (5.42) with initial condition I_2 then the product $G(x)M$ satisfies the same ODE with initial condition M. Here M is an arbitrary constant 2×2 matrix.

Exercise 5.8 Check carefully the representation formula (5.43).

Using (5.43) one can write

$$\|G(x)\| \leq \|G_{j_1}(x)\| \, \|G_{j_1-1}(+\infty)\| \cdots \|G_2(+\infty)\|.$$

Clearly, since the potential function of each individual W_j is an L^1 function, we know already from the previous section that $\|G_j(z)\| \leq C$ for some universal constant C. Using this for every $j \in \{j_0 - 3, j_0 + 3\}$ we conclude that

$$\|G(x)\| \leq C \prod_{j \notin \{j_0-3, j_0+3\}} \|G_j(+\infty)\|.$$

Lemma 5.6 *The following estimate holds:*

$$\|G_j(+\infty)\| = 1 + O\left(\frac{1}{|j - j_0|^2}\right)$$

for every $j \notin \{j_0 - 3, j_0 + 3\}$.

Using this result in the previous inequality completes the proof. We are therefore left with proving this lemma. Let us now recall (5.26) and (5.27) and use them in our particular case. First, we observe that

$$\widetilde{T}_1^\alpha = \int_{\mathbb{R}} \varphi_j(x_1) e^{ijx_1} e^{-i\lambda x_1}\, dx_1 = \int_{\mathbb{R}} \varphi_j(x_1) e^{i(j-\lambda)x_1}\, dx_1$$

and a simple integration by parts shows that this term is at least $O(1/|j - j_0|^2)$.

Now let us consider the next term, $\widetilde{T}_2^\alpha$, and write

$$\begin{aligned}\widetilde{T}_2^\alpha &= \int_{x_1 < x_2} \varphi_j(x_1)\varphi_j(x_2) e^{ij\xi_1} e^{-ij\xi_2} e^{-i\lambda x_1} e^{i\lambda x_2}\, dx_1 dx_2 \\ &= \int_{x_1 < x_2} \varphi_j(x_1)\varphi_j(x_2) e^{i(j-\lambda)x_1} e^{-i(j-\lambda)x_2}\, dx_1 dx_2.\end{aligned}$$

Fix x_1 and integrate by parts in x_2. This time there is a boundary term (when $x_2 = x_1$) where the two phase functions cancel. One obtains as a consequence that

$$\widetilde{T}_2^\alpha = \frac{iC}{j - \lambda} + O\left(\frac{1}{|j - j_0|^2}\right).$$

One can then treat every other single term similarly, eventually obtaining that the contribution of the rest of the sum is also $O(1/|j - j_0|^2)$. Putting everything together one obtains

$$G_j(+\infty) = \begin{bmatrix} 1 + \dfrac{iC_1}{j - \lambda} & 0 \\ 0 & 1 + \dfrac{iC_2}{j - \lambda} \end{bmatrix} + R_j$$

with $\|R_j\| = O(1/|j - j_0|^2)$. Since the operator norm of the first matrix is clearly of the type $1 + O(1/|j - j_0|^2)$ this concludes the proof of the lemma.

All this shows that the L^2 general case sometimes simply amounts to the obvious fact that e^{ix} is a bounded function even though the individual terms in its Taylor expansion are unbounded.

Notes

With the exception of some early parts of the chapter, most material here comes from Muscalu, Tao, and Thiele [92, 97]. See also Muscalu, Tao, and Thiele [91] for a related discussion. The counterexample in Theorem 5.4 is from [92]; see also Muscalu, Pipher, Tao *et al.* [93]. The transference theorem mentioned in the text can be found in Grafakos and Honzík [49]. The treatment of the bi-Carleson operator can be found in Muscalu, Tao, and Thiele [89, 95]. For more details about the L^p case for $1 \leq p < 2$ and its connections with the spectral theory of Schrödinger operators, see Christ and Kiselev [18, 19]. To learn about the relevance of AKNS systems to mathematical physics, consult the book by Ablowitz and Segur [1]. For connections with orthogonal polynomials, Riemann–Hilbert problems, etc. see the expository lecture notes by Tao and Thiele [110].

Problems

Problem 5.1 Let $\alpha_1, \alpha_2, \alpha_3$ be three real numbers all different from zero. Consider the trilinear expression

$$(f, f, f) \mapsto \int_{x_1 < x_2 < x_3} f(x_1) f(x_2) f(x_3) e^{ix(\alpha_1 x_1 + \alpha_2 x_2 + \alpha_3 x_3)} \, dx_1 dx_2 dx_3$$

and prove that it maps $L^p \times L^p \times L^p$ into $L^{p'}$ for any $1 \leq p < 2$.

Problem 5.2 Consider the function of two variables $(t, s) \mapsto e^{its}$. Construct a phase-space portrait for it similar to that described in Chapter 1 for $x \mapsto e^{ix^2}$.

6

The bilinear Hilbert transform

Let us start by recalling the first Calderón commutator, defined by

$$C_1 f(x) = p.v. \int_{\mathbb{R}} \frac{A(x) - A(y)}{(x - y)^2} f(y)\, dy,$$

where A is a Lipschitz function; in other words $A' \in L^\infty$. In Chapter 4 we observed that the above formula can be written as

$$\begin{aligned} C_1 f(x) &= -p.v. \int_{\mathbb{R}} \frac{A(x+t) - A(x)}{t} f(x+t) \frac{dt}{t} \\ &= -p.v. \int_{\mathbb{R}} \left(\int_0^1 a(x + \alpha t)\, d\alpha \right) f(x+t) \frac{dt}{t}, \end{aligned}$$

where $a = A'$. By switching the two integrations this expression can also be seen as an average over α of bilinear expressions in a and f, denoted $BHT_\alpha(a, f)$. As mentioned in Chapter 4, Calderón asked whether these bilinear operators are bounded from $L^\infty \times L^2$ into L^2. Notice that at the endpoints one has $BHT_0(a, f) = aHf$ and $BHT_1(a, f) = H(af)$, where H is the linear Hilbert transform; clearly, both satisfy the desired estimates.

Since all the operators BHT_α behave similarly for any real number α different from 0 or 1, for *symmetry* reasons the traditional approach was to consider the particular formula (which corresponds to $\alpha = -1$)

$$BHT(f, g)(x) = p.v. \int_{\mathbb{R}} f(x - t) g(x + t) \frac{dt}{t}, \tag{6.1}$$

which is known as the *bilinear Hilbert transform*. The main goal of the present chapter is to prove the following theorem of Lacey and Thiele, which in particular answers Calderón's original question.

Theorem 6.1 *The bilinear Hilbert transform maps $L^p \times L^q$ into L^r for any $1 < p, q \leq \infty$ with the property that $1/p + 1/q = 1/r$ and $2/3 < r < \infty$.*

The constraint $2/3 < r < \infty$ comes from the method of proof. It is not yet known whether the theorem holds for every $1/2 < r < \infty$. We will come back to this question at the end of the chapter.

Before starting to describe the proof of the theorem, we would like to make several observations. First, we recall that the meaning of (6.1) is

$$\lim_{\varepsilon \to 0} \int\limits_{|t|>\varepsilon} f(x-t)g(x+t)\,\frac{dt}{t} \tag{6.2}$$

whenever the limit exists. Assume now that both f and g are Schwartz functions on $\mathbb{R}$. We will show that in this case the limit (6.2) exists and we can rewrite it in a more convenient way.

Indeed, (6.2) can be written as

$$\lim_{\substack{\varepsilon \to 0 \\ N \to \infty}} \int\limits_{\varepsilon<|t|<N} f(x-t)g(x+t)\,\frac{dt}{t}, \tag{6.3}$$

which can be rewritten further as

$$\begin{aligned} &\lim_{\substack{\varepsilon \to 0 \\ N \to \infty}} \int\limits_{\varepsilon<|t|<N} \int\limits_{\mathbb{R}} \int\limits_{\mathbb{R}} \widehat{f}(\xi)e^{2\pi i\xi(x-t)}\,\widehat{g}(\eta)e^{2\pi i\eta(x+t)}\frac{1}{t}\,d\xi d\eta dt \\ &= \lim_{\substack{\varepsilon \to 0 \\ N \to \infty}} \int\limits_{\varepsilon<|t|<N} \int\limits_{\mathbb{R}} \int\limits_{\mathbb{R}} \widehat{f}(\xi)\widehat{g}(\eta)e^{2\pi ix(\xi+\eta)}\frac{1}{t}e^{-2\pi it(\xi-\eta)}\,d\xi d\eta dt. \end{aligned}$$

The above integrand, viewed as a function depending on ξ, η, t, is clearly integrable and, by applying Fubini's theorem together with dominated convergence, we see that the last formula becomes equal to

$$-i\pi \int\limits_{\mathbb{R}^2} \operatorname{sgn}(\xi-\eta)\,\widehat{f}(\xi)\widehat{g}(\eta)e^{2\pi ix(\xi+\eta)}\,d\xi d\eta, \tag{6.4}$$

which clearly makes perfect sense since both $\widehat{f}$ and $\widehat{g}$ are also Schwartz functions. Given also the fact that the product of f and g can be written as

$$\int\limits_{\mathbb{R}^2} \widehat{f}(\xi)\widehat{g}(\eta)e^{2\pi ix(\xi+\eta)}\,d\xi d\eta,$$

we see that in order to understand (6.4) it is enough to understand the *bilinear multiplier operator* given by

$$(f, g) \mapsto \int_{\xi<\eta} \widehat{f}(\xi)\widehat{g}(\eta)e^{2\pi i x(\xi+\eta)}\, d\xi d\eta. \tag{6.5}$$

Notice that the symbol of this operator is $1_{\{\xi<\eta\}}$, which is singular along the line $\xi = \eta$. As a consequence one cannot apply the paraproduct theory developed earlier in this case since, in particular, paraproduct symbols are smooth away from the origin. This fact is closely related to the symmetries of the operator. To be more specific, let us denote by T_y, D_λ, and M_a the *translation*, *dilation*, and *modulation* operators, defined by

$$T_y f(x) := f(x - y),$$
$$D_\lambda f(x) := f\left(\frac{x}{\lambda}\right)$$

and

$$M_a f(x) := f(x)e^{2\pi i x a},$$

for every $y, a \in \mathbb{R}$ and $\lambda > 0$.

Then, it is easy to observe that, from (6.1),

$$BHT\left(T_y f, T_y g\right)(x) = T_y(BHT(f, g))(x),$$
$$BHT(D_\lambda f, D_\lambda g)(x) = D_\lambda(BHT(f, g))(x)$$

and also

$$\left|BHT(M_a f, M_a g)(x)\right| = \left|M_{2a}(BHT(f, g))(x)\right|. \tag{6.6}$$

While the first two relations are (as we already know) available for paraproducts as well, the last relation is new and is the main source of new difficulty for the bilinear Hilbert transform; this one-dimensional *modulation invariance* is clearly related to the one-dimensional singularity of its symbol.

6.1. Discretization

As explained above, from now on the goal is to prove the desired estimates for the bilinear operator (6.5). The first task is to decompose its symbol in a natural way. Instead of using a paraproduct decomposition based on Littlewood–Paley projections (which, as mentioned earlier, cannot succeed), the idea is to decompose the region $\{\xi < \eta\}$ using *Whitney squares* (see Figure 6.1). These squares have the property that their side length is comparable with their distance from

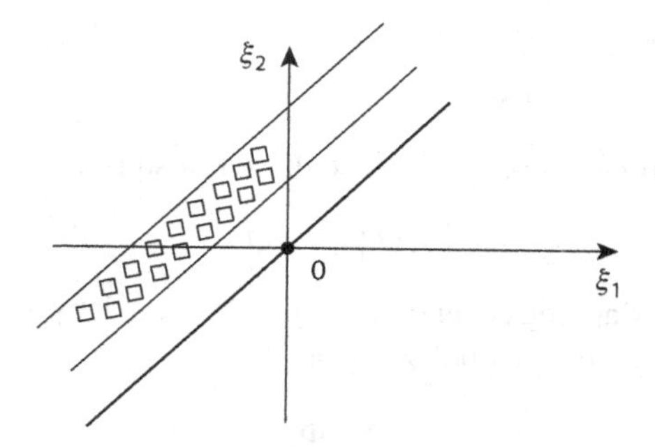

Figure 6.1. Decomposition of the symbol of the bilinear Hilbert transform.

the *singularity line* $\{\xi = \eta\}$. We claim that this is a very natural point of view. After all, as the reader may remember, even in the paraproduct case we used Whitney squares but there they were Whitney with respect to the *singularity point* $\{\xi = \eta = 0\}$.

To be able to describe our discretization fully, we need to set up some notation and definitions. Since our operator is bilinear, it has a trilinear form associated with it. This explains why our terminology will be three dimensional most of the time.

By a *shifted dyadic interval* we simply mean any interval of the form $2^j(k + (0, 1) + \alpha)$ for any $k, j \in \mathbb{Z}$ and $\alpha \in \left\{0, \frac{1}{3}, -\frac{1}{3}\right\}$. In other words, a shifted dyadic interval is either a dyadic interval or a translation to the right or left of a dyadic interval with one third the length of the latter. A *shifted dyadic cube* is a set of the form $Q = Q_1 \times Q_2 \times Q_3$ where each Q_j is a shifted dyadic interval and they all have the same length. Finally, a *shifted dyadic quasi-cube* is a set $Q = Q_1 \times Q_2 \times Q_3$ where again each Q_j is a shifted dyadic interval but this time we require only the more relaxed condition $|Q_1| \simeq |Q_2| \simeq |Q_3|$. It is a simple geometric observation that any arbitrary cube $Q \subseteq \mathbb{R}^3$ is contained in $\frac{7}{10}\widetilde{Q}$ (the cube having the same center as $\widetilde{Q}$ but with side length $7/10$ that of $\widetilde{Q}$) for a certain shifted dyadic cube $\widetilde{Q}$ of comparable side length.

The same terminology will be used in the plane $\mathbb{R}^2$. The only difference is that the previous cubes now become squares.

Denote by Γ the singularity line

$$\Gamma := \left\{(\xi_1, \xi_2) \in \mathbb{R}^2 \,\middle|\, \xi_1 = \xi_2\right\}$$

and consider the collection $\mathbf{Q}$ of all shifted dyadic squares $Q = Q_1 \times Q_2$ having the property that

$$Q \subseteq \{\xi_1 < \xi_2\}$$

and also satisfying

$$\mathrm{dist}(Q, \Gamma) \simeq C_0 \, \mathrm{diam}(Q),$$

where C_0 is a fixed large constant. Since the set of squares

$$\{\tfrac{7}{10}Q \mid Q \in \mathbf{Q}\}$$

forms a finitely overlapping cover of the region $\{\xi_1 < \xi_2\}$, by a standard partition of unity we can write the symbol $\chi_{\xi_1<\xi_2}$ as

$$\chi_{\xi_1<\xi_2} = \sum_Q \Phi_Q(\xi_1, \xi_2),$$

where each Φ_Q is a smooth bump function adapted to Q and supported in $\frac{8}{10}Q$. This is the only reason why we have chosen to work with *shifted* dyadic intervals and not simply with dyadic intervals. This extra freedom allows us to obtain the above decomposition. By splitting each Φ_Q further as a double Fourier series in ξ_1, ξ_2 we can rewrite the above expression as

$$\sum_{n\in\mathbb{Z}^2} C_n \sum_Q \Phi_{Q_1,n,1}(\xi_1)\Phi_{Q_2,n,2}(\xi_2), \tag{6.7}$$

where $(C_n)_n$ is a rapidly decreasing sequence and $\Phi_{Q_j,n,j}$ is a bump function adapted to Q_j and supported in $\frac{9}{10}Q_j$ uniformly in n for every $j = 1, 2$. The rapid decay of $(C_n)_n$ allows one to obtain the above uniformity in n.

Exercise 6.1 Convince yourself that our symbol can indeed be written in the form (6.7).

Since $\xi_1 \in \frac{9}{10}Q_1$ and $\xi_2 \in \frac{9}{10}Q_2$ it follows that $\xi_1 + \xi_2 \in \frac{9}{10}Q_1 + \frac{9}{10}Q_2$ and, as a consequence, one can find a shifted dyadic interval Q_3 with the property that $\frac{9}{10}Q_1 + \frac{9}{10}Q_2 \subseteq \frac{7}{10}Q_3$ and also satisfying $|Q_1| = |Q_2| \simeq |Q_3|$. In particular, there exist bump functions $\Phi_{Q_3,n,3}$ adapted to Q_3 uniformly in n and supported in $\frac{9}{10}Q_3$ such that $\Phi_{Q_3,n,3} \equiv 1$ on $\frac{9}{10}Q_1 + \frac{9}{10}Q_2$. This means that the expression (6.7) can also be written as

$$\sum_{n\in\mathbb{Z}^2} C_n \sum_Q \Phi_{Q_1,n,1}(\xi_1)\Phi_{Q_2,n,2}(\xi_2)\Phi_{Q_3,n,3}(\xi_1 + \xi_2), \tag{6.8}$$

where this time Q runs over a collection of shifted dyadic quasi-cubes in $\mathbb{R}^3$. We now claim that we can assume that the sum in (6.8) runs over a collection of *sparse* dyadic quasi-cubes. More precisely, we can assume that whenever Q and $\widetilde{Q}$ belong to this collection and $|Q_j| < |\widetilde{Q}_j|$ then $C_1|Q_j| \leq |\widetilde{Q}_j|$ and also that whenever $|Q_j| = |\widetilde{Q}_j|$ one has $C_2 Q_j \cap C_2 \widetilde{Q}_j = \emptyset$ for some fixed large constants C_1 and C_2 (much larger than the previous C_0) and every $j = 1, 2, 3$.

Indeed, it is not difficult to see the our sum (6.8) can be split into a sum of finitely many such sparse corresponding expressions.

Assuming this we then see that, for any Q in such a sparse collection, there exists a unique shifted dyadic cube $\widetilde{Q}$ in $\mathbb{R}^3$ such that $Q \subseteq \frac{7}{10}\widetilde{Q}$ and with $\mathrm{diam}(Q) \simeq \mathrm{diam}(\widetilde{Q})$. This allows us in particular to assume that the sum (6.8) runs over a sparse collection of shifted dyadic cubes (so that now $|Q_1| = |Q_2| = |Q_3|$).

The advantage of such a decomposition is that multipliers of the type $m(\xi_1, \xi_2) = \Phi_1(\xi_1)\Phi_2(\xi_2)\Phi_3(\xi_1 + \xi_2)$ are well localized and allow us to simplify the corresponding bilinear expression $T_m(f_1, f_2)$. More precisely, as in the case of paraproducts one can write the trilinear form associated with such a T_m as follows:

$$\int_{\mathbb{R}} T_m(f_1, f_2)(x) f_3(x)\, dx$$

$$= \int_{\mathbb{R}} \left(\int_{\mathbb{R}^2} \widehat{f_1}(\xi_1)\widehat{f_2}(\xi_2)\Phi_1(\xi_1)\Phi_2(\xi_2)\Phi_3(\xi_1 + \xi_2)e^{2\pi i x(\xi_1+\xi_2)}\, d\xi_1 d\xi_2 \right) f_3(x)\, dx$$

$$= \int_{\mathbb{R}^2} \widehat{f_1}(\xi_1)\widehat{f_2}(\xi_2)\Phi_1(\xi_1)\Phi_2(\xi_2)\Phi_3(\xi_1 + \xi_2)\widehat{f_3}(-\xi_1 - \xi_2)\, d\xi_1 d\xi_2$$

$$=: \int_{\mathbb{R}^2} \widehat{f_1}(\xi_1)\widehat{f_2}(\xi_2)\Phi_1(\xi_1)\Phi_2(\xi_2)\widetilde{\Phi}_3(-\xi_1 - \xi_2)\widehat{f_3}(-\xi_1 - \xi_2)\, d\xi_1 d\xi_2$$

$$= \int_{\lambda_1+\lambda_2+\lambda_3=0} \widehat{f_1 * \check{\Phi}_1}(\lambda_1)\widehat{f_2 * \check{\Phi}_2}(\lambda_2)\widehat{f_3 * \check{\widetilde{\Phi}}_3}(\lambda_3)\, d\lambda_1 d\lambda_2 d\lambda_3$$

$$= \int_{\mathbb{R}} (f_1 * \check{\Phi}_1)(x)\,(f_2 * \check{\Phi}_2)(x) \left(f_3 * \check{\widetilde{\Phi}}_3 \right)(x)\, dx, \tag{6.9}$$

where $\check{h}$ denotes the inverse Fourier transform of the function h.

Exercise 6.2 Justify the last equality in (6.9).

Now fix $n \in \mathbb{Z}$ and look at a particular term in the sum (6.8). Using (6.9), the trilinear form associated with it becomes

$$\int_{\mathbb{R}} (f_1 * \check{\Phi}_{Q_1,n,1})(x)(f_2 * \check{\Phi}_{Q_2,n,2})(x) \left(f_3 * \check{\widetilde{\Phi}}_{Q_3,n,3} \right)(x)\, dx. \tag{6.10}$$

Since Q is a shifted dyadic cube, there exists an integer $k \in \mathbb{Z}$ such that $|Q_1| = |Q_2| = |Q_3| = 2^k$. By setting $x = 2^{-k}y$, (6.10) becomes

$$2^{-k} \int_{\mathbb{R}} (f_1 * \check{\Phi}_{Q_1,n,1})(2^{-k}y)(f_2 * \check{\Phi}_{Q_2,n,2})(2^{-k}y)\left(f_3 * \check{\tilde{\Phi}}_{Q_3,n,3}\right)(2^{-k}y)\,dy$$

$$= 2^{-k} \int_0^1 \sum_{\ell \in \mathbb{Z}} (f_1 * \check{\Phi}_{Q_1,n,1})(2^{-k}(\ell+\alpha))\,(f_2 * \check{\Phi}_{Q_2,n,2})(2^{-k}(\ell+\alpha))$$
$$\times \left(f_3 * \check{\tilde{\Phi}}_{Q_3,n,3}\right)(2^{-k}(\ell+\alpha))\,d\alpha.$$

Now, a generic term of the type $f_j * \check{\Phi}_{Q_j,n,j}(2^{-k}(\ell+\alpha))$ can be written as

$$\begin{aligned}(f_j * \check{\Phi}_{Q_j,n,j})(2^{-k}(\ell+\alpha)) &= \int_{\mathbb{R}} f_j(y)\check{\Phi}_{Q_j,n,j}(2^{-k}\ell + 2^{-k}\alpha - y)\,dy \\ &=: \int_{\mathbb{R}} f_j(y)\tilde{\check{\Phi}}_{Q_j,n,j}(y - 2^{-k}\ell - 2^{-k}\alpha)\,dy \\ &= \int_{\mathbb{R}} f_j(y)\,\overline{\overline{\tilde{\check{\Phi}}}_{Q_j,n,j}(y - 2^{-k}\ell - 2^{-k}\alpha)}\,dy \\ &=: 2^{k/2}\langle f_j, \Phi_{P_j}^{j,n,\alpha}\rangle \end{aligned} \tag{6.11}$$

where P_j, $j = 1, 2$, denotes the rectangle $2^{-k}[\ell, \ell+1] \times Q_j$ and $P_3 := 2^{-k}[\ell, \ell+1] \times (-Q_3)$. Putting these together we see that (6.10) becomes an average over α of expressions of the type

$$\frac{1}{|I_P|^{1/2}}\langle f_1, \Phi_{P_1}^{1,n,\alpha}\rangle\langle f_2, \Phi_{P_2}^{2,n,\alpha}\rangle\langle f_3, \Phi_{P_3}^{3,n,\alpha}\rangle, \tag{6.12}$$

where $P := (P_1, P_2, P_3)$ and $I_P := I_{P_1} = I_{P_2} = I_{P_3} := 2^{-k}[\ell, \ell+1]$. The rectangles P_1, P_2, P_3 are called *tiles* (notice that they all have area 1) while P is called a *tri-tile*. The intervals Q_j for $j = 1, 2$ and $-Q_3$ are called the *frequency intervals* of P_j for $j = 1, 2, 3$ and will be denoted ω_{P_j} from now on, while I_{P_j} is the spatial (or time) interval of P_j. One should also observe the important fact that all the functions $\Phi_{P_j}^{j,n,\alpha}$ are L^2-normalized and are wave packets associated with the Heisenberg boxes P_j uniformly with respect to the parameter n (as a consequence of (6.7)). Also, we now have supp $\widehat{\Phi_{P_j}^{j,n,\alpha}} \subseteq \omega_{P_j}$ for $j = 1, 2, 3$. If one treats every term in (6.8) in a similar manner then one obtains a *discretized model operator* for the bilinear Hilbert transform.

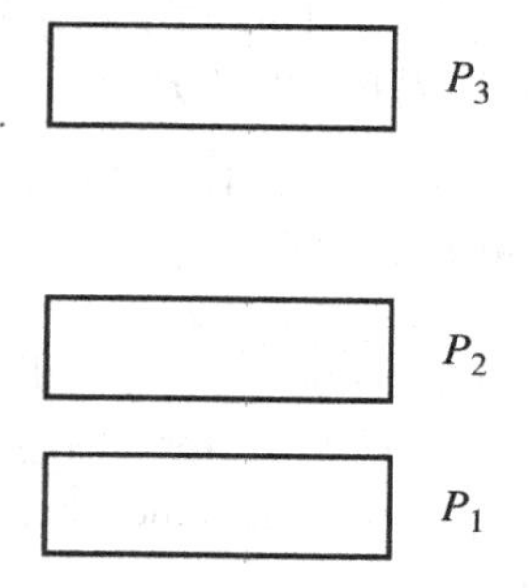

Figure 6.2. Tiles in phase space.

It is important at this point to realize the similarity between the generic terms in (6.12) and the corresponding terms that appeared in the case of the discretized paraproducts studied earlier, which were of the form

$$\frac{1}{|I|^{1/2}}\langle f_1, \Phi_I^1\rangle\langle f_2, \Phi_I^2\rangle\langle f_3, \Phi_I^3\rangle. \tag{6.13}$$

However, the fundamental difference now is that while the Heisenberg boxes of the wave packets Φ_I^j are uniquely determined by the spatial dyadic intervals I, in (6.12) for every given I there is potentially a whole column of tri-tiles P for which $I_P = I$. We also point out two other important properties of the tri-tiles in (6.12). First, if one knows the position of P_1, P_2, or P_3 then one knows precisely the positions of the other two as well. For instance, given that P_1 lies in the phase plane, P_2 must lie a certain number of steps above it (a number comparable with the constant C_0 used before), while the frequency coordinate of P_3 is essentially the sum of the frequency coordinates of P_1 and P_2, as in Figure 6.2.

Second, if one assumes for instance that all the frequency intervals of the P_1 tiles intersect each other (say, they all contain a fixed frequency ξ_0) then the frequency intervals of the corresponding P_2 tiles are disjoint and lacunarily aligned away from ξ_0 (remember that each P_2 has to be about C_0 steps above its corresponding P_1). A similar conclusion can be drawn for the P_3 tiles. This means that in such a situation, modulo certain harmless translations in frequency, (6.12) becomes a discretized paraproduct. This observation motivates the introduction of *trees* later on.

We will collect these properties in the definition below.

Definition 6.2 Let P and P' be tiles. Then

(i) we write $P' < P$ if $I_{P'} \subsetneq I_P$ and $\omega_P \subseteq 3\omega_{P'}$;
(ii) we write $P' \leq P$ if $P' < P$ or $P' = P$;

(iii) we write $P' \lesssim P$ if $I_{P'} \subseteq I_P$ and $\omega_P \subseteq 100C_0\omega_{P'}$;
(iv) we write $P' \lesssim' P$ if $P' \lesssim P$ but $P' \not\leq P$.

Definition 6.3 A collection $\mathbb{P}$ of tri-tiles is said to have rank 1 if the following properties are satisfied for all P, $P' \in \mathbb{P}$.

(i) If $P \neq P'$ then $P_j \neq P'_j$ for $1 \leq j \leq 3$.
(ii) If $\omega_{P_j} = \omega_{P'_j}$ for some j then $\omega_{P_j} = \omega_{P'_j}$ for all $1 \leq j \leq 3$.
(iii) If $P'_j \leq P_j$ for some j then $P'_j \lesssim P_j$ for all $1 \leq j \leq 3$.
(iv) If in addition to $P'_j \leq P_j$ one also assumes that $|I_{P'}| \ll |I_P|$ then one has $P'_i \lesssim' P_i$ for every $i \neq j$.

We also say that a collection $\mathbb{P}$ of tri-tiles is called *sparse* if the corresponding collection of frequency cubes $(\omega_{P_1} \times \omega_{P_2} \times \omega_{P_3})_P$ forms a sparse collection of shifted dyadic cubes (in the sense discussed earlier) and the set of dyadic time intervals $(I_P)_P$ is sparse as well. In addition, from now on we will assume, without loss of generality, that for every $j = 1, 2, 3$ the corresponding shifting parameter α is constant. It is not difficult to observe that the collection of tri-tiles that appear in the formula (6.12) can be written as a finite union of sparse collections of rank 1. The term rank 1 comes from the one-dimensional degree of freedom in frequency. Discretized paraproducts may be said to be associated with collections of rank 0. There are also multilinear operators defined by collections of *rank k* for any $k \geq 2$, but they will not be discussed here.

Now let $\mathbb{P}$ be a finite collection of tri-tiles that is sparse and has rank 1. Consider wave packets $(\Phi^j_{P_j})_{P\in\mathbb{P}}$ for $j = 1, 2, 3$, adapted to the tiles P_j respectively, as before. Assume also that they are L^2-normalized. Denote by $T_{\mathbb{P}}$ the discrete bilinear operator given by

$$T_{\mathbb{P}}(f_1, f_2) = \sum_{P\in\mathbb{P}} \frac{1}{|I_P|^{1/2}} \langle f_1, \Phi^1_{P_1}\rangle\langle f_2, \Phi^2_{P_2}\rangle\Phi^3_{P_3}. \tag{6.14}$$

We now claim that in order to prove Theorem 6.1 it is enough to prove corresponding *restricted weak type* estimates for the model operators (6.14). We need to recall a few facts from interpolation theory to be able to state this properly.

Definition 6.4 Let $1 < p_1, p_2 \leq \infty$ and $0 < p < \infty$ be such that $1/p_1 + 1/p_2 = 1/p$. An arbitrary bilinear operator T is said to be of the *restricted weak type* (p_1, p_2, p) if and only if for all measurable sets E_1, E_2, E of finite measure there exists $E' \subseteq E$ with $|E'| \simeq |E|$ such that

$$\left|\int_{\mathbb{R}} T(f_1, f_2)(x)f(x)\,dx\right| \lesssim |E_1|^{1/p_1}|E_2|^{1/p_2}|E'|^{1/p'}$$

for every $|f_1| \leq \chi_{E_1}$, $|f_2| \leq \chi_{E_2}$, and $|f| \leq \chi_{E'}$.

Notice that, as a consequence of Lemma 2.6, this implies in particular that

$$\|T(f_1, f_2)\|_{p,\infty} \lesssim |E_1|^{1/p_1}|E_2|^{1/p_2}$$

whenever f_1 and f_2 are as above.

The following theorem will be proved in detail during this chapter.

Theorem 6.5 *Let p and q be such that p is strictly larger than 1 and arbitrarily close to 1 and q is strictly smaller than 2 and arbitrarily close to 2 and such that for $1/r := 1/p + 1/q$ one has $2/3 < r < 1$. Then the operator $T_{\mathbb{P}}$ is of the restricted weak type (p, q, r). Moreover, the implicit constants do not depend on the particular finite rank-1 collection $\mathbb{P}$.*

Using interpolation arguments and the symmetry of $T_{\mathbb{P}}$ (more precisely, the fact that its two adjoints are similar operators), one can deduce from this the following.

Theorem 6.6 *If $\mathbb{P}$ is as before then $T_{\mathbb{P}}$ is bounded from $L^p \times L^q \to L^r$ for any $1 < p, q \leq \infty$ satisfying $1/p + 1/q = 1/r$ and $2/3 < r < \infty$.*

Indeed, first observe that if p, q, r are as in Theorem 6.6 then the point $(1/p, 1/q, 1/r')$ lies in the interior of the convex hull of the following six extremal points: $(-1/2, 1/2, 1)$, $(-1/2, 1, 1/2)$, $(1/2, -1/2, 1)$, $(1, -1/2, 1/2)$, $(1/2, 1, -1/2)$, and $(1, 1/2, -1/2)$. Then one can use multilinear interpolation between the restricted weak-type estimates of Theorem 6.5 for $T_{\mathbb{P}}$ and its adjoints to deduce the strong-type estimates of Theorem 6.6.

The claim now is that this is sufficient to obtain the same result for the bilinear Hilbert transform itself.

Our previous calculations, the rapid decay of the Fourier coefficients C_n, and the facts that our dualization procedure commutes with the average over α and that one can choose the implicit exceptional sets Ω independently of the averaging parameter α (as was the case for paraproducts), together with standard limiting arguments, allow us to conclude that whenever f and g are Schwartz functions one has

$$\|BHT(f, g)\|_r \lesssim \|f\|_p \|g\|_q \tag{6.15}$$

for any p, q, r as above. Then, if both p and q are strictly smaller than ∞, there exists a unique way of extending BHT by density considerations so that (6.15) is available for every $f \in L^p$ and $g \in L^q$. Finally, in the case where one of p or q equals ∞ one can use instead a duality argument to extend BHT rigorously to all the functions in L^∞, as we showed in Chapter 4.

This discussion proves that in order to complete the proof of Theorem 6.1 it is enough to prove the particular discrete variant of it, Theorem 6.5.

6.2. The particular scale-1 case of Theorem 6.5

As in the case of paraproducts, it is instructive at this point to prove Theorem 6.5 in a simpler situation in order to motivate the general approach to be described later. More precisely, consider the discrete model (6.14) for the bilinear Hilbert transform,

$$T_{\mathbb{P}}(f_1, f_2) = \sum_{P\in\mathbb{P}} \frac{1}{|I_P|^{1/2}} \langle f_1, \Phi^1_{P_1}\rangle \langle f_2, \Phi^2_{P_2}\rangle \Phi^3_{P_3},$$

and assume that here the sum is over a collection of tri-tiles having the property that the length $|I_P|$ is the same for every $P \in \mathbb{P}$ rather than over a generic collection of tri-tiles.

Fix measurable sets E_1, E_2 of finite measure and a set E of measure 1. Our goal is to construct a subset $E' \subseteq E$ with $|E'| \simeq 1$ and such that

$$\left| \sum_{P\in\mathbb{P}} \frac{1}{|I_P|^{1/2}} \langle f_1, \Phi^1_{P_1}\rangle \langle f_2, \Phi^2_{P_2}\rangle \langle f_3, \Phi^3_{P_3}\rangle \right| \lesssim |E_1|^{1/p} |E_2|^{1/q} \tag{6.16}$$

for every $|f_1| \leq \chi_{E_1}$, $|f_2| \leq \chi_{E_2}$, and $|f| \leq \chi_{E'}$, with indices p, q as in the hypothesis of Theorem 6.5. As in the paraproduct case, the fact that one can assume that $|E| = 1$ is a consequence of the *scaling invariance* of the operator $T_{\mathbb{P}}$. As a result one has to work with a different scale $|I_P|$, but the argument will be independent of that.

Define first an exceptional set

$$\Omega := \{x \mid M\chi_{E_1}(x) > C|E_1|\} \cup \{x \mid M\chi_{E_2}(x) > C|E_2|\},$$

where M is the usual Hardy–Littlewood maximal operator. Clearly, if C is a sufficiently large constant, one has $|\Omega| < \frac{1}{2}$. In particular, the set $E' := E \setminus \Omega$ now satisfies $|E'| \simeq 1$ and we claim that this is the set we were looking for.

To be able to estimate (6.16) we first split our collection of tri-tiles $\mathbb{P}$ as follows:

$$\mathbb{P} = \bigcup_{d\geq 0} \mathbb{P}_d,$$

where $\mathbb{P}_d$ contains all the tri-tiles in $\mathbb{P}$ having the property that

$$1 + \frac{\operatorname{dist}(I_P, \Omega^c)}{|I_P|} \simeq 2^d.$$

We will prove that for every $d \geq 0$ one has

$$\left| \sum_{P\in\mathbb{P}_d} \frac{1}{|I_P|^{1/2}} \langle f_1, \Phi^1_{P_1}\rangle \langle f_2, \Phi^2_{P_2}\rangle \langle f_3, \Phi^3_{P_3}\rangle \right| \lesssim |E_1|^{1/p} |E_2|^{1/q} 2^{-10d}, \tag{6.17}$$

and this will be enough to conclude (6.16). To be able to show this, we introduce the decomposition

$$\mathbb{P}_d = \bigcup_{n_1} \mathbb{P}_{d,1}^{n_1}, \tag{6.18}$$

where $\mathbb{P}_{d,1}^{n_1}$ contains the tri-tiles in $\mathbb{P}_d$ with the property that the average

$$\frac{|\langle f_1, \Phi_{P_1}^1 \rangle|}{|I_P|^{\frac{1}{2}}} \simeq 2^{-n_1}. \tag{6.19}$$

Similarly, using the functions f_2 and f_3 one obtains

$$\mathbb{P}_d = \bigcup_{n_2} \mathbb{P}_{d,2}^{n_2} \tag{6.20}$$

and also

$$\mathbb{P}_d = \bigcup_{n_3} \mathbb{P}_{d,3}^{n_3}. \tag{6.21}$$

In particular, the left-hand side of (6.17) can be decomposed naturally as

$$\left| \sum_{n_1,n_2,n_3} \sum_{P \in \mathbb{P}_d^{n_1,n_2,n_3}} \frac{1}{|I_P|^{1/2}} \langle f_1, \Phi_{P_1}^1 \rangle \langle f_2, \Phi_{P_2}^2 \rangle \langle f_3, \Phi_{P_3}^3 \rangle \right|, \tag{6.22}$$

where $\mathbb{P}_d^{n_1,n_2,n_3}$ is defined to be the intersection $\mathbb{P}_{d,1}^{n_1} \cap \mathbb{P}_{d,2}^{n_2} \cap \mathbb{P}_{d,3}^{n_3}$. Using these, the absolute value of (6.22) can be estimated by

$$\sum_{n_1,n_2,n_3} \sum_{P \in \mathbb{P}_d^{n_1,n_2,n_3}} \frac{|\langle f_1, \Phi_{P_1}^1 \rangle|}{|I_P|^{\frac{1}{2}}} \frac{|\langle f_2, \Phi_{P_2}^2 \rangle|}{|I_P|^{\frac{1}{2}}} \frac{|\langle f_3, \Phi_{P_3}^3 \rangle|}{|I_P|^{\frac{1}{2}}} |I_P|$$

$$\lesssim \sum_{n_1,n_2,n_3} 2^{-n_1} 2^{-n_2} 2^{-n_3} \sum_{P \in \mathbb{P}_d^{n_1,n_2,n_3}} |I_P|. \tag{6.23}$$

So far, everything is identical to the paraproduct case studied in Chapter 2. The difference comes at the next step, when we need to estimate the sum of the lengths $|I_P|$. Note that this time there are potentially many tri-tiles P with the property that $I_P = I_0$ for a certain fixed dyadic interval I_0 of the same length as I_P. However, if such a situation occurs, notice that for every $1 \leq j \leq 3$ the frequency intervals ω_{P_j} are disjoint. This fact suggests that one should take advantage of the orthogonality in L^2 of the wave packets corresponding to the tiles P_j. Using this intuition, on the one hand we can write down the Bessel-type estimate

$$\sum_{P \in \mathbb{P}_d^{n_1,n_2,n_3}} |I_P| \leq \sum_{P \in \mathbb{P}_d^{n_1}} |I_P| \lesssim 2^{2n_1} \sum_{P \in \mathbb{P}_d^{n_1}} |\langle f_1, \Phi_{P_1}^1 \rangle|^2$$

$$\lesssim 2^{2n_1} \|f_1\|_2^2 \leq 2^{2n_1} |E_1|.$$

Similarly

$$\sum_{P \in \mathbb{P}_d^{n_1,n_2,n_3}} |I_P| \lesssim 2^{2n_2}|E_2|$$

and also

$$\sum_{P \in \mathbb{P}_d^{n_1,n_2,n_3}} |I_P| \lesssim 2^{2n_3},$$

given that $|E'| \simeq 1$. By interpolating between these estimates one obtains

$$\sum_{P \in \mathbb{P}_d^{n_1,n_2,n_3}} |I_P| \lesssim 2^{2n_1\theta_1} 2^{2n_2\theta_2} 2^{2n_3\theta_3} |E_1|^{\theta_1} |E_2|^{\theta_2} \tag{6.24}$$

for every $0 \leq \theta_1, \theta_2, \theta_3 < 1$ with $\theta_1 + \theta_2 + \theta_3 = 1$.

On the other hand, as in the case of paraproducts, since every P has the property that it is in $\mathbb{P}_d$ we must have

$$2^{-n_1} \lesssim 2^d |E_1|, \quad 2^{-n_2} \lesssim 2^d |E_2|, \quad 2^{-n_3} \lesssim 2^{-Md}, \tag{6.25}$$

where M is an arbitrarily large constant. Using all these facts one can estimate (6.23) by

$$\sum_{n_1,n_2,n_3} 2^{-n_1(1-2\theta_1)} 2^{-n_2(1-2\theta_2)} 2^{-n_3(1-2\theta_3)} |E_1|^{\theta_1} |E_2|^{\theta_2} \lesssim 2^{-\widetilde{M}d} |E_1|^{1-\theta_1} |E_2|^{1-\theta_2},$$

assuming that for every $1 \leq j \leq 3$ one has $0 \leq \theta_j < 1/2$ to ensure that the geometric series are convergent. Finally, one chooses θ_1 and θ_2 such that $1/p = 1 - \theta_1$ and $1/q = 1 - \theta_2$ and observes that p becomes close to 1 if we choose θ_1 close to 0 and q becomes close to 2 if θ_2 is chosen close to $1/2$. Since M can be as large as we want, this indeed proves (6.17), as desired.

It is also important to notice that, since $1/r = 1/p + 1/q$, $1/r$ must be equal to $1 + \theta_3$ and, in particular, $\theta_3 < 1/2$ is equivalent to $r > 2/3$ as assumed in the hypothesis of Theorem 6.5.

6.3. Trees, L^2 sizes, and L^2 energies

To prove the general case of Theorem 6.5 we have to be much more careful, since the tri-tiles can now have different scales. The reader may remember that we faced a somewhat similar difficulty in Chapter 2, but the situation is more complex here because of the *extra degree of freedom* in frequency. Recall that for paraproducts the lacunary frequency intervals are already organized, and

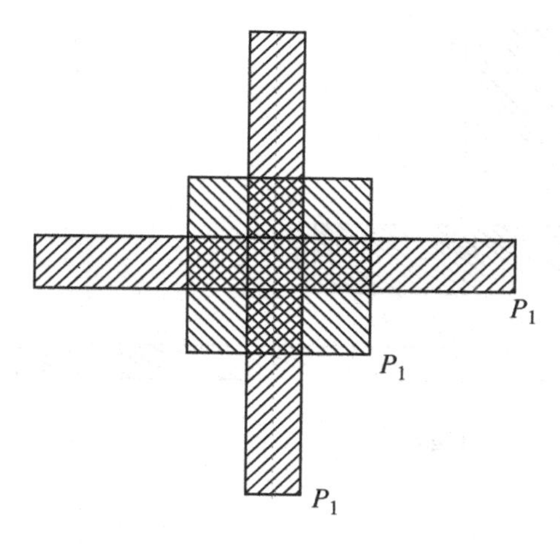

Figure 6.3. P_1-tiles in a 1-tree.

we could use square-function estimates to handle the corresponding terms quite easily, at least in the Banach case. There are no easy estimates this time.

As we mentioned and briefly explained earlier, the trees that will be introduced in the next definition play an important role in the proof.

Definition 6.7 Let $\mathbb{P}$ be a sparse rank-1 collection of tri-tiles. Let $j \in \{1, 2, 3\}$. A subcollection $T \subseteq \mathbb{P}$ is called a *j-tree* if and only if there exists a tri-tile P_T (called the top of the tree) such that

$$P_j \leq P_{T,j}$$

for every $P \in T$.

Notice that a tree does not necessarily have to contain the corresponding top. We will write I_T and $\omega_{T,j}$ for I_{P_T} and $\omega_{P_T,j}$ respectively. Then, we simply say that T is a tree if it is a j-tree for some $j = 1, 2, 3$. Trees are fundamental objects since, as we mentioned, if one restricts the general sum in the definition of $T_{\mathbb{P}}$ (see (6.14)) to a tree then essentially one sees a discrete paraproduct. This should be clear from the Heisenberg boxes shown in Figures 6.3 and 6.4. Thus the difficulty comes from the fact that potentially there are infinitely many paraproducts in the general sum of $T_{\mathbb{P}}$. The idea is to find a way of organizing and selecting these trees one at a time, estimating each of them separately, and then adding all these estimates together, using orthogonality arguments for distinct trees. Suppose that two tiles P and Q do not intersect. Then either $\omega_P \cap \omega_Q = \emptyset$ or $I_P \cap I_Q = \emptyset$. In the first case the wave packets associated

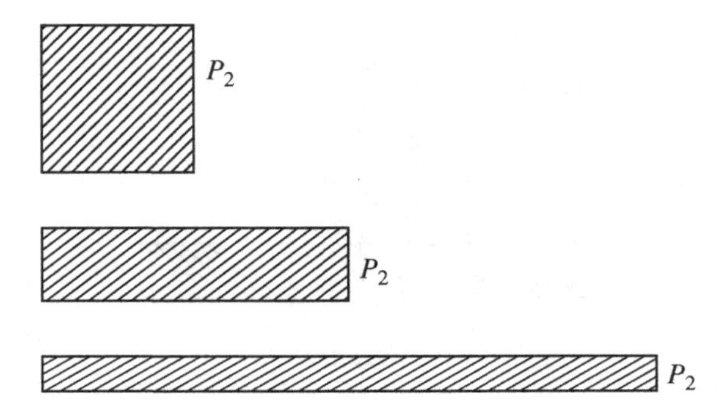

Figure 6.4. P_2-tiles in a 1-tree.

with the two tiles are orthogonal since their Fourier transforms are supported in ω_P and ω_Q. In the second case it is true that their tails intersect each other, so perfect orthogonality no longer holds; however, their scalar product becomes smaller when the spatial intervals are farther away. This suggests that if the tiles of two distinct trees are *sufficiently far* from each other then the functions naturally associated with them will be *almost orthogonal*. This fact motivates the next definition.

Definition 6.8 Let $1 \leq i \leq 3$. A finite sequence of trees $T_1, \ldots, T_M$ is said to be a *chain of strongly i-disjoint trees* if and only if

(i) $P_i \neq P'_i$ for every $P \in T_{\ell_1}$ and $P' \in T_{\ell_2}$ with $\ell_1 \neq \ell_2$;
(ii) whenever $P \in T_{\ell_1}$ and $P' \in T_{\ell_2}$ with $\ell_1 \neq \ell_2$ are such that $2\omega_{P_i} \cap 2\omega_{P'_i} \neq \emptyset$ then if $|\omega_{P_i}| < |\omega_{P'_i}|$ one has $I_{P'} \cap I_{T_{\ell_1}} = \emptyset$ and if $|\omega_{P'_i}| < |\omega_{P_i}|$ one has $I_P \cap I_{T_{\ell_2}} = \emptyset$;
(iii) whenever $P \in T_{\ell_1}$ and $P' \in T_{\ell_2}$ with $\ell_1 < \ell_2$ are such that $2\omega_{P_i} \cap 2\omega_{P'_i} \neq \emptyset$ then if $|\omega_{P_i}| = |\omega_{P'_i}|$ one has $I_{P'} \cap I_{T_{\ell_1}} = \emptyset$.

Given these definitions then to be able to achieve and quantify our plan, as in the case of paraproducts, one needs also to define two quantities called the *size* and the *energy*. The first allows us to estimate each individual tree while the second adds all the tree estimates together. They are both L^2 quantities rather than the $L^{1,\infty}$ quantities used for paraproducts. This should not be at all surprising since, as we have seen already in the scale-1 case, L^2 Bessel-type inequalities such as that following equation (6.23) are more natural for counting the trees this time.

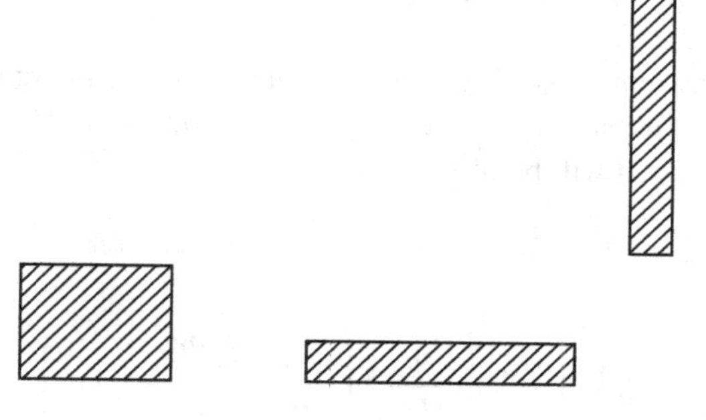

Figure 6.5. Disjoint tiles.

Definition 6.9 Let $\mathbb{P}$ be a finite collection of tri-tiles, let $j \in \{1, 2, 3\}$, and let f be an arbitrary function. We define the *size* of the sequence $\langle f, \Phi^j_{P_j} \rangle_P$ by

$$\text{size}\left(\left\langle f, \Phi^j_{P_j} \right\rangle_P\right) := \sup_{T \subseteq \mathbb{P}} \left(\frac{1}{|I_T|} \sum_{P \in T} \left| \left\langle f, \Phi^j_{P_j} \right\rangle \right|^2 \right)^{1/2},$$

where T ranges over all trees in $\mathbb{P}$ that are i-trees for some $i \neq j$.

Definition 6.10 Let $\mathbb{P}$ be a finite collection of tri-tiles as before, let $j \in \{1, 2, 3\}$, and let f again be a fixed function. We define the *energy* of the sequence $\langle f, \Phi^j_{P_j} \rangle_P$ by

$$\text{energy}\left(\left\langle f, \Phi^j_{P_j} \right\rangle_P\right) := \sup_{n \in \mathbb{Z}} \sup_{\mathbb{T}} 2^n \left(\sum_{T \in \mathbb{T}} |I_T| \right)^{1/2},$$

where now $\mathbb{T}$ ranges over all chains of strongly j-disjoint trees (see Definition 6.8) in $\mathbb{P}$ (which are i-trees for some $i \neq j$) having the property that

$$\left(\sum_{P \in T} \left| \langle f, \Phi^j_{P_j} \rangle \right|^2 \right)^{1/2} \geq 2^n \, |I_T|^{1/2}$$

for all $T \in \mathbb{T}$ and such that

$$\left(\sum_{P \in T'} \left| \langle f, \Phi^j_{P_j} \rangle \right|^2 \right)^{1/2} \leq 2^{n+1} \, |I_{T'}|^{1/2}$$

for all subtrees $T' \subseteq T \in \mathbb{T}$.

We emphasize that the j-tiles of the trees in $\mathbb{T}$ are all disjoint, as in Figure 6.5.

The following lemma is another form of the John–Nirenberg inequality and has been proved already in the context of paraproducts in Corollary 2.8. As was the case there, it will be of great help later.

Lemma 6.11 *Let $\mathbb{P}$ be a finite collection of tri-tiles and let $j \in \{1, 2, 3\}$. Then*

$$\text{size}\left(\left(\langle f, \Phi^j_{P_j}\rangle\right)_P\right) \simeq \sup_{T \subseteq \mathbb{P}} \frac{1}{|I_T|} \left\| \left(\sum_{P \in T} \frac{\left|\langle f, \Phi^j_{P_j}\rangle\right|^2}{|I_P|} 1_{I_P} \right)^{1/2} \right\|_{1,\infty},$$

where T ranges over all trees in $\mathbb{P}$ which are i-trees for some $i \neq j$.

The following proposition provides a way of estimating a generic trilinear form associated with $T_{\mathbb{P}}(f_1, f_2)$. It is clearly analogous to Proposition 2.12. First we write

$$\Lambda_{\mathbb{P}}(f_1, f_2, f_3) := \int_{\mathbb{R}} T_{\mathbb{P}}(f_1, f_2)(x) f_3(x)\, dx.$$

Proposition 6.12 *Let $\mathbb{P}$ be a finite collection of tri-tiles. Then*

$$|\Lambda_{\mathbb{P}}(f_1, f_2, f_3)| \lesssim \prod_{j=1}^{3} \left(\text{size}\left(\left(\langle f, \Phi^j_{P_j}\rangle\right)_P\right)\right)^{\theta_j} \left(\text{energy}\left(\left(\langle f, \Phi^j_{P_j}\rangle\right)_P\right)\right)^{1-\theta_j} \tag{6.26}$$

for any $0 \leq \theta_1, \theta_2, \theta_3 < 1$ with $\theta_1 + \theta_2 + \theta_3 = 1$; the implicit constants depend on the θ_j but are independent of the other parameters.

This inequality will play an import role in the proof of Theorem 6.5. In order to be able to use it effectively we need ways to estimate further the sizes and energies appearing on the right-hand side of (6.26).

Lemma 6.13 *Let $j \in \{1, 2, 3\}$ and let E be a set of finite measure. Then for every $|f| \leq \chi_E$ one has*

$$\text{size}\left(\left(\langle f, \Phi^j_{P_j}\rangle\right)_P\right) \lesssim \sup_{P \in \mathbb{P}} \frac{1}{|I_P|} \int_E \widetilde{\chi}^M_{I_P}\, dx$$

for all $M > 0$, with implicit constants depending on M.

Lemma 6.14 *Let* $j \in \{1, 2, 3\}$ *and* $f \in L^2(\mathbb{R})$. *Then*

$$\text{energy}\left(\left\langle f, \Phi^j_{P_j} \right\rangle_P\right) \lesssim \|f\|_2.$$

The proofs of Proposition 6.12 and Lemma 6.14 will be presented at the end of the chapter. Lemma 6.13 is a simple consequence of Lemmas 6.11 and 2.13. We are now going to use these statements to finally complete the proof of Theorem 6.5.

6.4. Proof of Theorem 6.5

Fix indices p, q, r as in the hypothesis of Theorem 6.5. Write $\alpha := 1/p$, $\beta := 1/q$, and $\gamma := 1/r'$. Fix sets of finite measure E_1, E_2, E_3. The goal is to find $E_3' \subseteq E_3$ with $|E_3'| \simeq |E_3|$ such that, for every $|f_1| \leq \chi_{E_1}$, $|f_2| \leq \chi_{E_2}$, and $|f_3| \leq \chi_{E_3'}$, one has

$$|\Lambda_{\mathbb{P}}(f_1, f_2, f_3)| \lesssim |E_1|^\alpha |E_2|^\beta |E_3|^\gamma. \tag{6.27}$$

Note that α is smaller than but close to 1, while β is greater than but close to 1/2. Using the dilation invariance of the bilinear Hilbert transform, which transfers naturally to $T_{\mathbb{P}}$, one can clearly assume without loss of generality that $|E_3| = 1$.

Define the set

$$\Omega := \left\{x \,\middle|\, M\left(\frac{\chi_{E_1}}{|E_1|}\right)(x) > C\right\} \cup \left\{x \,\middle|\, M\left(\frac{\chi_{E_2}}{|E_2|}\right)(x) > C\right\}$$

and observe that $|\Omega| \ll 1$ if C is a large enough constant. Then set $E_3' := E_3 \backslash \Omega$ and note that $|E_3'| \simeq 1$ as desired. For any positive integer d, define the collection of tri-tiles

$$\mathbb{P}_d := \left\{P \in \mathbb{P} \,\middle|\, 2^d \leq 1 + \frac{\text{dist}(I_P, \Omega^c)}{|I_P|} < 2^{d+1}\right\}$$

with the convention that $\mathbb{P}_0$ contains also the tri-tiles that intersect Ω^c. Clearly, $\bigcup_{d\geq 0} \mathbb{P}_d = \mathbb{P}$. In particular, one has that

$$\Lambda_{\mathbb{P}}(f_1, f_2, f_3) = \sum_{d=0}^{\infty} \int_{\mathbb{R}} T_{\mathbb{P}_d}(f_1, f_2)(x) f_3(x)\, dx. \tag{6.28}$$

Fix $d \geq 0$ and consider the corresponding trilinear form in (6.28). It can be estimated using Proposition 6.12, as we have seen. Moreover for $i = 1, 2$ we also have, thanks to Lemmas 6.13 and 6.14 that

$$\operatorname{size}\left(\left\langle f_i, \Phi^i_{P_i}\right\rangle_{P\in\mathbb{P}_d}\right) \lesssim \sup_{P\in\mathbb{P}_d} \frac{1}{|I_P|} \int_{E_i} \widetilde{\chi}^M_{I_P}(x)\,dx \lesssim 2^d |E_i|$$

and also

$$\operatorname{energy}\left(\langle f_i, \Phi^i_{P_i}\rangle_{P\in\mathbb{P}_d}\right) \lesssim |E_i|^{1/2}.$$

However, since $|E_3| = 1$ and $E_3' \subseteq \Omega^c$, one has

$$\operatorname{size}\left(\left\langle f_3, \Phi^3_{P_3}\right\rangle_{P\in\mathbb{P}_d}\right) \lesssim 2^{-Md}$$

for some large $M > 0$ and also

$$\operatorname{energy}\left(\langle f_3, \Phi^3_{P_3}\rangle_{P\in\mathbb{P}_d}\right) \lesssim 1.$$

Putting all these together, Proposition 6.12 allows us to bound the corresponding trilinear form in (6.28) (for a fixed $d \geq 0$) by

$$\begin{aligned} &2^{-\widetilde{M}d} |E_1|^{\theta_1} |E_1|^{(1-\theta_1)/2} |E_2|^{\theta_2} |E_2|^{(1-\theta_2)/2} \\ &= 2^{-\widetilde{M}d} |E_1|^{(1+\theta_1)/2} |E_2|^{(1+\theta_2)/2}. \end{aligned}$$

One can clearly make the exponent $(1+\theta_1)/2$ arbitrarily close to 1 by taking θ_1 close to 1; similarly one can make the second exponent $(1+\theta_2)/2$ arbitrarily close to $1/2$ by taking θ_2 close to 0. In the end, the series over $d \geq 0$ is summable if we choose the original M large enough.

This completes the proof of Theorem 6.5. We are therefore left with the proofs of Lemma 6.14 and Proposition 6.12.

6.5. Bessel-type inequalities

We start with the proof of Lemma 6.14, which can be considered as a Bessel-type inequality.[1] Fix $j \in \{1, 2, 3\}$ and $f \in L^2(\mathbb{R})$. Let n and $\mathbb{T}$ be as in Definition 6.10, in which the supremum of the $\operatorname{energy}\left(\left\langle f, \Phi^j_{P_j}\right\rangle_P\right)$ is attained. We want to show that

$$2^n \left(\sum_{T\in\mathbb{T}} |I_T|\right)^{1/2} \lesssim \|f\|_2. \tag{6.29}$$

[1] The terms "Bessel-type" inequalities and "Bessel methods" are sometimes used in connection with L^2 estimates involving orthogonal functions.

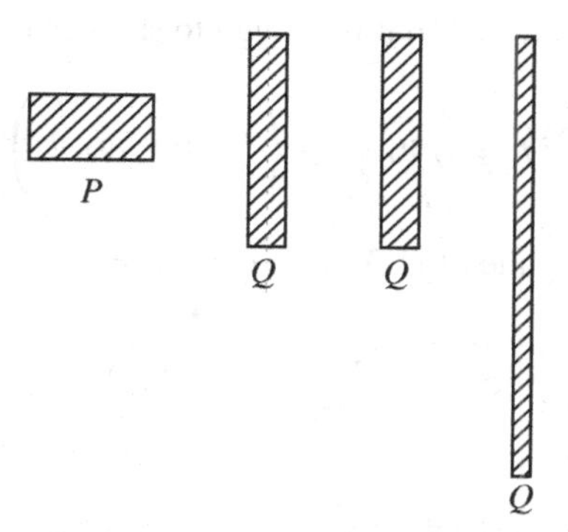

Figure 6.6. Relative positions of the P and Q tiles.

If we square the left-hand side of (6.29) and use the properties of the trees in $\mathbb{T}$ we can write

$$\begin{aligned}(\text{LHS})^2 &= 2^{2n}\sum_{T\in\mathbb{T}}|I_T| \lesssim 2^{2n}2^{-2n}\sum_{T\in\mathbb{T}}\left(\sum_{P\in T}\left|\left\langle f, \Phi^j_{P_j}\right\rangle\right|^2\right)\\ &= \sum_{T\in\mathbb{T}}\left(\sum_{P\in T}\left|\left\langle f, \Phi^j_{P_j}\right\rangle\right|^2\right),\end{aligned}$$

and we will show that this expression is smaller than $\|f\|_2{}^2$.

From now on, we will suppress the index j for simplicity. Assume first that we are in the simpler situation when all the trees T in $\mathbb{T}$ are one tri-tile trees. In this case we have to show that, given any collection of disjoint tiles $P \in \mathbb{D}$ and L^2-normalized wave packets $(\Phi_P)_P$ adapted to them with the property that

$$\frac{1}{|I_P|^{1/2}}\,|\langle f, \Phi_P\rangle| \simeq 2^n,$$

one has that

$$\sum_{P\in\mathbb{D}}|\langle f, \Phi_P\rangle|^2 \lesssim \|f\|_2^2. \tag{6.30}$$

One can write the left-hand side of (6.30) as

$$\begin{aligned}\left|\sum_{P\in\mathbb{D}}\langle f, \Phi_P\rangle\overline{\langle f, \Phi_P\rangle}\right| &= \left|\left\langle\sum_{P\in\mathbb{D}}\langle f, \Phi_P\rangle\Phi_P, f\right\rangle\right|\\ &\leq \|f\|_2\left\|\sum_{P\in\mathbb{D}}\langle f, \Phi_P\rangle\Phi_P\right\|_2\end{aligned}$$

and observe that to prove (6.30) it is enough to show that

$$\left\| \sum_{P \in \mathbb{D}} \langle f, \Phi_P \rangle \Phi_P \right\|_2 \lesssim \left(\sum_{P \in \mathbb{D}} |\langle f, \Phi_P \rangle|^2 \right)^{1/2}. \tag{6.31}$$

Squaring the left-hand side of (6.31), we can write

$$\begin{aligned}
(\text{LHS})^2 &= \left| \left\langle \sum_{P \in \mathbb{D}} \langle f, \Phi_P \rangle \Phi_P, \sum_{Q \in \mathbb{D}} \langle f, \Phi_Q \rangle \Phi_Q \right\rangle \right| \\
&= \left| \sum_{P,Q} \langle f, \Phi_P \rangle \overline{\langle f, \Phi_Q \rangle} \langle \Phi_P, \Phi_Q \rangle \right| \\
&\le \sum_{\substack{P,Q: \\ \omega_Q \cap \omega_P \neq \emptyset;\, |\omega_Q| < |\omega_P|}} |\langle f, \Phi_P \rangle| \, |\langle f, \Phi_Q \rangle| \, |\langle \Phi_P, \Phi_Q \rangle| \\
&\quad + \sum_{\substack{P,Q: \\ \omega_P \cap \omega_Q \neq \emptyset;\, |\omega_P| < |\omega_Q|}} |\langle f, \Phi_P \rangle| \, |\langle f, \Phi_Q \rangle| \, |\langle \Phi_P, \Phi_Q \rangle| \\
&\quad + \sum_{\substack{P,Q: \\ \omega_Q = \omega_P}} |\langle f, \Phi_P \rangle| \, |\langle f, \Phi_Q \rangle| \, |\langle \Phi_P, \Phi_Q \rangle| =: I + II + III.
\end{aligned}$$

Since I and II are symmetric it is enough to discuss, say, II. For this case, the generic relative positions of a tile P and the tiles Q are represented in Figure 6.6. Since $|\langle f, \Phi_P \rangle| \simeq 2^n |I_P|^{1/2}$ and $|\langle f, \Phi_Q \rangle| \simeq 2^n |I_Q|^{1/2}$, one can estimate II as

$$\sum_P |\langle f, \Phi_P \rangle|^2 \left(\sum_{\substack{Q: \\ \omega_P \cap \omega_Q \neq \emptyset;\, |\omega_P| < |\omega_Q|}} \frac{|I_Q|^{1/2}}{|I_P|^{1/2}} \, |\langle \Phi_P, \Phi_Q \rangle| \right) \tag{6.32}$$

and the goal is to show that the contents of the large parentheses in (6.32) are $O(1)$. Fix $P \in \mathbb{D}$. Then, we estimate the inner sum in (6.32) by

$$\sum_{\substack{Q: \\ \omega_P \cap \omega_Q \neq \emptyset;\, |\omega_P| < |\omega_Q|}} \frac{1}{|I_P|} \left| \langle \widetilde{\chi}_{I_P}^M, \widetilde{\chi}_{I_Q}^M \rangle \right| = \frac{1}{|I_P|} \sum_{\substack{Q: \\ \omega_P \cap \omega_Q \neq \emptyset;\, |\omega_P| < |\omega_Q|}} \left| \langle \widetilde{\chi}_{I_P}^M, \widetilde{\chi}_{I_Q}^M \rangle \right|. \tag{6.33}$$

Clearly, one also has

$$\left| \langle \widetilde{\chi}_{I_P}^M, \widetilde{\chi}_{I_Q}^M \rangle \right| \lesssim \left(1 + \frac{\text{dist}(I_P, I_Q)}{|I_P|} \right)^{-M} |I_Q|,$$

and so one can write

$$\sum_{\substack{Q:\\ \omega_P\cap\omega_Q\neq\emptyset;\, |\omega_P|<|\omega_Q|}} \left|\langle \widetilde{\chi}_{I_P}^M, \widetilde{\chi}_{I_Q}^M\rangle\right| \lesssim \sum_{\substack{Q:\\ \omega_P\cap\omega_Q\neq\emptyset;\, |\omega_P|<|\omega_Q|}} \left(1+\frac{\operatorname{dist}(I_P, I_Q)}{|I_P|}\right)^{-M} |I_Q|$$
$$\lesssim \left\| \left(1+\frac{\operatorname{dist}(x, I_P)}{|I_P|}\right)^{-M} \right\|_{L^1_x} \lesssim |I_P|,$$

since the I_Q are disjoint.

This inequality together with (6.33) allows one to estimate (6.32) by $\sum_{P\in\mathbb{D}} |\langle f, \Phi_P\rangle|^2$, as desired. Now we need to estimate III. One writes

$$\sum_{\omega_P=\omega_Q} |\langle f, \Phi_P\rangle|\, |\langle f, \Phi_Q\rangle|\, |\langle \Phi_P, \Phi_Q\rangle|$$
$$\lesssim \sum_P |\langle f, \Phi_P\rangle|^2 \left(\sum_{\substack{Q:\\ \omega_Q=\omega_P}} \frac{1}{|I_P|} \left|\langle \widetilde{\chi}_{I_P}^M, \widetilde{\chi}_{I_Q}^M\rangle\right| \right)$$
$$\lesssim \sum_P |\langle f, \Phi_P\rangle|^2 \frac{1}{|I_P|} \left(\sum_{\substack{Q:\\ \omega_Q=\omega_P}} \langle \widetilde{\chi}_{I_P}^M, \widetilde{\chi}_{I_Q}^M\rangle \right)$$
$$\lesssim \sum_P |\langle f, \Phi_P\rangle|^2,$$

also as desired.

Let us now treat the general case, when the trees in $\mathbb{T}$ are generic. Then, as before, one can write

$$\sum_{T\in\mathbb{T}} \sum_{P\in T} |\langle f, \Phi_P\rangle|^2 = \left| \left\langle \sum_T \sum_{P\in T} \langle f, \Phi_P\rangle \Phi_P, f \right\rangle \right|$$
$$\lesssim \|f\|_2 \left\| \sum_T \sum_{P\in T} \langle f, \Phi_P\rangle \Phi_P \right\|_2$$

and, also as before, it is enough to prove that

$$\left\| \sum_T \sum_{P\in T} \langle f, \Phi_P\rangle \Phi_P \right\|_2 \lesssim \left(\sum_T \sum_{P\in T} |\langle f, \Phi_P\rangle|^2 \right)^{1/2}. \tag{6.34}$$

The square of the left-hand side of (6.34) is smaller than

$$\sum_{T,T'} \sum_{\substack{P\in T\\ Q\in T'}} |\langle f, \Phi_P\rangle|\, |\langle f, \Phi_Q\rangle|\, |\langle \Phi_P, \Phi_Q\rangle| := I + II \tag{6.35}$$

where I contains the term where $T \neq T'$ while II contains the $T = T'$ term.

We first estimate I. Observe that if $P \in T$ and $Q \in T'$ then, in order for $\langle \Phi_P, \Phi_Q \rangle$ to be nonzero, one must have $\omega_P \cap \omega_Q \neq \emptyset$. Because of the symmetry, we can also assume that one always has $|\omega_P| \leq |\omega_Q|$. There are two subcases, $|\omega_P| < |\omega_Q|$ and $|\omega_P| = |\omega_Q|$. Consider the first case. Then, since T and T' are *strongly disjoint*, this means in particular that $I_Q \cap I_T = \emptyset$ for any $Q \in T'$.

Fix T, T', $P \in T$, and $Q \in T'$ such that $|\omega_P| < |\omega_Q|$. Clearly, using the properties of the trees in $\mathbb{T}$, we can write

$$\frac{1}{|I_P|^{1/2}} |\langle f, \Phi_P \rangle| \lesssim 2^n \lesssim \frac{1}{|I_T|^{1/2}} \left(\sum_{P' \in T} |\langle f, \Phi_{P'} \rangle|^2 \right)^{1/2},$$

from which we deduce that

$$|\langle f, \Phi_P \rangle| \lesssim \frac{|I_P|^{1/2}}{|I_T|^{1/2}} \left(\sum_{P' \in T} |\langle f, \Phi_{P'} \rangle|^2 \right)^{1/2}. \tag{6.36}$$

Similarly, one has

$$|\langle f, \Phi_Q \rangle| \lesssim \frac{|I_Q|^{1/2}}{|I_T|^{1/2}} \left(\sum_{P' \in T} |\langle f, \Phi'_P \rangle|^2 \right)^{1/2}. \tag{6.37}$$

Using (6.36) and (6.37) one can estimate the corresponding sum in (6.35) by

$$\begin{aligned}
&\sum_{T \neq T'} \sum_{\substack{P \in T \\ Q \in T' \\ |\omega_P| < |\omega_Q|}} \left(\frac{|I_P|^{1/2}}{|I_T|^{1/2}} \left(\sum_{P' \in T} |\langle f, \Phi_{P'} \rangle|^2 \right)^{1/2} \right) \\
&\qquad \times \left(\frac{|I_Q|^{1/2}}{|I_T|^{1/2}} \left(\sum_{P' \in T} |\langle f, \Phi_{P'} \rangle|^2 \right)^{1/2} \right) |\langle \Phi_P, \Phi_Q \rangle| \\
&= \sum_{T \in \mathbb{T}} \left(\sum_{P' \in T} |\langle f, \Phi_{P'} \rangle|^2 \right) \\
&\qquad \times \sum_{P \in T} \sum_{T' \neq T} \sum_{\substack{Q \in T' \\ \omega_P \cap \omega_Q \neq \emptyset;\, |\omega_P| < |\omega_Q|}} \frac{1}{|I_T|} |I_P|^{1/2} |I_Q|^{1/2} |\langle \Phi_P, \Phi_Q \rangle| \qquad (6.38) \\
&\lesssim \sum_{T \in \mathbb{T}} \left(\sum_{P' \in T} |\langle f, \Phi_{P'} \rangle|^2 \right) \\
&\qquad \times \sum_{P \in T} \sum_{T' \neq T} \sum_{\substack{Q \in T' \\ \omega_P \cap \omega_Q \neq \emptyset;\, |\omega_P| < |\omega_Q|}} \frac{1}{|I_T|} \langle \widetilde{\chi}_{I_P}^M, \widetilde{\chi}_{I_Q}^M \rangle. \qquad (6.39)
\end{aligned}$$

Fix T and consider the inner sum in (6.39),

$$\sum_{P\in T}\sum_{T'\neq T}\sum_{\substack{Q\in T'\\ \omega_P\cap\omega_Q\neq\emptyset;\,|\omega_P|<|\omega_Q|}}\langle\widetilde{\chi}_{I_P}^M,\widetilde{\chi}_{I_Q}^M\rangle. \tag{6.40}$$

It is clearly enough to show that this expression is $O(|I_T|)$ in order to complete the contribution of the first part of I in (6.39). Also fix $P\in T$ and recall that

$$\left|\langle\widetilde{\chi}_{I_P}^M,\widetilde{\chi}_{I_Q}^M\rangle\right|\lesssim\left(1+\frac{\operatorname{dist}(I_P,I_Q)}{|I_P|}\right)^{-M}|I_Q|.$$

Since every I_Q has the property that $I_Q\cap I_T=\emptyset$, one can see that the contribution of all possible Q in (6.40) is bounded by

$$|I_p|\left(1+\frac{\operatorname{dist}(I_P,I_T{}^c)}{|I_P|}\right)^{-M}. \tag{6.41}$$

Finally, one has

$$\begin{aligned}
&\sum_{P\in T}|I_p|\left(1+\frac{\operatorname{dist}(I_P,I_T{}^c)}{|I_P|}\right)^{-M}\\
&\quad=\sum_{\substack{k:\\ 2^k\leq|I_T|}}\sum_{\substack{P\in T\\ |I_P|=2^k}}|I_P|\left(1+\frac{\operatorname{dist}(I_P,I_T^c)}{|I_P|}\right)^{-M}\\
&\quad=\sum_{\substack{k:\\ 2^k\leq|I_T|}}2^k\sum_{\substack{P\in T\\ |I_P|=2^k}}\left(1+\frac{\operatorname{dist}(I_P,I_T^c)}{|I_P|}\right)^{-M}\lesssim\sum_{\substack{k:\\ 2^k\leq|I_T|}}2^k\lesssim|I_T|,
\end{aligned}$$

as desired. Then, in the second subcase, $|\omega_P|=|\omega_Q|$, we rewrite the sum $\sum_{T\neq T'}\sum_{P\in T}\sum_{Q\in T'}$, using symmetry, as

$$2\sum_{\ell_1<\ell_2}\sum_{P\in T_{\ell_1}}\sum_{Q\in T_{\ell_2}}$$

and observe that one still has $I_Q\cap I_{T_{\ell_1}}=\emptyset$ as a consequence of the fact that all these trees are part of a *chain of strongly disjoint trees*. The rest of the argument is completely identical to that for the first subcase.

We are now left with the *diagonal term* II, when the sum runs over $T=T'$. In this case one observes that if $P,Q\in T$ are such that $\omega_P\cap\omega_Q\neq\emptyset$ then one must have $\omega_P=\omega_Q$. Hence, the main contribution comes from the situation when $P=Q$ and it is not difficult to see that this gives the desired bound.

Exercise 6.3 Complete the details in the estimate of the diagonal term II in (6.35).

6.6. Stopping-time decompositions

Finally, we are ready to describe the proof of Proposition 6.12. Fix a collection $\mathbb{P}$ of tri-tiles and the functions f_1, f_2, f_3 as in the proposition. As discussed, we assume also that $\mathbb{P}$ is sparse. We will write for simplicity

$$S_j := \operatorname{size}(\langle f_j, \Phi^j_{P_j}\rangle_{P\in\mathbb{P}})$$

and

$$E_j := \operatorname{energy}(\langle f_j, \Phi^j_{P_j}\rangle_{P\in\mathbb{P}})$$

for $j \in \{1, 2, 3\}$. We first need to prove the following.

Proposition 6.15 *Let $j \in \{1, 2, 3\}$, $\mathbb{P}' \subseteq \mathbb{P}$ and $n \in \mathbb{Z}$ be such that*

$$\operatorname{size}\left(\langle f_j, \Phi^j_{P_j}\rangle_{P\in\mathbb{P}'}\right) \leq 2^{-n} E_j. \tag{6.42}$$

Then, one can decompose $\mathbb{P}' = \mathbb{P}'' \cup \mathbb{P}'''$ in such a way that

$$\operatorname{size}\left(\langle f_j, \Phi^j_{P_j}\rangle_{P\in\mathbb{P}''}\right) \leq 2^{-n-1} E_j \tag{6.43}$$

and $\mathbb{P}'''$ can be written as a disjoint union of trees $T \in \mathbb{T}$ such that

$$\sum_{T\in\mathbb{T}} |I_T| \lesssim 2^{2n}. \tag{6.44}$$

Proof This is the analogue of Lemma 2.10. As we will see, the selection process will be more complicated this time, because trees can lie not only near one another, as in the paraproducts case, but also one on top of the other.

The basic idea is to remove trees from $\mathbb{P}'$ when their sizes are as large as possible and place their tiles in $\mathbb{P}'''$ until (6.43) is satisfied. There are two kinds of tree, to be defined rigorously below: upward trees and downward trees.

If P is a tile, we denote by ξ_P the center of ω_P. Then, if P' is also a tile, we write $P' \lesssim^+ P$ if $P' \lesssim' P$ and $\xi_{P'} > \xi_P$ hold and $P' \lesssim^- P$ if $P' \lesssim' P$ and $\xi_{P'} < \xi_P$ hold. Then we perform the following algorithm. We first consider the set of all possible-trees T (see Definition 6.7) for $i \neq j$ that are *upward trees* (see Figure 6.7) in the sense that

$$P_j \lesssim^+ P_{T,j} \tag{6.45}$$

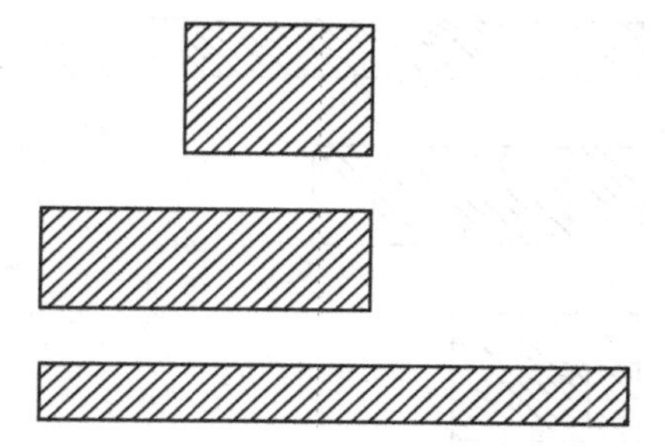

Figure 6.7. Upward tree.

for all $P \in T$ and which satisfy

$$\sum_{P \in T} \left|\left\langle f, \Phi^j_{P_j} \right\rangle\right|^2 > 2^{-2n-2} |I_T| |E_j|^2. \tag{6.46}$$

If there are no such trees, we terminate the algorithm. Otherwise, we choose a T that is maximal with respect to the set inclusion and such that the center $\xi_{T,j}$ of $\omega_{P_T,j}$ is as large as possible. Then, once we have selected such a tree T, we also define

$$T' = \{P \in \mathbb{P}' \backslash T | P_j \le P_{T,j}\},$$

which by definition is a j-tree. We remove both T and T' from $\mathbb{P}'$ and add them to our new collection of trees $\mathbb{T}$ and their tiles to the new set $\mathbb{P}'''$. Then, we continue to select other trees in the same manner. Since $\mathbb{P}'$ is also finite the algorithm will end after finitely many steps, producing pairs of trees $T_1, T_1', T_2, T_2', \ldots, T_M, T_M'$, as before. We claim now that the trees $T_1, T_2, \ldots, T_M$ constructed in this way form a chain of strongly j-disjoint trees.

To see why, consider two indices s and s', with $s \neq s'$, and tri-tiles $P \in T_s$, $P' \in T_{s'}$ such that $2\omega_{P_j} \cap 2\omega_{P_j'} \neq \emptyset$. Assume first that $|\omega_{P_j}| \neq |\omega_{P_j'}|$ or more precisely, that $|\omega_{P_j}| < |\omega_{P_j'}|$. We would like to show that in this case one has $I_{P'} \cap I_{T_s} = \emptyset$. If not then clearly the only alternative is that $I_{P'} \subseteq I_{T_s}$. By the sparseness assumption, we know that $|\omega_{P_j}| \ll |\omega_{P_j'}|$ and this implies in particular that $\xi_{P_{T_{s'}},j} < \xi_{P_{T_s},j}$, which, by our selection algorithm, implies that $s < s'$. In other words, T_s is selected before $T_{s'}$. Since $|\omega_{P'_j}|$ is much larger than $|\omega_{P_j}|$ we also deduce that $P_j' \le P_{T_s,j}$. This means that the tri-tile P' should have been selected earlier, in T_s', which is a contradiction with the fact that $P' \in T_{s'}$.

If, however, $2\omega_{P_j} \cap 2\omega_{P_j'} \neq \emptyset$ but $|\omega_{P_j}| = |\omega_{P_j'}|$ then (again by the sparseness assumption) we have $\omega_{P_j} = \omega_{P_j'}$. Assume for instance that $s < s'$. We need to

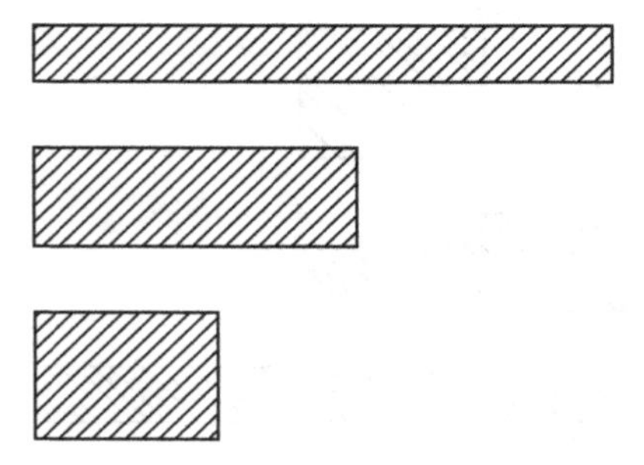

Figure 6.8. Downward tree.

check that $I_{P'} \cap I_{T_s} = \emptyset$. If this were not the case then the only possibility is $I_{P'} \subseteq I_{T_s}$, but then, using the maximality condition, the tri-tile P' should have been selected in T_s, which again is a contradiction with the fact that $P' \in T_{s'}$. This proves our *strong disjointness* claim above.

Now that we have clarified this let us observe, from the definition of the energies and our construction, that

$$\sum_{s=1}^{M} |I_{T_s}| \lesssim 2^{2n}. \tag{6.47}$$

Since T_s and T_s' have the same top, this means that

$$\sum_{T \in \mathbb{T}} |I_T| \lesssim 2^{2n}$$

as well. Consider now the set of the remaining tiles in $\mathbb{P}'$. Clearly, we must have

$$\sum_{\substack{P \in T: \\ P_j \lesssim^{+} P_{T,j}}} \left|\langle f, \Phi^j_{P_j} \rangle\right|^2 < 2^{-2n-2} |I_T| \, |E_j|^2$$

for all trees $T \in \mathbb{P}' \setminus \mathbb{P}'''$, otherwise our algorithm will continue. Now we simply repeat the above algorithm but replace the $\lesssim^{+}$ condition by the similar one $\lesssim^{-}$ (so now the trees we are considering are *downward trees*, see Figure 6.8) and we select maximal trees T such that $\xi_{T,j}$ is minimized instead. It is then clear that in the end the new total collection of $T \in \mathbb{T}$ (containing both upward and downward trees) satisfies (6.44) and that (6.43) is satisfied as well.

This ends the proof of the proposition. □

By iterating this result carefully for all the indices $j = 1, 2, 3$ simultaneously, one obtains the following consequence.

Corollary 6.16 *Let $\mathbb{P}$ be a finite and sparse collection of tri-tiles. Then, one can split $\mathbb{P}$ as*

$$\mathbb{P} = \bigcup_{n\in\mathbb{Z}} \mathbb{P}_n,$$

where for each $n \in \mathbb{Z}$ we have

$$\text{size}\left(\langle f, \Phi^j_{P_j}\rangle_{P\in\mathbb{P}_n}\right) \leq \min\left(2^{-n}E_j, S_j\right)$$

for every $j = 1, 2, 3$ and, for at least one index j, we also have

$$2^{-n-1}E_j \leq \text{size}\left(\langle f, \Phi^j_{P_j}\rangle_{P\in\mathbb{P}_n}\right) \leq \min\left(2^{-n}E_j, S_j\right).$$

Also, one can cover $\mathbb{P}_n$ by a collection of trees $T \in \mathbb{T}_n$ for which

$$\sum_{T\in\mathbb{T}_n} |I_T| \lesssim 2^{2n}.$$

This decomposition of $\mathbb{P}$ plays an important role in the proof of Proposition 6.12, and we describe this role in what follows. First, let us point out the following simple but important observation.

Lemma 6.17 *Let T be a tree in $\mathbb{P}$ and f_1, f_2, f_3 fixed functions. Then*

$$\sum_{P\in T} \frac{1}{|I_P|^{\frac{1}{2}}} |\langle f_1, \Phi^1_{P_1}\rangle| \, |\langle f_2, \Phi^2_{P_2}\rangle| \, |\langle f_3, \Phi^3_{P_3}\rangle| \leq |I_T| \prod_{j=1}^{3} \text{size}\left(\langle f_j, \Phi^j_{P_j}\rangle_{P\in T}\right).$$

Proof Suppose that T is a 3-tree. Then the left-hand side of the inequality can be estimated by

$$\left(\sum_{P\in T} |\langle f_1, \Phi^1_{P_1}\rangle|^2\right)^{1/2} \left(\sum_{P\in T} |\langle f_2, \Phi^2_{P_2}\rangle|^2\right)^{1/2} \sup_{P\in T} \frac{|\langle f_3, \Phi^3_{P_3}\rangle|}{|I_P|^{1/2}}.$$

Since for $j = 1, 2$ one clearly has

$$\left(\sum_{p\in T} |\langle f_j, \Phi^j_{P_j}\rangle|^2\right)^{1/2} \leq |I_T|^{1/2} \, \text{size}\left(\langle f_j, \Phi^j_{P_j}\rangle_{P\in T}\right)$$

and since a singleton tree $\{P\}$ can be seen as a 1-tree or 2-tree with top P, we also have

$$\frac{|\langle f_3, \Phi^3_{P_3}\rangle|}{|I_P|^{1/2}} \leq \text{size}\left(\langle f_3, \Phi^3_{P_3}\rangle_{P\in T}\right),$$

which completes the argument. □

6.7. Generic estimate of the trilinear BHT form

In this section we give the proof of Proposition 6.12. First, we renormalize f_1, f_2, f_3 and rewrite the left-hand side of (6.26) as

$$E_1 E_2 E_3 \left| \Lambda_{\mathbb{P}} \left(\frac{f_1}{E_1}, \frac{f_2}{E_2}, \frac{f_3}{E_3} \right) \right|,$$

which means that we now need to prove that

$$\left| \Lambda_{\mathbb{P}} \left(\frac{f_1}{E_1}, \frac{f_2}{E_2}, \frac{f_3}{E_3} \right) \right| \lesssim \left(\frac{S_1}{E_1} \right)^{\theta_1} \left(\frac{S_2}{E_2} \right)^{\theta_2} \left(\frac{S_3}{E_3} \right)^{\theta_3} \tag{6.48}$$

for every $\theta_1, \theta_2, \theta_3$ as specified in Proposition 6.12.

By applying Corollary 6.16 together with the tree estimate in Lemma 6.17, one can decompose $\mathbb{P}$ as there and estimate the left-hand side of (6.48) by

$$\sum_{n \in \mathbb{Z}} \text{size} \left(\left\langle \frac{f_1}{E_1}, \Phi^1_{P_1} \right\rangle_{P \in \mathbb{P}_n} \right) \text{size} \left(\left\langle \frac{f_2}{E_2}, \Phi^2_{P_2} \right\rangle_{P \in \mathbb{P}_n} \right)$$

$$\times \text{size} \left(\left\langle \frac{f_3}{E_3}, \Phi^3_{P_3} \right\rangle_{P \in \mathbb{P}_n} \right) \sum_{T \in \mathbb{T}_n} |I_T|$$

$$\lesssim \sum_{n \in \mathbb{Z}} \left(\prod_{j=1}^{3} \text{size} \left(\left\langle \frac{f_j}{E_j}, \Phi^j_{P_j} \right\rangle_{P \in \mathbb{P}_n} \right) \right) 2^{2n}. \tag{6.49}$$

However, we also know that

$$\text{size} \left(\left\langle \frac{f_j}{E_j}, \Phi^j_{P_j} \right\rangle_{P \in \mathbb{P}_n} \right) \lesssim \min \left(2^{-n}, \frac{S_j}{E_j} \right)$$

and that, as long as 2^{-n} is larger than $\max(S_1/E_1, S_2/E_2, S_3/E_3)$, the corresponding $\mathbb{P}_n$ is an empty collection. Assume for example that one has

$$\frac{S_1}{E_1} \leq \frac{S_2}{E_2} \leq \frac{S_3}{E_3}. \tag{6.50}$$

Since 2^{-n} must be smaller than S_3/E_3 we have three possible cases, which will be analyzed one by one.

Case A: $S_2/E_2 \leq 2^{-n} \leq S_3/E_3$ ***or equivalently*** $E_3/S_3 \leq 2^n \leq E_2/S_2$. In this case one can estimate (6.49) by

$$\sum_{n \in \mathbb{Z}} \frac{S_1}{E_1} \frac{S_2}{E_2} 2^n \lesssim \frac{S_1}{E_1} \frac{S_2}{E_2} \frac{E_2}{S_2} = \frac{S_1}{E_1},$$

which is clearly smaller than

$$\left(\frac{S_1}{E_1}\right)^{\theta_1}\left(\frac{S_2}{E_2}\right)^{\theta_2}\left(\frac{S_3}{E_3}\right)^{\theta_3},$$

given (6.50).

Case B: $S_1/E_1 \leq 2^{-n} \leq S_2/E_2$ ***or equivalently*** $E_2/S_2 \leq 2^n \leq E_1/S_1$. This time the expression (6.49) can estimated by

$$\sum_{n\in\mathbb{Z}} \frac{S_1}{E_1} 2^{-n}2^{-n}2^{2n} = \frac{S_1}{E_1}\sum_n 1, \tag{6.51}$$

where the sum now runs over those integers satisfying

$$\frac{E_2}{S_2} \leq 2^n \leq \frac{E_1}{S_1}.$$

This means that there are at most

$$\log\frac{E_1}{S_1} - \log\frac{E_2}{S_2} = \log\frac{(E_1/S_1)}{(E_2/S_2)}$$

such indices. In particular, (6.51) is smaller than

$$\frac{S_1}{E_1}\log\frac{(E_1/S_1)}{(E_2/S_2)}$$

which is even smaller than

$$\frac{S_1}{E_1}\frac{(E_1/S_1)^{\varepsilon}}{(E_2/S_2)^{\varepsilon}} = \left(\frac{S_1}{E_1}\right)^{1-\varepsilon}\left(\frac{S_2}{E_2}\right)^{\varepsilon} \tag{6.52}$$

for every $\varepsilon > 0$, with an implicit constant depending on ε. Now we just need to recognize that

$$\left(\frac{S_1}{E_1}\right)^{1-\varepsilon}\left(\frac{S_2}{E_2}\right)^{\varepsilon} \leq \left(\frac{S_1}{E_1}\right)^{\theta_1}\left(\frac{S_2}{E_2}\right)^{\theta_2}\left(\frac{S_3}{E_3}\right)^{\theta_3},$$

if ε, depending on θ_i for $i = 1, 2, 3$, is chosen suitably again by recalling (6.50).

Case C: $2^{-n} \leq S_1/E_1$. In this case we see that (6.49) is smaller than

$$\sum_n 2^{-n}2^{-n}2^{-n}2^{2n} = \sum_n 2^{-n} \lesssim \frac{S_1}{E_1},$$

which is clearly smaller than

$$\left(\frac{S_1}{E_1}\right)^{\theta_1}\left(\frac{S_2}{E_2}\right)^{\theta_2}\left(\frac{S_3}{E_3}\right)^{\theta_3}$$

for any particular choice of θ_1, θ_2 θ_3, as in the hypothesis of Proposition 6.12.

This ends the proof of the boundedness of the bilinear Hilbert transform.

6.8. The $1/2 < r < 2/3$ counterexample

The goal of this section is to show that (modulo the endpoint case) the result for the model operators in Theorem 6.6 is sharp. More precisely, we will prove that if $1/2 < r < 2/3$ it is impossible for discrete operators of the type $T_{\mathbb{P}}$ to map $L^p \times L^q$ into L^r with bounds that are uniform with respect to the rank-1 collection $\mathbb{P}$ and the wave packets attached to it. This counterexample is due to Lacey.

The construction is very simple and, as we shall see, it will be enough to work with scale-1 tiles. Let us consider an L^2-normalized Schwartz function $\Phi(x)$ adapted to the unit interval $[0, 1]$ and such that $\operatorname{supp}\widehat{\Phi} \subseteq [1/3, 2/3]$. Define families of wave packets $\Phi_n^j(x)$, for $j = 1, 2, 3$ and $n \in \mathbb{Z}$, by $\Phi_n^1(x) := \Phi(x)e^{2\pi i n x}$, $\Phi_n^2(x) := \Phi(x)e^{2\pi i(n+3)x}$, and $\Phi_n^3(x) := \Phi(x)e^{2\pi i(2n+3)x}$. Let N be a large fixed positive integer and consider the discrete model operator T_N^t given by

$$T_N^t(f, g) := \sum_{n=1}^{N} r_n(t)\langle f, \Phi_n^1\rangle\langle g, \Phi_n^2\rangle\Phi_n^3, \tag{6.53}$$

where $(r_n(t))_n$ are the Rademacher functions defined on the interval $[0, 1]$. Notice that this model corresponds to tri-tiles $P = (P_1, P_2, P_3)$ that are vertically contiguous, have scale 1 and the same time interval, $[0, 1]$. Now define the functions $f = \sum_{n=1}^{N} \Phi_n^1$ and $g = \sum_{n=1}^{N} \Phi_n^2$ and observe that, in particular, one has

$$T_N^t(f, g) = \sum_{n=1}^{N} r_n(t)\Phi_n^3. \tag{6.54}$$

Suppose that the inequality

$$\|T_N^t(f, g)\|_r \lesssim \|f\|_p\|g\|_q, \tag{6.55}$$

holds with bounds that are uniform with respect to N and $t \in [0, 1]$. If we integrate over t and use Khinchine's inequality, the left-hand side becomes comparable with $N^{1/2}$. However, it is not difficult to see that $\|f\|_p \lesssim N^{1-1/p}$ and $\|g\|_q \lesssim N^{1-1/q}$, which means that one must have $N^{1/2} \lesssim N^{1-1/p}N^{1-1/q}$ for every large integer N; clearly this is possible only if $r \geq 2/3$.

Exercise 6.4 Check that $\|f\|_p \lesssim N^{1-1/p}$ and $\|g\|_q \lesssim N^{1-1/q}$ for the particular functions f and g defined earlier.

What are then the chances that the bilinear Hilbert transform itself satisfies L^p estimates all the way down to $1/2$? One might ask this, given the above counterexample. Briefly, we claim that one should still remain optimistic about this possibility. Let us explain why. If we recall the way in which we arrived at the model operator (6.14), we will realize that the part of the bilinear Hilbert transform that gives rise to scale-1 model operators can be written as

$$(f, g) \mapsto \int_{\mathbb{R}} f(x-t)g(x+t)\Phi(t)\,dt$$

with Φ a Schwartz function adapted to the interval $[0, 1]$ and having the property that $\operatorname{supp}\widehat{\Phi} \subseteq [1, 2]$. The reason is that the symbol of the above operator is given by $\widehat{\Phi}(\xi_2 - \xi_1)$, which is *smoothly supported* on the scale-1 strip

$$\{(\xi_1, \xi_2) \in \mathbb{R}^2 | 1 \leq \xi_2 - \xi_1 \leq 2\}.$$

It is also not difficult to see that, in order to understand the above operator, it is enough to understand a sharper variant, given by

$$(f, g) \mapsto \int_0^1 f(x-t)g(x+t)\,dt. \tag{6.56}$$

We will prove now that this operator (6.56) is bounded even from $L^1 \times L^1$ into $L^{1/2}$; this result appeared implicitly in a previous work of Kenig and Stein. To see this, assume that both f and g are positive functions and observe first that for every interval $I \subseteq \mathbb{R}$ with $|I| = 1$ one has

$$\int_I \int_0^1 f(x-t)g(x+t)\,dt\,dx \lesssim \left(\int_{\widetilde{I}} f(y)\,dy\right)\left(\int_{\widetilde{I}} g(y)\,dy\right), \tag{6.57}$$

where $\widetilde{I} = 3I$ is the interval having the same center as I but three times as long. This is just a simple consequence of a natural two-dimensional change of variables.

Using (6.57) one then decomposes the real line $\mathbb{R}$ as a disjoint union of intervals of length 1 and writes

$$\begin{aligned}
&\left\| \int_0^1 f(x-t)g(x+t)\,dt \right\|_{1/2} \\
&= \left(\sum_{|I|=1} \left\| \left(\int_0^1 f(x-t)g(x+t)\,dt \right) \chi_I(x) \right\|_{1/2}^{1/2} \right)^2 \\
&\lesssim \left(\sum_{|I|=1} \left\| \left(\int_0^1 f(x-t)g(x+t)\,dt \right) \chi_I(x) \right\|_{1}^{1/2} \right)^2
\end{aligned}$$

$$\lesssim \left(\sum_{|I|=1} \left(\int_{\widetilde{I}} f(y)\,dy \right)^{1/2} \left(\int_{\widetilde{I}} g(y)\,dy \right)^{1/2} \right)^2$$

$$\lesssim \left(\sum_{|I|=1} \int_{\widetilde{I}} f(y)\,dy \right) \left(\sum_{|I|=1} \int_{\widetilde{I}} g(y)\,dy \right)$$

$$\lesssim \|f\|_1 \|g\|_1,$$

given that the intervals $\widetilde{I}$ have bounded overlaps.

6.9. The bilinear Hilbert transform on polydisks

Given the previous chapters on paraproducts it is of interest to ask whether there is a way to generalize the results on the bilinear Hilbert transform to the multiparameter settings of polydisks. The natural (and simplest) candidate, formally given by $BHT \otimes BHT$, is defined by the formula

$$BHT \otimes BHT(f, g)(x, y) = p.v. \int_{\mathbb{R}^2} f(x - t, y - s) g(x + t, y + s) \frac{dt}{t} \frac{ds}{s}. \tag{6.58}$$

We will prove here the surprising (at least at first glance) fact that the operator defined in (6.58) does not satisfy *any* L^p estimates of the Hölder inequality type. This is a result of Muscalu, Pipher, Tao, and Thiele. It is based on the following observation, similar to that described in Chapter 5. Let $f(x, y) = g(x, y) = e^{ixy}$ and observe that formally one can write

$$\begin{aligned} BHT \otimes BHT(f, g)(x, y) &= e^{2ixy} \int_{\mathbb{R}^2} \frac{e^{2its}}{ts}\, dtds \\ &= Ce^{2ixy} \int_{\mathbb{R}} \frac{\operatorname{sgn}(t)}{t}\, dt = 2Ce^{2ixy} \int_0^\infty \frac{1}{t}\, dt. \end{aligned}$$

In other words we have

$$BHT \otimes BHT(f, g)(x, y) = Cf(x, y)g(x, y) \int_0^\infty \frac{1}{t}\, dt$$

for the particular functions f and g. Then, as in Chapter 5, we pick an arbitrarily large positive integer N and define the localized functions $f_N(x, y) = g_N(x, y) = e^{ixy} \chi_{[-N,N]}(x) \chi_{[-N,N]}(y)$. As there, one may observe that

$$BHT \otimes BHT(f_N, g_N)(x, y) \geq \left| \int_{-N/100}^{N/100} \int_{-N/100}^{N/100} \frac{e^{2its}}{ts} \, dtds \right| + O(1)$$

for every $x, y \in [-N/1000, N/1000]$. Since we know from Lemma 5.5 that the above expression is larger than $C \log N$, this proves that no L^p Hölder-type estimates are available for the bilinear operator in (6.58).

Notes

The original proof of the main result, Theorem 6.1, appeared in the two papers by Lacey and Thiele [70, 71]. The presentation here follows Muscalu, Tao and Thiele [94] closely. For a Walsh analogue of these results see Thiele [112, 113]. A *maximal* variant of the theorem on the bilinear Hilbert transform was proved by Lacey [69]. In the same paper one can also find the $1/2 < p \leq 2/3$ counterexample described in the text. Uniform estimates can be found in Thiele [114], Grafakos and Li [51], and Li [74]. The result in Problem 6.3 is found in Gilbert and Nahmod [47]. Generalizations of this to multilinear multipliers of arbitrary rank appeared in Muscalu, Tao, and Thiele [90].

The paper of Kenig and Stein mentioned in the text is [65]. The result in Problem 6.4 is found in Bernicot [6]. A particular case was proved independently by Bényi, Demeter, Nahmod *et al.* [5]. The statements contained in Problems 6.5 and 6.6 together with their heuristical proofs go back to some unpublished notes by Muscalu [83]. Rigorous arguments (based on similar ideas) appeared later on in Demeter [34]. Regarding the n-linear Hilbert transform for $n \geq 3$, no positive results are presently known. The counterexample in Section 6.9 was inspired by the previous work of Fefferman [39]. Various other generalizations can be found in Demeter, Pramanik, and Thiele [36] and in Demeter and Thiele [35], Bernicot [7], Grafakos and Li [52], Muscalu [81], Christ, Li, Tao *et al.* [21], Christ [16], and Palsson [99]. See also the expository papers by Lacey [68] and Thiele [115].

Problems

Problem 6.1 Denote by $BHT_{\alpha,\beta}$ the bilinear operator defined by

$$BHT_{\alpha,\beta}(f, g)(x) := p.v. \int_{\mathbb{R}^2} f(x + t + s) g(x + \alpha t + \beta s) \frac{dt}{t} \frac{ds}{s},$$

where α and β are real numbers both different from 0 and 1. Prove that this operator maps $L^{p_1} \times L^{p_2}$ into L^p provided that $1/p_1 + 1/p_2 = 1/p$, $1 < p_1, p_2 < \infty$, and $2/3 < p < \infty$. Show that the result can be extended naturally to an arbitrary number of kernels.

Problem 6.2 Extend Theorem 6.1 in a natural way to Euclidean spaces of arbitrary dimension.

Problem 6.3 Consider the one-dimensional subspace

$$\Gamma = \{(\xi_1, \xi_2) \in \mathbb{R}^2 | \xi_1 = \xi_2\}.$$

Consider a symbol $m(\xi_1, \xi_2)$ that is smooth away from Γ and satisfies

$$|\partial^\alpha m(\xi)| \lesssim \frac{1}{\mathrm{dist}(\xi, \Gamma)^{|\alpha|}}$$

for every $\xi = (\xi_1, \xi_2) \in \mathbb{R}^2 \setminus \Gamma$ and sufficiently many multi-indices α. Prove that the bilinear operator T_m associated with this symbol satisfies the same estimates as the bilinear Hilbert transform.

Hint: Show that one can reduce this to the same model operator as that in (6.14).

Problem 6.4 Consider the smooth symbol $a(x, \xi_1, \xi_2)$ satisfying

$$|\partial_x^\beta \partial_\xi^\alpha a(x, \xi)| \lesssim \frac{1}{(1 + \mathrm{dist}(\xi, \Gamma))^{|\alpha|}}$$

for every $x \in \mathbb{R}$, every $\xi = (\xi_1, \xi_2) \in \mathbb{R}^2$, and sufficiently many multi-indices α and β. Prove that the bilinear operator T_a associated with this symbol satisfies the same estimates as the bilinear Hilbert transform.

Hint: Apply the same ideas as those used in the paraproduct version of Theorem 2.16.

Problem 6.5 Let $1 < p_1, p_2, p_3 \leq \infty$ and $1/3 < p < \infty$ be such that $1/p_1 + 1/p_2 + 1/p_3 = 1/p$. Consider the following scale-1 trilinear operator

$$(f_1, f_2, f_3) \mapsto \int_0^1 f_1(x + t) f_2(x + 2t) f_3(x + 3t)\, dt.$$

Prove that it is impossible for this operator to be bounded from $L^{p_1} \times L^{p_2} \times L^{p_3}$ into L^p for every p arbitrarily close to $1/3$.

Hint: Try to understand why the Kenig–Stein bilinear approach used before in Section 6.1 no longer works. Then, use scale-1 functions to find a counterexample.

Problem 6.6 Let $1 < p_1, p_2, p_3 \leq \infty$, and $1/3 < p < \infty$ be such that $1/p_1 + 1/p_2 + 1/p_3 = 1/p$. Consider the *trilinear Hilbert transform* defined by

$$(f_1, f_2, f_3) \mapsto p.v. \int_{\mathbb{R}} f_1(x + t) f_2(x + 2t) f_3(x + 3t) \frac{dt}{t}.$$

Use Problem 6.5 to prove that it is impossible for this operator to be bounded from $L^{p_1} \times L^{p_2} \times L^{p_3}$ into L^p for every p arbitrarily close to $1/3$.

Hint: Observe first that if f_1, f_2, f_3 are supported on compact intervals then both x and t must belong to compact intervals. Use this to show that one can then reduce this problem to the previous problem.

7

Almost everywhere convergence of Fourier series

The goal of Chapter 7 is to prove the classical theorem of Carleson and Hunt, Theorem 7.1, on the almost everywhere convergence of Fourier series. The reader should also recall the related topics discussed in detail in the first volume of the book, in particular, convergence in L^p, which can be found in Chapters I.1 and I.3 and the divergence of L^1 Fourier series, which can be found in Chapter I.6.

Theorem 7.1 *Let $1 < p < \infty$ and let f be a 2π-periodic function on the real line such that $f \in L^p([-\pi, \pi])$. Then*

$$\sum_{n=-N}^{N} \widehat{f}(n)e^{in\theta} \xrightarrow[N\to\infty]{} f(\theta) \tag{7.1}$$

for almost every $\theta \in [-\pi, \pi]$.

First we remark that (7.1) is indeed true if f is a smooth and 2π-periodic function, since the Fourier series of such functions are absolutely convergent, as we saw in the first volume of the book.

Now consider f, an arbitrary 2π-periodic function in $L^p([-\pi, \pi])$. To prove (7.1) it is clearly enough to show that

$$\limsup_{N\to\infty} |(S_N f - f)(\theta)| = 0 \tag{7.2}$$

for almost every $\theta \in [-\pi, \pi]$, where $S_N f$ denotes the partial sum of order N of the Fourier series of f, given by the left hand side of (7.1). This fact would follow if one could prove that

$$\left\| \limsup_{N\to\infty} |S_N f - f| \right\|_p = 0. \tag{7.3}$$

Let $\varepsilon > 0$ be an arbitrarily small number and g a smooth 2π-periodic function such that $\|f - g\|_p < \varepsilon$. Then, we can write

$$\left\| \limsup_N |S_N f - f| \right\|_p \leq \|f - g\|_p + \left\| \limsup_N |S_N g - g| \right\|_p$$
$$+ \left\| \limsup_N |S_N f - S_N g| \right\|_p \leq \|f - g\|_p + \left\| \sup_N |S_N(f - g)| \right\|_p . \quad (7.4)$$

Thus it can be seen that in order to show (7.3) one needs to prove an inequality of the type

$$\| C_{\mathbb{T}} h \|_p \lesssim \| h \|_p, \quad (7.5)$$

where $C_{\mathbb{T}} h$ is given by

$$C_{\mathbb{T}} h(\theta) := \sup_N |S_N h(\theta)|.$$

The operator $C_{\mathbb{T}}$ is the *discrete maximal Carleson operator*. The rest of the chapter will be devoted to the proof of the inequality (7.5).

First, there are several reductions that need to be made.

7.1. Reduction to the continuous case

If f is now an arbitrary L^p function on the real line for some $1 < p < \infty$, we define $C_{\mathbb{R}} f$ by the formula

$$C_{\mathbb{R}} f(x) := \sup_N \left| \int_{\xi < N} \widehat{f}(\xi) e^{2\pi i x \xi} \, d\xi \right| . \quad (7.6)$$

The operator $C_{\mathbb{R}}$ is the *continuous maximal Carleson operator*. The following inequality represents the continuous analogue of (7.5):

$$\|C_{\mathbb{R}} f\|_p \lesssim \|f\|_p. \quad (7.7)$$

In this section we will prove that (7.5) can in fact be reduced to (7.7). First, we observe that

$$C_{\mathbb{T}} f(\theta) \lesssim |f(\theta)| + \sup_N |\widetilde{H}(M_N f)(\theta)|, \quad (7.8)$$

where $M_N f(\theta) := f(\theta) e^{iN\theta}$ is the modulation operator from the previous chapter and $\widetilde{H}$ is the harmonic conjugate, given by

$$\widetilde{H} f(\theta) = \frac{1}{2\pi} \int_{-\pi}^{\pi} \cot\left(\frac{\varphi}{2}\right) f(\theta - \varphi) \, d\varphi.$$

Of course, the above integral should be understood in a principal value sense.

This is a consequence of the fact that the symbol of $\widetilde{H}$ is a multiple of the sign function, as we learned in Chapter I.3.

Exercise 7.1 Check the estimate (7.8) above.

We will now relate the harmonic conjugate $\widetilde{H}$ to the classical Hilbert transform H on the real line; see again Chapter I.3. Generically, one can rewrite any $\widetilde{H}f$ as follows:

$$\begin{aligned}\widetilde{H}f(\theta) &= \frac{1}{2\pi}\int_{\mathbb{R}} \cot\left(\frac{\varphi}{2}\right)\chi_{[-\pi,\pi]}(\varphi)f(\theta-\varphi)\,d\varphi\\ &= \frac{1}{2\pi}\int_{\mathbb{R}} \cot\left(\frac{\varphi}{2}\right)\chi_{[-\pi,\pi]}(\varphi)\widetilde{f}(\theta-\varphi)\,d\varphi,\end{aligned} \tag{7.9}$$

where

$$\widetilde{f}(\theta-\varphi) := f(\theta-\varphi)\chi_{[-2\pi,2\pi]}(\theta-\varphi)$$

for $\theta, \varphi \in [-\pi,\pi]$.

The relation (7.9) can be rewritten as

$$\begin{aligned}&\frac{1}{2\pi}\int_{\mathbb{R}} \frac{2}{\varphi}\widetilde{f}(\theta-\varphi)\,d\varphi + \frac{1}{2\pi}\int_{\mathbb{R}}\left(\cot\left(\frac{\varphi}{2}\right)\chi_{[-\pi,\pi]}(\varphi) - \frac{2}{\varphi}\right)\widetilde{f}(\theta-\varphi)\,d\varphi\\ &= \frac{1}{\pi}H(\widetilde{f})(\theta) + \frac{1}{2\pi}\int_{\mathbb{R}}\left(\cot\left(\frac{\varphi}{2}\right)\chi_{[-\pi,\pi]}(\varphi) - \frac{2}{\varphi}\right)\widetilde{f}(\theta-\varphi)\,d\varphi.\end{aligned} \tag{7.10}$$

Now observe that

$$\cot\left(\frac{\varphi}{2}\right)\chi_{[-\pi,\pi]}(\varphi) - \frac{2}{\varphi} = K_1(\varphi) - \frac{2}{\varphi}\chi_{[-\pi,\pi]^c}(\varphi),$$

K_1 being defined by the previous equality, and this allows us to write the second term in (7.10) as

$$\frac{1}{2\pi}\int_{\mathbb{R}} K_1(\varphi)\widetilde{f}(\theta-\varphi)\,d\varphi - \frac{1}{2\pi}\int_{\mathbb{R}} \frac{2}{\varphi}\chi_{[-\pi,\pi]^c}(\varphi)\widetilde{f}(\theta-\varphi)\,d\varphi.$$

Observe also that K_1 is bounded and has compact support, which implies in particular that it is an L^1 function. Consider now the other kernel,

$$\frac{1}{\varphi}\chi_{[-\pi,\pi]^c}(\varphi).$$

Let K_2 also be a smooth, positive, and even function, compactly supported inside $[-\pi,\pi]$ and such that

$$\int_{\mathbb{R}} K_2(x)\,dx = 1.$$

One can then decompose this kernel as

$$\frac{1}{\varphi}\chi_{[-\pi,\pi]^c}(\varphi) = \frac{1}{\varphi} * K_2 + \left(\frac{1}{\varphi}\chi_{[-\pi,\pi]^c}(\varphi) - \frac{1}{\varphi} * K_2\right). \tag{7.11}$$

Notice that

$$K_3 := \frac{1}{\varphi}\chi_{[-\pi,\pi]^c}(\varphi) - \frac{1}{\varphi} * K_2$$

is bounded on $\mathbb{R}$ and decays quadratically near infinity, which shows in particular that it is an L^1 function.

Putting all this together, one has the pointwise inequality

$$|\widetilde{H} f(\theta)| \lesssim |H\widetilde{f}(\theta)| + |(\widetilde{f} * K_1)(\theta)| + |(H\widetilde{f} * K_2)(\theta)| + |(\widetilde{f} * K_3)(\theta)| \tag{7.12}$$

for every $\theta \in [-\pi, \pi]$. Modulating this and also using (7.8) one immediately deduces that for the discrete maximal Carleson operator we have

$$\begin{aligned} C_{\mathbb{T}} f(\theta) \lesssim \big|\widetilde{f}(\theta)\big| &+ \sup_N \big|H\big(M_N \widetilde{f}\big)(\theta)\big| + \big(\big|\widetilde{f}\big| * |K_1|\big)(\theta) \\ &+ \big(\sup_N \big|H\big(M_N \widetilde{f}\big)\big| * K_2\big)(\theta) + \big(\big|\widetilde{f}\big| * |K_3|\big)(\theta). \end{aligned}$$

Since we also observe that the L^p norms of $C_{\mathbb{R}} g$ and $\sup_N |H(M_N g)|$ are comparable for any function g, this proves that (7.5) can indeed be reduced to (7.7). Here we have also used the fact that convolutions with L^1 functions are bounded linear operations in L^q for any $1 \leq q \leq \infty$; see Lemma I.1.1.

Let us say now a few words about the symmetries of the continuous maximal Carleson operator. They are described by the three identities

$$\begin{gathered} C_{\mathbb{R}}(T_y f)(x) = T_y(C_{\mathbb{R}} f)(x), \qquad C_{\mathbb{R}}(D_\lambda f)(x) = D_\lambda(C_{\mathbb{R}} f)(x), \\ C_{\mathbb{R}}(M_a f)(x) = C_{\mathbb{R}} f(x), \end{gathered}$$

which are valid for every $y, a \in \mathbb{R}$ and $\lambda > 0$, where T_y, D_λ, and M_a are the translation, dilation, and modulation operators introduced in the previous chapter. Notice the similarity to the corresponding symmetries for the bilinear Hilbert transform. While the first two are typical of Calderón–Zygmund operators, the modulation symmetry is new and is the main source of difficulty. Because of the symmetries that it possesses, it will be more natural to study the continuous maximal operator $C_{\mathbb{R}}$ rather than its periodic analogue $C_{\mathbb{T}}$.

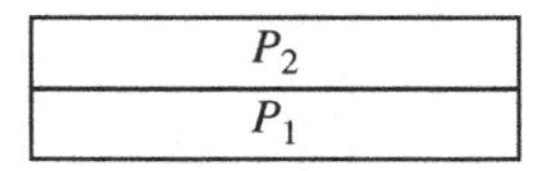

Figure 7.1. A bi-tile.

7.2. Discrete models

As in the case of the bilinear Hilbert transform, the plan for this section is to show that (7.7) can be reduced to a similar estimate for some discretized model operators. We need to introduce some notation and definitions first. Denote by $\mathcal{D}$ the collection of all dyadic intervals on the real line:

$$\mathcal{D} := \{\omega \mid \omega \text{ dyadic interval}\}.$$

For $k \in [0, 1]$ and $\eta \in \mathbb{R}$ denote by $\mathcal{D}_{k,\eta}$ the set

$$\mathcal{D}_{k,\eta} := 2^k\mathcal{D} + \eta := \{2^k\omega + \eta \mid \omega \in \mathcal{D}\}.$$

Here, if $\omega = 2^\ell[n, n+1]$ then $2^k\omega$ denotes the interval $2^{k+\ell}[n, n+1]$. By a generic *dyadic grid* we simply mean one of these collections, corresponding to fixed k and η.

Fix such a dyadic grid. A *bi-tile* $P = I_P \times \omega_P$ is a rectangle in the *phase plane* of area 2, where $I_P \in \mathcal{D}$ and ω_P belongs to the fixed dyadic grid. Every P splits naturally as a disjoint union of the tiles P_1 and P_2 of area 1, as in Figure 7.1. More specifically, $P_1 = I_P \times \omega_{P_1}$ and $P_2 = I_P \times \omega_{P_2}$, where ω_{P_1} is the left half of ω_P and ω_{P_2} is the right half of ω_P.

Let $\mathbb{P}$ be a finite collection of bi-tiles. Define the linear operator $C_{\mathbb{P}}f$ by

$$C_{\mathbb{P}}f(x) := \sum_{P\in\mathbb{P}} \langle f, \widetilde{\Phi}_{P_1}\rangle \widetilde{\widetilde{\Phi}}_{P_1} \chi_{\{x \mid N(x)\in\omega_{P_2}\}}, \tag{7.13}$$

where $N : \mathbb{R} \to \mathbb{R}$ is an arbitrary function while $\widetilde{\Phi}_{P_1}$ and $\widetilde{\widetilde{\Phi}}_{P_1}$ are wave packets adapted to the Heisenberg box P_1 that are L^2-normalized (recall that the support of their Fourier transform lies within ω_{P_1}).

We will argue in what follows that the desired L^p estimates for $C_{\mathbb{R}}$ can be reduced to the following theorem.

Theorem 7.2 *The linear operator $C_{\mathbb{P}}$ is bounded from $L^p(\mathbb{R})$ into $L^p(\mathbb{R})$ with bounds that are independent of the cardinality of $\mathbb{P}$, the wave packets associated to it, and the fixed dyadic grid.*

To see this we need to discretize the operator $C_{\mathbb{R}}$. Before we start to do this, we will make a few comments. If $N(x) = N$ in (7.13) then intuitively one expects

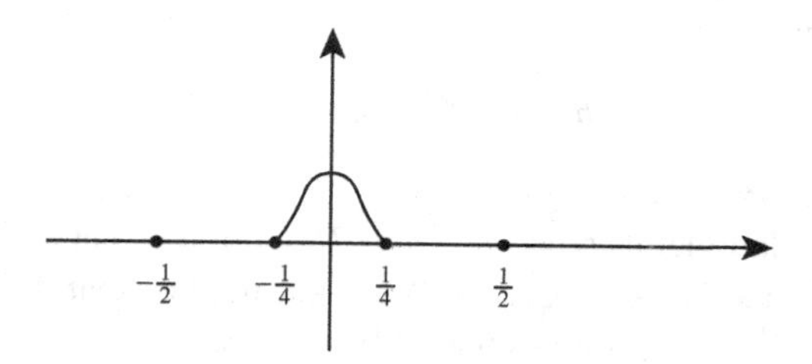

Figure 7.2. The function $\widehat{\Phi}$.

that the corresponding expression will come from the discretization of the linear multiplier operator defined by the symbol $\chi_{(-\infty,N]}(\xi)$. However, since the sum in (7.13) is over those bi-tiles P with the property that $N \in \omega_{P_2}$, it should be clear that the symbol decomposition that will be used cannot come from that obtained directly by translating the Littlewood–Paley decomposition of $\chi_{(-\infty,0]}(\xi)$. Instead, what one needs to do is to construct a translation-invariant Littlewood–Paley decomposition for the original symbol $\chi_{(-\infty,0]}(\xi)$ by simply averaging over all possible (classical) Littlewood–Paley decompositions naturally associated with arbitrary dyadic grids.

The details are as follows. Pick a smooth compactly supported positive function $\widehat{\Phi}$ that $\operatorname{supp}\widehat{\Phi} \subseteq [-\frac{1}{4}, \frac{1}{4}]$, say, as in Figure 7.2. If ω is any interval, denote by $\widehat{\Phi}_\omega$ the function defined by

$$\widehat{\Phi}_\omega(\xi) := \widehat{\Phi}\left(\frac{\xi - c_\omega}{|\omega|}\right),$$

where c_ω is the center of ω. For any ω, denote by ω_ℓ and ω_r its left-hand and right-hand halves respectively. Define $m(\xi)$ by

$$m(\xi) := \mathrm{ave}_{k,\eta}\Bigg\{\sum_{\omega\in\mathcal{D}_{k,\eta}} \widehat{\Phi}_{\omega_\ell}(\xi)\chi_{\omega_r}(0)\Bigg\}, \tag{7.14}$$

where the average on the right-hand side is given by

$$\lim_{M\to\infty}\frac{1}{2M}\int_{-M}^{M}\int_0^1\left(\sum_{\omega\in\mathcal{D}_{k,\eta}} \widehat{\Phi}_{\omega_\ell}(\xi)\chi_{\omega_r}(0)\right)dkd\eta.$$

To see that the limit exists one has to observe that different scales for the integrand correspond to essentially disjoint (in support) functions and, if one restricts the summation to $|\omega| = \text{constant}$, one sees that the integrand, as a function of η, is periodic with period $|\omega|$.

We claim now that

$$m(\xi) = C\chi_{(-\infty,0]}(\xi). \tag{7.15}$$

for some constant $C > 0$.

To prove the claim let us recall that $\omega \in \mathcal{D}_{k,\eta}$ implies that $\omega = 2^k\widetilde{\omega} + \eta$ for some $\widetilde{\omega} \in \mathcal{D}$. In particular, $\xi \in \omega_\ell = 2^k\widetilde{\omega}_\ell + \eta$ implies that $2^{-k}(\xi - \eta) \in \widetilde{\omega}_\ell$ while $0 \in \omega_r = 2^k\widetilde{\omega}_r + \eta$ implies that $2^{-k}(-\eta) \in \widetilde{\omega}_r$.

As a consequence, we have that

$$m(\xi) = \text{ave}_{k,\eta}\left\{\sum_{\omega\in\mathcal{D}} \widehat{\Phi}_{\omega_\ell}\left(2^{-k}(\xi+\eta)\right)\chi_{\omega_r}\left(2^{-k}\eta\right)\right\}. \tag{7.16}$$

Let $\lambda > 0$. Since the support of $m(\xi)$ obviously lies inside $(-\infty, 0]$, it is enough to prove that

$$m(2^\lambda\xi) = m(\xi) \tag{7.17}$$

for every $\xi \in \mathbb{R}$. Using (7.16) one has

$$m(2^\lambda\xi) = \text{ave}_{k,\eta}\left\{\sum_{\omega\in\mathcal{D}} \widehat{\Phi}_{\omega_\ell}\left(2^k\left(2^\lambda\xi+\eta\right)\right)\chi_{\omega_r}(2^k\eta)\right\}$$

after changing variables ($k \to -k$) from k to $-k$; the average over k is now an integral from -1 to 0. Then the above equality can be written as

$$\begin{aligned}
&\text{ave}_{k,\eta}\left\{\sum_{\omega\in\mathcal{D}} \widehat{\Phi}_{\omega_\ell}\left(2^{k+\lambda}\xi+2^k\eta\right)\chi_{\omega_r}(2^k\eta)\right\}\\
&\quad= \text{ave}_{k,\eta}\left\{\sum_{\omega\in\mathcal{D}} \widehat{\Phi}_{\omega_\ell}\left(2^k\xi+2^{k-\lambda}\eta\right)\chi_{\omega_r}(2^{k-\lambda}\eta)\right\},
\end{aligned}$$

where the average over k becomes an integral from $\lambda - 1$ to λ. We change variables again (this time with respect to η) and rewrite the expression as

$$\text{ave}_{k,\eta}\left\{\sum_{\omega\in\mathcal{D}} \widehat{\Phi}_{\omega_\ell}\left(2^k\xi+2^k\eta\right)\chi_{\omega_r}\left(2^k\eta\right)\right\}, \tag{7.18}$$

where the average over η has become an average over the interval $2^{-\lambda}[-M, M]$ as $M \to \infty$.

At this point one needs to realize that (7.18) is the same as (7.16), since the inner function is periodic of period 1 in k and, as a consequence, the integral from $\lambda - 1$ to λ is the same as the integral from -1 to 0. This proves our claim (7.15).

Using this, one then observes that

$$\chi_{(-\infty,N]}(\xi) = \chi_{(-\infty,0]}(\xi - N) = \text{ave}_{k,\eta}\left\{ \sum_{\omega \in \mathcal{D}_{k,\eta}} \widehat{\Phi}_{\omega_\ell}(\xi - N)\chi_{\omega_r}(0) \right\}, \quad (7.19)$$

where we have renamed Φ as the previous one divided by the constant C in (7.15).

Now $\xi - N \in \omega_\ell$ means that $\xi \in \omega_\ell + N$ and similarly $0 \in \omega_r$ means that $N \in \omega_r + N$. Because of the average with respect to η, it is not difficult to realize that (7.19) is equal to

$$\text{ave}_{k,\eta}\left\{ \sum_{\omega \in \mathcal{D}_{k,\eta}} \widehat{\Phi}_{\omega_\ell}(\xi)\chi_{\omega_r}(N) \right\}. \quad (7.20)$$

This is a very useful fact, since the collection of intervals $D_{k,\eta}$ is independent of N.

As a consequence, for every fixed well-behaved function f, one can write

$$\int_{\xi<N} \widehat{f}(\xi)e^{2\pi i x\xi}\, d\xi = \text{ave}_{k,\eta}\left\{ \sum_{\omega \in \mathcal{D}_{k,\eta}} (f * \Phi_{\omega_\ell})(x)\, \chi_{\omega_r}(N) \right\}. \quad (7.21)$$

As we have observed several times before, one can further discretize every term $f * \Phi_{\omega_\ell}$ as

$$f * \Phi_{\omega_\ell} = \text{ave}_\alpha \left\{ \sum_{|I|=|\omega_\ell|^{-1}} \langle f, \Phi^1_{I,\alpha}\rangle \Phi^2_{I,\alpha} \right\}$$

where the average over α means an average over the interval $[0, 1]$ and where $\Phi^1_{I,\alpha}$ and $\Phi^2_{I,\alpha}$ are both L^2-normalized bump functions adapted to I and whose Fourier supports lie inside ω_ℓ. Taking all these facts into account and realizing also that the supremum over N in the definition of $C_{\mathbb{R}}$ will produce the function $N(x)$ (where the supremum is attained), we see that $C_{\mathbb{R}}$ applied to the given function f gives the absolute value of an average over the parameters k, η, α of discrete model operators of the type (7.13). This (together with the usual limiting arguments) proves that it is indeed enough to demonstrate Theorem 7.2, in order to complete the proof of the Carleson–Hunt theorem.

Exercise 7.2 Fill in the details of the last few statements above.

Regarding (7.13), the fact that we have two distinct families of wave packets, $(\widetilde{\Phi}_{P_1})_P$ and $(\widetilde{\widetilde{\Phi}}_{P_1})_P$, in the definition of $C_{\mathbb{P}}$ is irrelevant. Because of this, we will prove the estimates for the cleaner expression (which we also denote

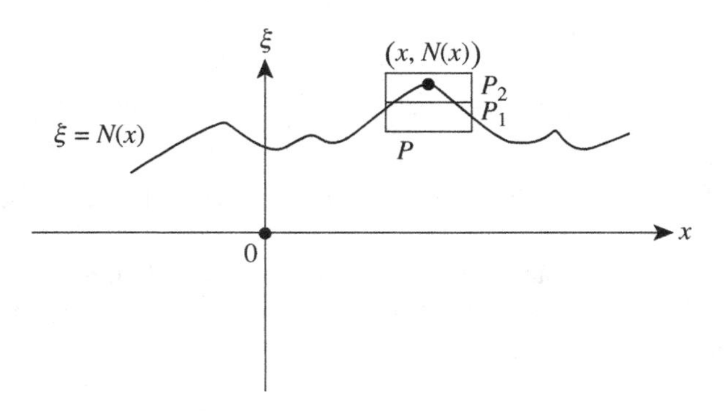

Figure 7.3

by $C_{\mathbb{P}}$)

$$C_{\mathbb{P}} f(x) = \sum_{P \in \mathbb{P}} \langle f, \Phi_{P_1} \rangle \Phi_{P_1} \chi_{\{x | N(x) \in \omega_{P_2}\}}. \tag{7.22}$$

This simplifies the notation, although an identical argument works in both cases. A diagrammatic description of the summation in (7.22) is given in Figure 7.3.

Let us consider (7.22) and assume that $N(x)$ is constant. Then one can see that all the tiles ω_{P_2} intersect each other, and this implies that the corresponding ω_{P_1} are all disjoint. This is a simple observation, but it is of fundamental importance. It is a consequence of the fact that our rectangles P of area 2 are bi-tiles (see Figure 7.4). It is also important for the reader to realize that, because of this fact, our collections of bi-tiles here are similar to the collections of rank-1 *tri-tiles* used in the study of the bilinear Hilbert transform in Chapter 6. In particular, one can also see that, modulo a translation in *frequency*, the corresponding operator becomes in this case a discretized variant of a classical linear singular integral (or, equivalently, of a classical multiplier operator).

Exercise 7.3 Let m be a classical Marcinkiewicz–Mikhlin–Hörmander symbol that is a function of one variable, and consider the linear multiplier operator defined by

$$T_m f(x) := \int_{\mathbb{R}} m(\xi) \widehat{f}(\xi) e^{2\pi i x \xi} \, d\xi.$$

Show that the discretized model of this operator corresponds to formula (7.13) in the case $N(x) = 0$.

The fact that the ω_{P_2} tiles intersect should remind the reader of the trees of the previous chapter. The latter will play a similarly important role here.

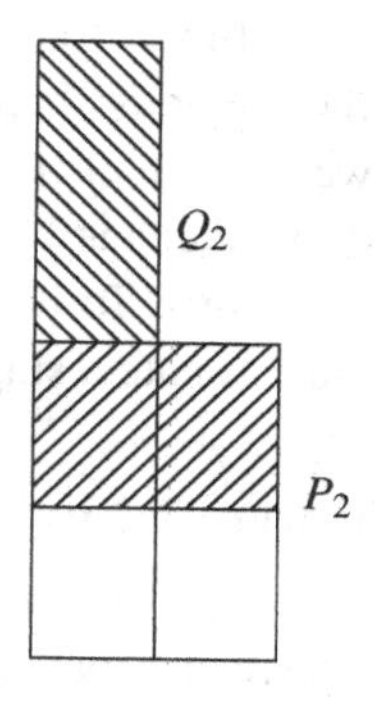

Figure 7.4 Generic bi-tiles P and Q intersecting each other. Notice that P_2 and Q_2 are disjoint in this case.

Definition 7.3 We define (as before) a partial ordering $\leq$ between distinct bi-tiles and say that $P \leq P'$ if and only if $I_P \subseteq I_{P'}$ and $\omega_{P'} \subseteq \omega_P$. Then, a collection of bi-tiles T is called a *tree* if there exists a bi-tile $P_T = I_T \times \omega_T$ (called the top of the tree) having the property that $P \leq P_T$ for every $P \in T$.

As in the case of the bilinear Hilbert transform, the top corresponding to the tree does not necessarily have to belong to the tree. Clearly, since $\omega_{P_T} \subseteq \omega_P$ it follows that one either has $\omega_{P_T} \subseteq \omega_{P_1}$ or $\omega_{p_T} \subseteq \omega_{P_2}$, where $P_1 \cup P_2 = P$ are the tiles whose union equals the bi-tile P. We say that the tree is a j-tree if and only if one has $\omega_{P_T,j} \subseteq \omega_{P_j}$ for any $P \in T$ and $j \in \{1, 2\}$. It is easy to observe that every tree of bi-tiles can be written as a disjoint union of an 1-tree and a 2-tree.

We emphasize again that the reason for which one considers these clusters of bi-tiles called *trees* is that (as we will see rigorously later on) the corresponding Carleson operator $C_{\mathbb{T}}$ becomes essentially a discrete version of the Hilbert transform, for which classical arguments can be used. Our finite collection of bi-tiles $\mathbb{P}$ can in principle contain many such trees and the idea is to find a way of selecting these trees, to estimate each of them, and then to add these estimates using *orthogonality* arguments for distinct trees. In other words, the plan is *identical* to that used in the previous chapter. However, as we will see, the Carleson operator is technically more challenging than the bilinear Hilbert transform. Let us briefly explain why.

As the reader might have expected already, soon we will start estimating bilinear forms associated with $C_{\mathbb{P}}$ of the type

$$\sum_{P \in \mathbb{P}} \langle f, \Phi_{P_1} \rangle \langle \Phi_{P_1} \chi_{\{x | N(x) \in \omega_{P_2}\}}, g \rangle.$$

Notice that now we have two types of scalar product that we need to handle, namely $\langle f, \Phi_{P_1} \rangle$ and $\langle \Phi_{P_1} \chi_{\{x | N(x) \in \omega_{P_2}\}}, g \rangle$. The first type is similar to those that we met in the case of the bilinear Hilbert transform, but the second type is new. Everything we learned in the previous chapter can be used to handle the first type of scalar product, but we need an analogous treatment for the second type. More specifically, we need to understand which are the natural sizes and energies that will help us to achieve this. To be able to guess them correctly, we need first of all to learn how to estimate a single tree.

7.3. Proof of Theorem 7.2 in the scale-1 case

Before doing that, let us first treat the much easier case of Theorem 7.2 when all the bi-tiles have the same scale; the reader may remember that this is precisely what we did when we treated the paraproducts case in Chapter 2 and the bilinear Hilbert transform case in Chapter 6. The understanding of this situation will help us to understand what happens in the general case, later on.

Let us consider discrete model of the Carleson operator defined by

$$C_{\mathbb{P}} f(x) := \sum_{P \in \mathbb{P}} \langle f, \Phi_{P_1} \rangle \Phi_{P_1} \chi_{\{x | N(x) \in \omega_{P_2}\}}$$

where we assume that the sum is over a collection of bi-tiles having the property that their length $|I_P|$ is the same for every $P \in \mathbb{P}$.

Fix measurable sets E_1, E_2 of finite measure such that $|E_2| = 1$; our goal is to construct a subset $E_2' \subseteq E_2$ with $|E_2'| \simeq 1$ and such that

$$\left| \sum_{P \in \mathbb{P}} \langle f_1, \Phi_{P_1} \rangle \langle \Phi_{P_1} \chi_{\{x | N(x) \in \omega_{P_2}\}}, f_2 \rangle \right| \lesssim |E_1|^{1/p} \tag{7.23}$$

for every $|f_1| \leq \chi_{E_1}$ and $|f_2| \leq \chi_{E_2'}$ where p is a number strictly larger than 1 but arbitrarly close to 1. If one could prove this then by interpolation $C_{\mathbb{P}}$ would be bounded on every L^p space for p larger than but arbitrarily close to 1.

Define first an exceptional set Ω by

$$\Omega := \{x \, | M\chi_{E_1}(x) > C|E_1|\},$$

where M is the Hardy–Littlewood maximal operator. As usual, if C is a sufficiently large constant, one has $|\Omega| < \frac{1}{2}$ and so one can define $E_2' := E_2 \setminus \Omega$ to satisfy $|E_2'| \simeq 1$. To be able to estimate (7.23) we first split our collection of bi-tiles $\mathbb{P}$ as follows:

$$\mathbb{P} = \bigcup_{d \geq 0} \mathbb{P}_d,$$

where, as before, $\mathbb{P}_d$ contains the bi-tiles in $\mathbb{P}$ with the property that

$$1 + \frac{\text{dist}(I_P, \Omega^c)}{|I_P|} \simeq 2^d.$$

We will show that for every $d \geq 0$ one has

$$\left| \sum_{P \in \mathbb{P}_d} \langle f_1, \Phi_{P_1} \rangle \langle \Phi_{P_1} \chi_{\{x | N(x) \in \omega_{P_2}\}}, f_2 \rangle \right| \lesssim |E_1|^{1/p} 2^{-10d}, \tag{7.24}$$

and this will be clearly enough to conclude (7.23). To be able to prove (7.24), we decompose further $\mathbb{P}_d$ as follows:

$$\mathbb{P}_d = \bigcup_{n_1} \mathbb{P}_{d,1}^{n_1}, \tag{7.25}$$

where $\mathbb{P}_{d,1}^{n_1}$ contains the bi-tiles in $\mathbb{P}_d$ having the property that

$$\frac{|\langle f_1, \Phi_{P_1} \rangle|}{|I_P|^{\frac{1}{2}}} \simeq 2^{-n_1}. \tag{7.26}$$

Similarly, using the function f_2 one obtains a decomposition of $\mathbb{P}_d$ as follows:

$$\mathbb{P}_d = \bigcup_{n_2} \mathbb{P}_{d,2}^{n_2}, \tag{7.27}$$

where $\mathbb{P}_{d,2}^{n_2}$ contains the bi-tiles for which one has

$$\frac{|\langle \Phi_{P_1} \chi_{\{x | N(x) \in \omega_{P_2}\}}, f_2 \rangle|}{|I_P|^{\frac{1}{2}}} \simeq 2^{-n_2}. \tag{7.28}$$

The left-hand side of (7.24) can be split naturally as

$$\left| \sum_{n_1, n_2} \sum_{P \in \mathbb{P}_d^{n_1, n_2}} \langle f_1, \Phi_{P_1} \rangle \langle \Phi_{P_1} \chi_{\{x | N(x) \in \omega_{P_2}\}}, f_2 \rangle \right|, \tag{7.29}$$

where $\mathbb{P}_d^{n_1,n_2}$ is defined to be the intersection $\mathbb{P}_{d,1}^{n_1} \cap \mathbb{P}_{d,2}^{n_2}$. Using all this the absolute value of (7.29) can be estimated by

$$\sum_{n_1, n_2} 2^{-n_1} 2^{-n_2} \sum_{P \in \mathbb{P}_d^{n_1, n_2}} |I_P|. \tag{7.30}$$

Notice that so far everything is identical to the paraproduct case studied in Chapter 2 and to the bilinear Hilbert transform case studied in Chapter 6. The difference comes at the next step, when we estimate the sum of the lengths $|I_P|$. As we will see, to estimate this sum we need to use the same L^2 Bessel method in the case of f_1 (the reader should remember it from the corresponding section for the bilinear Hilbert transform, Section 6.9), but a completely different argument is needed in the case of f_2.

So, on the one hand, as in the discussion after (6.23) we observe that

$$\sum_{P \in \mathbb{P}_d^{n_1, n_2}} |I_P| \leq \sum_{P \in \mathbb{P}_d^{n_1}} |I_P| \lesssim 2^{2n_1} |E_1|$$

and on the other hand we note the stronger estimate

$$\sum_{P\in\mathbb{P}_d^{n_1,n_2}} |I_P| \leq \sum_{P\in\mathbb{P}_d^{n_2}} |I_P| \lesssim 2^{n_2}|E_2| = 2^{n_2}. \tag{7.31}$$

To see why (7.31) is true, one can write

$$\begin{aligned}\sum_{P\in\mathbb{P}_d^{n_2}} |I_P| &\lesssim 2^{n_2} \sum_{P\in\mathbb{P}_d^{n_2}} \left|\left\langle |I_P|^{1/2}\Phi_{P_1}\chi_{\{x|N(x)\in\omega_{P_2}\}}, f_2\right\rangle\right| \\ &=: 2^{n_2} \sum_{P\in\mathbb{P}_d^{n_2}} |\langle \Phi_{P_1}^{\infty}\chi_{\{x|N(x)\in\omega_{P_2}\}}, f_2\rangle| \\ &= 2^{n_2}\sum_{\omega} \sum_{P\in\mathbb{P}_d^{n_2}:\,\omega_{P_2}=\omega} |\langle \Phi_{P_1}^{\infty}\chi_{\{x|N(x)\in\omega_{P_2}\}}, f_2\rangle| \\ &\lesssim 2^{n_2}\sum_{\omega} \|f_2\chi_{\{x|N(x)\in\omega\}}\|_1 \lesssim 2^{n_2}\|f_2\|_1 \leq 2^{n_2}|E_2|,\end{aligned}$$

as claimed. In the above we used that fact that all the bi-tiles have the same length $|I_P|$ and that the intervals ω are all disjoint.

By interpolating between these estimates one obtains

$$\sum_{P\in\mathbb{P}_d^{n_1,n_2}} |I_P| \lesssim 2^{2n_1\theta_1}2^{n_2\theta_2}|E_1|^{\theta_1} \tag{7.32}$$

for every $0 \leq \theta_1, \theta_2 < 1$ with $\theta_1 + \theta_2 = 1$. However, as in the case of the paraproducts and the bilinear Hilbert transform, since every P is in $\mathbb{P}_d$ one must have

$$\begin{aligned}2^{-n_1} &\lesssim 2^d|E_1|,\\ 2^{-n_2} &\lesssim 2^{-Md},\end{aligned}$$

where M is an arbitrarily large constant. Using all these relations together, one can estimate (7.30) by

$$\sum_{n_1,n_2} 2^{-n_1(1-2\theta_1)}2^{-n_2(1-\theta_2)}|E_1|^{\theta_1} \lesssim 2^{-\widetilde{M}d}|E_1|^{1-\theta_1},$$

assuming that $0 \leq \theta_1 < 1/2$ in order to ensure that the first geometric series is convergent. Finally one simply has to choose θ_1 such that $1-\theta_1 = 1/p$ and large enough M to conclude our desired estimate (7.24).

7.4. Estimating a single tree

We now come back to the general case of Theorem 7.2 and, as promised, consider the contribution of a single tree. As we will see, estimating a single tree is a rather more difficult task than it was in the bilinear Hilbert transform case.

Lemma 7.4 *Let $T \subseteq \mathbb{P}$ be a tree and f, g be two fixed functions. Then one has*

$$\sum_{P \in T} |\langle f, \Phi_{P_1} \rangle| \, |\langle \Phi_{P_1} \chi_{\{x | N(x) \in \omega_{P_2}\}}, g \rangle|$$

$$\lesssim \left(\sup_{\substack{\widetilde{T} \subseteq T \\ \widetilde{T}\ 2-tree}} \left(\frac{1}{|I_{\widetilde{T}}|} \sum_{P \in \widetilde{T}} |\langle f, \Phi_{P_1} \rangle|^2 \right)^{1/2} \right)$$

$$\times \left(\sup_{P \in T} \sup_{\substack{P' \in \overline{\mathbb{P}} \\ P' \geq P}} \frac{1}{|I_{P'}|} \int_{\mathbb{R}} \widetilde{\chi}^M_{I_{P'}}(x) \chi_{\{x | N(x) \in \omega_{P'}\}} |g(x)| \, dx \right) |I_T|. \tag{7.33}$$

In the above, M is an arbitrarily large constant while $\overline{\mathbb{P}}$ denotes the collection of *all* possible dyadic bi-tiles in the plane.

Proof Fix a tree $T \subseteq \mathbb{P}$. First, let us denote by $\mathcal{J}$ the set of all dyadic intervals $J \subseteq \mathbb{R}$ satisfying $3J \not\supseteq I_P$ for any $P \in T$ and such that J is maximal with this property. Observe that such a collection forms a disjoint partition of $\mathbb{R}$. Then, the left-hand side of (7.33) can be estimated by

$$\left\| \sum_{P \in T} \varepsilon_P \langle f, \Phi_{P_1} \rangle \Phi_{P_1} \chi_{\{x | N(x) \in \omega_{P_2}\}} g \right\|_{L^1}$$

$$= \sum_{J \in \mathcal{J}} \left\| \sum_{P \in T} \varepsilon_P \langle f, \Phi_{P_1} \rangle \Phi_{P_1} \chi_{\{x | N(x) \in \omega_{P_2}\}} g \right\|_{L^1(J)}$$

$$\leq \sum_{J \in \mathcal{J}} \left\| \sum_{\substack{P \in T \\ |I_P| \leq |J|}} \varepsilon_P \langle f, \Phi_{P_1} \rangle \Phi_{P_1} \chi_{\{x | N(x) \in \omega_{P_2}\}} g \right\|_{L^1(J)}$$

$$+ \sum_{J \in \mathcal{J}} \left\| \sum_{\substack{P \in T \\ |I_P| > |J|}} \varepsilon_P \langle f, \Phi_{P_1} \rangle \Phi_{P_1} \chi_{\{x | N(x) \in \omega_{P_2}\}} g \right\|_{L^1(J)} := I + II. \tag{7.34}$$

Here, $(\varepsilon_P)_P$ are complex numbers of modulus 1.

Estimates for term I in (7.34) Fix J and look at the corresponding term. From the definition of J it is obvious that $I_P \cap J = \emptyset$ for any $P \in T$ in this case. In particular, one can write

$$\left\| \sum_{\substack{P\in T\\ |I_P|\le|J|}} \varepsilon_P \langle f, \Phi_{P_1}\rangle \Phi_{P_1} \chi_{\{x|N(x)\in\omega_{P_2}\}} g \right\|_{L^1(J)}$$
$$\lesssim \sum_{\substack{P\in T\\ |I_P|\le|J|}} \left(\sup_{P\in T} \frac{|\langle f, \Phi_{P_1}\rangle|}{|I_{P_1}|^{1/2}}\right)\left(1+\frac{\operatorname{dist}(I_P, J)}{|I_P|}\right)^{-100}$$
$$\times \left(\sup_{P\in T} \frac{1}{|I_P|}\int_{\mathbb{R}} \widetilde{\chi}_{I_P}^M(x)\chi_{\{x|N(x)\in\omega_{P_2}\}}|g(x)|\,dx\right)|I_P|. \qquad (7.35)$$

Clearly, the first and third factors in (7.35) are bounded by the corresponding expressions on the right-hand side of (7.33), while the rest of the sum can be estimated by

$$\sum_{\substack{P\in T\\ |I_P|\le|J|}} \left(1+\frac{\operatorname{dist}(I_P, J)}{|I_P|}\right)^{-100} |I_P|$$
$$= \sum_{\substack{k:2^k\le|J|\\ 2^k\le|I_T|}} 2^k \sum_{\substack{P\in T\\ |I_P|=2^k}} \left(1+\frac{\operatorname{dist}(I_P, J)}{|I_P|}\right)^{-100}$$
$$\lesssim \sum_{\substack{k:2^k\le|J|\\ 2^k\le|I_T|}} 2^k \left(1+\frac{\operatorname{dist}(J, I_T)}{|I_T|}\right)^{-50}$$
$$= \min(|I_T|, |J|)\left(1+\frac{\operatorname{dist}(J, I_T)}{|I_T|}\right)^{-50}.$$

Then, one just needs to observe that the sum over J of the above expression is smaller than $|I_T|$.

Estimates for term II in (7.34) Here the argument depends on the type of tree. We have two cases.

Case 1: T is type 1. In this situation, the frequency intervals of $(\Phi_{P_1})_P$ intersect. In particular, we can estimate term II by

$$\sum_{J\in\mathcal{J}} \left(\sup_{P\in T} \frac{|\langle f, \Phi_{P_1}\rangle|}{|I_P|^{1/2}}\right) \sum_{\substack{P\in T\\ |I_P|>|J|}} \int_J \widetilde{\chi}_{I_P}^M(x)\chi_{\{x|N(x)\in\omega P_2\}}|g(x)|\,dx. \qquad (7.36)$$

We also observe that now the intervals J must have the property that $J \subseteq 3I_T$, otherwise there would be no bi-tiles P in T such that $|I_P| > |J|$. Fix $J \in \mathcal{J}$. Pick a dyadic interval $\widetilde{J} \supseteq J$ such that $|\widetilde{J}| = 2|J|$. Because of the maximality of J there must exist $P \in T$ such that $I_P \subseteq 3\widetilde{J}$. Now, choose a bi-tile $P(J) \in \overline{\mathbb{P}}$ such that $|I_{P(J)}| = 2|\widetilde{J}|$ and also such that $P \leq P(J) \leq P_T$, and simply observe that (since the sets $\{x | N(x) \in \omega_{P_2}\}_P$ are disjoint for different scales) (7.36) can be further estimated by

$$\left(\sup_{P \in P} \frac{|\langle f, \Phi_{P_1}\rangle|}{|I_P|^{1/2}}\right) \sum_{J \in \mathcal{J}} \int_{\mathbb{R}} \widetilde{\chi}^M_{P(J)}(x) \chi_{\{x|N(x) \in \omega_{P(J)}\}} |g(x)|\, dx$$
$$\lesssim \left(\sup_{P \in P} \frac{|\langle f, \Phi_{P_1}\rangle|}{|I_P|^{1/2}}\right)$$
$$\times \sum_{J \in \mathcal{J}} \left(\sup_{P \in T} \sup_{\substack{P' \in \overline{\mathbb{P}} \\ P' \geq P}} \frac{1}{|I_{P'}|} \int_{\mathbb{R}} \widetilde{\chi}^M_{I_{P'}}(x) \chi_{\{x|N(x) \in \omega_{P'}\}} |g(x)|\, dx \right) |I_{P(J)}|,$$

which is smaller than the right-hand side of (7.33) since this time the intervals J are within $3I_T$ and are also disjoint.

Case 2: T is type 2. This time the frequency intervals of $(\Phi_{P_1})_P$ are disjoint for different scales and we can take advantage of this. The corresponding term II can be written as

$$\sum_{\substack{J \in \mathcal{J} \\ J \subseteq 3I_T}} \int_J \left| \sum_{\substack{P \in T \\ |I_P| \geq |J|}} \varepsilon_P \langle f, \Phi_{P_1}\rangle \Phi_{P_1} \chi_{\{x|N(x) \in \omega_{P_2}\}} g(x) \right| dx. \tag{7.37}$$

Fix $J \subseteq 3I_T$. For every $x \in J$ observe that one has the pointwise estimate

$$\left| \sum_{\substack{P \in T \\ |I_P| \geq |J|}} \varepsilon_P \langle f, \Phi_{P_1}\rangle \Phi_{P_1} \chi_{\{x|N(x) \in \omega_{P_2}\}}(x) \right| \lesssim \sup_{I \supseteq J} \frac{1}{|I|} \int_I \left| \sum_P \varepsilon_P \langle f, \Phi_{P_1}\rangle \Phi_{P_1}(z) \right| dz, \tag{7.38}$$

which is a quantity independent of $x \in J$. To see this, one first has to realize that the pointwise expression

$$\sum_{\substack{P \in T \\ |I_P| \geq |J|}} \varepsilon_P \langle f, \Phi_{P_1}\rangle \Phi_{P_1} \chi_{\{x|N(x) \in \omega_{P_2}\}}(x)$$

can also be written as a convolution of the type

$$\left(\sum_{P \in T} \varepsilon_P \langle f, \Phi_{P_1}\rangle \Phi_{P_1}\right) * \Psi^x(x)$$

for a well-chosen bump function Ψ^x (depending on the fixed point $x \in J$). After that one observes that the absolute value of this convolution is smaller than the right-hand side of (7.38). Here the condition $|I_P| \geq |J|$ also plays an important role.

Exercise 7.4 Check the estimate in (7.38).

Recalling the definition of $P(J) \in \overline{\mathbb{P}}$ from case 1, we then see that (7.37) can be estimated by

$$\sum_{\substack{J \in \mathcal{J} \\ J \subseteq 3I_T}} \left(\sup_{I \supseteq J} \frac{1}{|I|} \int_I \left| \sum_P \varepsilon_P \langle f, \Phi_{P_1} \rangle \Phi_{P_1}(z) \right| dz \right)$$
$$\times \int_{\mathbb{R}} \widetilde{\chi}^M_{I_{P(J)}}(x) \widetilde{\chi}_{\{x | N(x) \in \omega_{P(J)}\}} |g(x)|\, dx$$
$$\lesssim \left(\sup_{P \in T} \sup_{\substack{P' \in \overline{\mathbb{P}} \\ P' > P}} \frac{1}{|I_{P'}|} \int_{\mathbb{R}} \widetilde{\chi}^M_{I_{P'}}(x) \chi_{\{x | N(x) \in \omega_{P'}\}} |g(x)|\, dx \right)$$
$$\times \sum_{J \subseteq 3I_T} \left(\sup_{I \supseteq J} \frac{1}{|I|} \int_I \left| \sum_P \varepsilon_P \langle f, \Phi_{P_1} \rangle \Phi_{P_1}(z) \right| dz \right) |I_{P(J)}|. \tag{7.39}$$

Since the first factor on the right-hand side of (7.39) coincides with an expression in (7.33), it is enough to estimate the second factor, namely

$$\sum_{\substack{J \in \mathcal{J} \\ J \subseteq 3I_T}} \left(\sup_{I \supseteq J} \frac{1}{|I|} \int_I \left| \sum_P \varepsilon_P \langle f, \Phi_{P_1} \rangle \Phi_{P_1}(z) \right| dz \right) |J|. \tag{7.40}$$

Remember that by construction we have $|I_{P(J)}| \simeq |J|$. The expression (7.40) can be further estimated by

$$\left\| \sum_{\substack{J \in \mathcal{J} \\ J \subseteq 3I_T}} \left(\sup_{I \supseteq J} \frac{1}{|I|} \int_I \left| \sum_P \varepsilon_P \langle f, \Phi_{P_1} \rangle \Phi_{P_1}(z) \right| dz \right) \chi_J \right\|_{L^1(3I_T)}$$
$$\lesssim \left\| M \left(\sum_P \varepsilon_P \langle f, \Phi_{P_1} \rangle \Phi_{P_1} \right) \right\|_{L^1(3I_T)}$$
$$\lesssim \left\| M \left(\sum_P \varepsilon_P \langle f, \Phi_{P_1} \rangle \Phi_{P_1} \right) \right\|_{L^2} |I_T|^{1/2}$$

$$\lesssim \left\| \sum_P \varepsilon_P \langle f, \Phi_{P_1}\rangle \Phi_{P_1} \right\|_2 |I_T|^{1/2}$$

$$\lesssim \frac{1}{|I_T|^{1/2}} \left(\sum_{P\in T} |\langle f, \Phi_{P_1}\rangle|^2 \right)^{1/2} |I_T|,$$

and this completes the *tree estimates* relating to (7.34). □

7.5. Additional sizes and energies

There are two averaging factors on the right-hand side of (7.33). The first appeared earlier in the bilinear Hilbert transform case. As a consequence, from now on we will denote by $\text{size}((\langle f, \Phi_{P_1}\rangle_{P\in\mathbb{P}})$ and $\text{energy}((\langle f, \Phi_{P_1}\rangle_{P\in\mathbb{P}})$ the same L^2 sizes and L^2 energies as those used in the previous chapter. The only difference is a minor one, namely that $\mathbb{P}$ now is a finite collection of bi-tiles, instead of a sparse rank-1 collection of tri-tiles. But the second averaging factor is new and it suggests the following definitions.

Definition 7.5 Let $\mathbb{P}$ be an arbitrary finite collection of bi-tiles and g a given function. We define

$$\text{size}((\langle \Phi_{P_1}\chi_{\{x|N(x)\in\omega_{P_2}\}}, g\rangle_{P\in\mathbb{P}})$$

$$:= \sup_{P\in\mathbb{P}} \sup_{\substack{P'\in\overline{\mathbb{P}}\\ P'\geq P}} \frac{1}{|I_{P'}|} \int_{\mathbb{R}} \widetilde{\chi}^M_{I_{P'}}(x)\chi_{\{x|N(x)\in\omega_{P'}\}}|g(x)|\,dx.$$

Similarly, we define a quantity that is easier to treat:

$$\text{size}^{\text{e}}((\langle \Phi_{P_1}\chi_{\{x|N(x)\in\omega_{P_2}\}}, g\rangle_{P\in\mathbb{P}}) := \sup_{P\in\mathbb{P}} \frac{1}{|I_P|} \int_{\mathbb{R}} \widetilde{\chi}^M_{I_P}(x)\chi_{\{x|N(x)\in\omega_P\}}|g(x)|\,dx.$$

Then, we have similarly

Definition 7.6 Let $\mathbb{P}$ and g be as before. We define

$$\text{energy}((\langle \Phi_{P_1}\chi_{\{x|N(x)\in\omega_{P_2}\}}, g\rangle_{P\in\mathbb{P}}) := \sup_{n\in\mathbb{Z}} \sup_{\mathbb{D}} 2^n \left(\sum_{P'\in\mathbb{D}} |I_{P'}| \right),$$

where $\mathbb{D}$ ranges over all collections of disjoint bi-tiles $P' \in \overline{\mathbb{P}}$ with the property that there exists $P \in \mathbb{P}$ with $P \leq P'$ and such that

$$\int_{\mathbb{R}} \widetilde{\chi}^M_{I_{P'}}(x)\chi_{\{x|N(x)\in\omega_{P'}\}}|g(x)|\,dx \geq 2^n|I_{P'}|.$$

As in the case of the bilinear Hilbert transform, the following generic proposition will play an important role.

Proposition 7.7 *Let $\mathbb{P}$ be a finite collection of bi-tiles and f and g measurable functions. Then*

$$\left|\int_{\mathbb{R}} C_{\mathbb{P}} f(x) g(x)\,dx\right|$$
$$\lesssim \left(\text{size}(\langle f, \Phi_{P_1}\rangle_{P\in\mathbb{P}})\right)^{\theta_1} \left(\text{size}(\langle \Phi_{P_1}\chi_{\{x|N(x)\in\omega_{P_2}\}}, g\rangle_{P\in\mathbb{P}})\right)^{\theta_2}$$
$$\times \left(\text{energy}(\langle f, \Phi_{P_1}\rangle_{P\in\mathbb{P}})\right)^{1-\theta_1} \left(\text{energy}(\langle \Phi_{P_1}\chi_{\{x|N(x)\in\omega_{P_2}\}}, g\rangle_{P\in\mathbb{P}})\right)^{1-\theta_2}$$

for any $0 \le \theta_1 < 1,\ 0 < \theta_2 \le 1/2$ with $\theta_1 + 2\theta_2 = 1$. Moreover, if the bi-tiles of $\mathbb{P}$ are disjoint then the inequality holds even if one replaces

$$\text{size}\big(\langle \Phi_{P_1}\chi_{\{x|N(x)\in\omega_{P_2}\}}, g\rangle_{P\in\mathbb{P}}\big)$$

by the smaller quantity

$$\text{size}^{\text{e}}\big(\langle \Phi_{P_1}\chi_{\{x|N(x)\in\omega_{P_2}\}}, g\rangle_{P\in\mathbb{P}}\big).$$

This result is analogous to Proposition 6.12 Briefly, the explanation for its lack of symmetry is related to the fact that the new energies are L^1 type while the previous energies were L^2 type. The proof of Proposition 7.7 will be presented later. The following size and energy estimates are clearly necessary for one to be able to use the above inequality.

Lemma 7.8 *Let E be a set of finite measure and $|f| \le \chi_E$. Then*

$$\text{size}\big(\langle \Phi_{P_1}\chi_{\{x|N(x)\in\omega_{P_2}\}}, f\rangle_{P\in\mathbb{P}}\big) \lesssim \sup_{P\in\mathbb{P}} \sup_{\substack{P'\in\overline{\mathbb{P}}\\ P'\ge P}} \frac{\int_E \widetilde{\chi}_{I_{P'}}^M\,dx}{|I_{P'}|}$$

and similarly

$$\text{size}^{\text{e}}\big(\langle \Phi_{P'}\chi_{\{x|N(x)\in\omega_{P_2}\}}, f\rangle_{P\in\mathbb{P}}\big) \lesssim \sup_{P\in\mathbb{P}} \frac{\int_E \widetilde{\chi}_{I_P}^M\,dx}{|I_P|}.$$

Lemma 7.9 *Let $f \in L^1(\mathbb{R})$ and let $\mathbb{P}$ be a finite collection of bi-tiles. Then*

$$\text{energy}\left(\left\langle \Phi_{P_1}\chi_{\{x|N(x)\in\omega_{P_2}\}}, f\right\rangle_{P\in\mathbb{P}}\right) \lesssim \|f\|_1.$$

The proof of Lemma 7.8 is straightforward. The proof of Lemma 7.9 requires a detailed analysis and will be postponed until the end of the chapter.

Using all this information, we will complete the proof of Theorem 7.2 in the next section.

7.6. Proof of Theorem 7.2

Let E_1, E_2 be sets of finite measure and let $1 < q < 2$. We will show that there exists $E_2' \subseteq E_2$, $|E_2'| \simeq |E_2|$, such that

$$\left| \int_{\mathbb{R}} C_{\mathbb{P}}(f_1)(x) f_2(x)\, dx \right| \lesssim |E_1|^{1/q} |E_2|^{1/q'} \tag{7.41}$$

for every $|f_1| \leq \chi_{E_1}$ and $|f_2| \leq \chi_{E_2'}$ and also that there exists $E_1' \subseteq E_1$, $|E_1'| \simeq |E_1|$, such that

$$\left| \int_{\mathbb{R}} C_{\mathbb{P}}(f_1)(x) f_2(x)\, dx \right| \lesssim |E_2| \tag{7.42}$$

for every $|f_1| \leq \chi_{E_1'}$ and $|f_2| \leq \chi_{E_2}$. If we can show both (7.41) and (7.42) then Theorem 7.2 follows immediately from Marcinkiewicz's interpolation theorem.

To prove (7.41), assume for simplicity and without loss of generality that $|E_2| = 1$ (by scaling invariance). Define Ω by

$$\Omega := \left\{ x \,\middle|\, M\left(\frac{\chi_{E_1}}{|E_1|}\right)(x) > C \right\},$$

where M is the Hardy–Littlewood maximal function.

Clearly, if $C > 0$ is a large constant then one has that $|\Omega| < 1/10$ and, as a consequence, $E_2' := E_2 \backslash \Omega$ has a measure comparable with the measure of E_2. Fix $|f_1| \leq \chi_{E_1}$ and $|f_2| \leq \chi_{E_2'}$. We need to prove that

$$\left| \int_{\mathbb{R}} C_{\mathbb{P}}(f_1)(x) f_2(x)\, dx \right| \lesssim |E_1|^{1/q}.$$

The left-hand side can be majorized by

$$\sum_{P \in \mathbb{P}} \left|\langle f_1, \Phi_{P_1} \rangle\right| \left|\langle \Phi_{P_1} \chi_{\{x | N(x) \in \omega_{P_2}\}}, f_2 \rangle\right|. \tag{7.43}$$

Then we decompose $\mathbb{P}$ as $\bigcup_{d \geq 0} \mathbb{P}_d$, where

$$\mathbb{P}_d := \left\{ P \in \mathbb{P} \,\middle|\, 1 + \frac{\operatorname{dist}(I_P, \Omega^c)}{|I_P|} \simeq 2^d \right\}.$$

In particular this allows one to rewrite (7.43) as

$$\sum_{d \geq 0} \sum_{P \in \mathbb{P}_d} \left|\langle f_1, \Phi_{P_1} \rangle\right| \left|\langle \Phi_{P_1} \chi_{\{x | N(x) \in \omega_{P_2}\}}, f_2 \rangle\right|. \tag{7.44}$$

Fix $d \geq 0$. For every $P \in \mathbb{P}_d$ observe that

$$\frac{\int_{E_1} \widetilde{\chi}_{I_P}^M \, dx}{|I_P|} \lesssim 2^d |E_1|$$

and as a consequence of the size and energy estimates from Lemmas 6.13 and 6.14 we know that

$$\text{size}(\langle f_1, \Phi_{P_1} \rangle_{P \in \mathbb{P}_d}) \lesssim 2^d |E_1| \tag{7.45}$$

and also that

$$\text{energy}(\langle f_1, \Phi_{P_1} \rangle_{P \in \mathbb{P}_d}) \lesssim |E_1|^{1/2}.$$

Then, for small $d \geq 0$ we have

$$\text{size}\big(\langle \Phi_{P_1} \chi_{\{x|N(x) \in \omega_{P_2}\}}, f_2 \rangle_{P \in \mathbb{P}_d}\big) \lesssim 1$$

while for large d one observes that

$$\text{size}^{\text{e}}\big(\langle \Phi_{P_1} \chi_{\{x|N(x) \in \omega_{P_2}\}}, f_2 \rangle_{P \in \mathbb{P}_d}\big) \lesssim 2^{-Md}$$

for any $M > 0$, taking advantage of the fact that f_2 is supported inside Ω^c. However, by Lemma 7.9 we also know that

$$\text{energy}\big(\langle \Phi_{P_1} \chi_{\{x|N(x) \in \omega_{P_2}\}}, f_2 \rangle_{P \in \mathbb{P}_d}\big) \lesssim 1.$$

Observing that when d is large the bi-tiles $P \in \mathbb{P}_d$ are essentially disjoint and also applying Proposition 7.7, one can estimate

$$\left| \int_{\mathbb{R}} C_{\mathbb{P}}(f_1)(x) f_2(x) \, dx \right|$$

by

$$\left(2^d |E_1|\right)^{\theta_1} (2^{-Md})^{\theta_2} |E_1|^{(1-\theta_1)/2} = 2^{-d(M\theta_2 - \theta_1)} |E_1|^{(1+\theta_1)/2}$$

for any $\theta_1 \in (0, 1)$. If we now choose θ_1 such that $(1 + \theta_1)/2 = 1/q$ and M large enough, we obtain an upper bound of the type $|E_1|^{1/q}$ as desired.

To prove the inequality (7.42) assume, again using scaling invariance, that $|E_1| = 1$ and define E_1' similarly by discarding an exceptional set depending on E_2. Split $\mathbb{P}$ as earlier and remark that this time one can estimate the sizes and energies as follows:

$$\begin{aligned}
\operatorname{size}\big((\langle f_1, \Phi_{P_1}\rangle_{P\in\mathbb{P}_d}\big) &\lesssim 2^{-Md},\\
\operatorname{energy}\big((\langle f_1, \Phi_{P_1}\rangle_{P\in\mathbb{P}_d}\big) &\lesssim 1,\\
\operatorname{size}\left(\left(\langle \Phi_{P_1}\chi_{\{x|N(x)\in\omega_{P_2}\}}, f_2\rangle_{P\in\mathbb{P}_d}\right)\right) &\lesssim 2^d|E_2|,\\
\operatorname{energy}\left(\left(\langle \Phi_{P_1}\chi_{\{x|N(x)\in\omega_{P_2}\}}, f_2\rangle_{P\in\mathbb{P}_d}\right)\right) &\lesssim |E_2|.
\end{aligned}$$

Finally, Proposition 7.7 allows us to estimate the left-hand side of (7.42) by

$$\sum_{d\geq 0}\left(2^{-Md}\right)^{\theta_1}\left(2^d|E_2|\right)^{\theta_2}|E_2|^{1-\theta_2} = \sum_{d\geq 0} 2^{-d(M\theta_1-\theta_2)}|E_2|,$$

which is summable and bounded by $|E_2|$, as desired, if M is once again chosen large enough.

We also observe that the above proof of (7.42) can be applied to any $f_2 \in L^1(\mathbb{R})$, not just to functions for which $|f_2| \leq \chi_{E_2}$. However, the difference is that now one defines the corresponding exceptional set Ω by

$$\Omega = \left\{x \,\middle|\, M\left(\frac{f_2}{\|f_2\|_1}\right)(x) > C\right\}.$$

In particular, we obtain the following result.

Theorem 7.10 *The adjoint Carleson operator $C^*_{\mathbb{P}}$ maps L^1 into $L^{1,\infty}$ boundedly.*

This is an interesting fact since it is well known that the operator $C_{\mathbb{P}}$ itself does not have this property. The reader may remember the L^1 counterexample of the almost everywhere convergence of Fourier series from the first volume of the book. These issues are clearly related since if $C_{\mathbb{P}}$ were bounded from L^1 into $L^{1,\infty}$ then Fourier series of L^1-periodic functions would converge almost everywhere.

We are left with the proofs of Lemma 7.9 and Proposition 7.7.

7.7. Estimates for Carleson energies

We start with the proof of Lemma 7.9. Fix $\mathbb{P}$ as finite and $f \in L^1(\mathbb{R})$. Pick $n \in \mathbb{Z}$ and $\mathbb{D}$ for which the supremum in Definition 7.6 is attained. Then

$$\begin{aligned}
\operatorname{energy}\big((\langle \Phi_{P_1}\chi_{\{x|N(x)\in\omega_{P_2}\}}, f\rangle_{P\in\mathbb{P}}\big) &= 2^n\left(\sum_{P'\in\mathbb{D}}|I_{P'}|\right)\\
&\leq \sum_{P'\in\mathbb{D}}\int_{\mathbb{R}}\widetilde{\chi}^M_{I_{P'}}(x)\chi_{\{x|N(x)\in\omega_{P'}\}}|f(x)|\,dx.
\end{aligned} \tag{7.46}$$

Now, if the bump functions $\widetilde{\chi}^M_{I_{P'}}$ were actually supported on $I_{P'}$ then the disjointness property of the bi-tiles in $\mathbb{D}$ would imply that the sets $\{x \in I_{P'} | N(x) \in \omega_{P'}\}$ are all disjoint, which would immediately allow us to estimate (7.46) by $\|f\|_1$, as desired. However, the $\widetilde{\chi}^M_{I_{P'}}$ are only adapted to $I_{P'}$ (there are tails decreasing rapidly away from $I_{P'}$) and so we need to exercise care.

Fix $P' \in \mathbb{D}$. Rewrite each $\widetilde{\chi}^M_{I_{P'}}$ as follows:

$$\widetilde{\chi}^M_{I_{P'}} = \sum_{k=0}^{\infty} 2^{-100k} \widetilde{\chi}_{2^k I_{P'}}, \tag{7.47}$$

where the $\widetilde{\chi}_{2^k I_{P'}}$ are still bump functions adapted to $I_{P'}$ but having the additional property that $\operatorname{supp}(\widetilde{\chi}_{2^k I_{P'}}) \subseteq 2^k I_{P'}$. Thus we can write

$$\begin{aligned}
&\frac{1}{|I_{P'}|} \int_{\mathbb{R}} \widetilde{\chi}^M_{I_{P'}}(x) \chi_{\{x|N(x) \in \omega_{P'}\}} |f(x)|\, dx \\
&= \sum_{k=0}^{\infty} 2^{-100k} \frac{1}{|I_{P'}|} \int_{\mathbb{R}} \widetilde{\chi}_{2^k I_{P'}}(x) \chi_{\{x|N(x) \in \omega_{P'}\}} |f(x)|\, dx \\
&= \sum_{k=0}^{\infty} 2^{-99k} \frac{1}{|2^k I_{P'}|} \int_{\mathbb{R}} \widetilde{\chi}_{2^k I_{P'}}(x) \chi_{\{x|N(x) \in \omega_{P'}\}} |f(x)|\, dx,
\end{aligned}$$

and we know that this is larger than 2^n. In particular, there must exist an integer k such that

$$\frac{1}{|2^k I_{P'}|} \int_{\mathbb{R}} \widetilde{\chi}_{2^k I_{P'}}(x) \chi_{\{x|N(x) \in \omega_{P'}\}} |f(x)|\, dx > c 2^{5k} 2^n \tag{7.48}$$

for a fixed but small $c > 0$, since otherwise the above statement would be contradicted.

Denote by $\mathbb{P}_k$ the set of all $P' \in \mathbb{D}$ for which (7.48) holds. Clearly, every $P' \in \mathbb{D}$ belongs to one of these sets $\mathbb{P}_k$. In particular,

$$2^n \left(\sum_{P' \in \mathbb{D}} |I_{P'}| \right) \le \sum_{k=0}^{\infty} 2^n \left(\sum_{P' \in \mathbb{P}_k} |I_{P'}| \right)$$

and so it would be enough to prove that, for each k, one has

$$2^n \left(\sum_{P' \in \mathbb{P}_k} |I_{P'}| \right) \lesssim 2^{-k} \|f\|_1. \tag{7.49}$$

To show (7.49) we need to reorganize the set $\mathbb{P}_k$ somewhat. For every $P' \in \mathbb{P}_k$, write $2^k P' := 2^k I_{P'} \times \omega_{P'}$ and call this the 2^k *enlargement* of P'.

We select $P' \in \mathbb{P}_k$ as follows. First, pick $P' \in \mathbb{P}_k$ such that $|I_{P'}|$ is as large as possible. If there are many bi-tiles having the same maximal length, we just pick one and place it in a subset of $\mathbb{P}_k$ called $\mathbb{P}_k^{\text{select}}$. Then, we look at the remaining bi-tiles and pick another one such that its 2^k enlargement is disjoint from the 2^k enlargements of all the previously selected bi-tiles. This process is continued until it ends (and it must end since $\mathbb{D}$ and $\mathbb{P}$ are finite sets).

Observe that by construction (and by using again the disjointness of the bi-tiles in $\mathbb{D}$) one has that

$$2^n \left(\sum_{P' \in \mathbb{P}_k} |I_{P'}| \right) \lesssim 2^n 2^k \left(\sum_{P' \in \mathbb{P}_k^{\text{select}}} |I_{P'}| \right) \tag{7.50}$$

and, since the bi-tiles $2^k P'$ are all disjoint for $P' \in \mathbb{P}_k^{\text{select}}$, one can use (7.48) and estimate the above expression by

$$2^{-5k} \sum_{P' \in \mathbb{P}_k^{\text{select}}} \int_{\mathbb{P}} \widetilde{\chi}_{2^k I_{P'}}(x) \chi_{\{x | N(x) \in \omega_{P'}\}} |f(x)| \, dx \lesssim 2^{-5k} \|f\|_1,$$

as desired.

Exercise 7.5 Check the estimate (7.50).

7.8. Stopping-time decompositions

To prove Proposition 7.7 we argue in essentially the same way as in Chapter 6 when we proved the analogous Proposition 6.12. We will treat only the case of a generic collection of bi-tiles $\mathbb{P}$ and leave the easier situation when all the bi-tiles in $\mathbb{P}$ are disjoint (and one can work with size^{e} instead) to the reader.

As in the proof of Proposition 6.12, the argument is based on the following result, which corresponds to Proposition 6.15.

Proposition 7.11 *Let $\mathbb{P}$ be a finite collection of bi-tiles and let $\mathbb{P}' \subseteq \mathbb{P}$. Also let $n \in \mathbb{Z}$ and f be a measurable function and assume that*

$$\text{size}\big(\langle \Phi_{P_1} \chi_{\{x | N(x) \in \omega_{P_2}\}}, f \rangle_{P \in \mathbb{P}'}\big) \leq 2^{-n} \, \text{energy}\big(\langle \Phi_{P_1} \chi_{\{x | N(x) \in \omega_{P_2}\}}, f \rangle_{P \in \mathbb{P}}\big).$$

Then there exists a decomposition of $\mathbb{P}'$ as $\mathbb{P}' = \mathbb{P}'' \cup \mathbb{P}'''$ such that

$$\text{size}\big(\langle \Phi_{P_1} \chi_{\{x | N(x) \in \omega_{P_2}\}}, f \rangle_{P \in \mathbb{P}''}\big) \leq 2^{-n-1} \, \text{energy}\big(\langle \Phi_{P_1} \chi_{\{x | N(x) \in \omega_{P_2}\}}, f \rangle_{P \in \mathbb{P}}\big) \tag{7.51}$$

and also such that $\mathbb{P}'''$ can be written as a disjoint union of trees $T \in \mathbb{T}$ for which

$$\sum_{T \in \mathbb{T}} |I_T| \lesssim 2^n. \tag{7.52}$$

Proof First, consider all bi-tiles $P' \in \overline{\mathbb{P}}$ for which there exists $P \in \mathbb{P}$ with $P \leq P'$ and such that

$$\begin{aligned} &\frac{1}{|I_{P'}|} \int_{\mathbb{R}} \widetilde{\chi}_{I_{P'}}^M(x) \chi_{\{x | N(x) \in \omega_{P'}\}} |f(x)| \, dx \\ &\geq \frac{2^{-n}}{2} \operatorname{energy} \left(\left\langle \Phi_{P_1} \chi_{\{x | N(x) \in \omega_{P_2}\}}, f \right\rangle_{P \in \mathbb{P}} \right). \end{aligned} \tag{7.53}$$

Among all these bi-tiles P', pick those that are also maximal with respect to the partial order $\leq$. Clearly, they will be in particular disjoint. Then, the set of tiles $P \in \mathbb{P}$ for which there is such a P' with $P' \geq P$ can be organized as a union of trees $T \in \mathbb{T}$ with tops P'. Define $\mathbb{P}'''$ to be precisely the collection of these bi-tiles belonging to some $T \in \mathbb{T}$. As a consequence, $\mathbb{P}''$ is defined to be the complement of $\mathbb{P}'''$ in $\mathbb{P}'$.

Clearly (7.51) is satisfied now, and we need only to check (7.52). Using (7.53) and the disjointness of the maximal bi-tiles P' satisfying (7.53), we can write

$$\begin{aligned} &\frac{2^{-n}}{2} \operatorname{energy} \left(\left\langle \Phi_{P_1} \chi_{\{x | N(x) \in \omega_{P_2}\}}, f \right\rangle_{P \in \mathbb{P}} \right) \left(\sum_T |I_T| \right) \\ &\leq \operatorname{energy} \left(\left\langle \varphi_{P_1} \chi_{\{x | N(x) \in \omega_{P_2}\}}, f \right\rangle_{P \in \mathbb{P}} \right) \end{aligned} \tag{7.54}$$

by also using the definition of the energy. However, (7.54) reduces to (7.52) and so the proof of Proposition 7.11 is complete. □

If we carefully iterate Proposition 6.15 (applied this time to the case of our bi-tiles) and also Proposition 7.11, we obtain the following consequence.

Corollary 7.12 *Let $\mathbb{P}$ be a finite collection of bi-tiles. Then there exists a partition*

$$\mathbb{P} = \bigcup_{n \in \mathbb{Z}} \mathbb{P}_n$$

with the property that, for every $n \in \mathbb{Z}$, one has

$$\operatorname{size} \left(\langle f_1, \Phi_{P_1} \rangle_{P \in \mathbb{P}_n} \right) \leq \min \left(2^{-n} E_1, S_1 \right)$$

and

$$\text{size}\left(\left\langle \Phi_{P_1}\chi_{\{x|N(x)\in\omega_{P_2}\}}, f_2\right\rangle_{P\in\mathbb{P}_n}\right) \le \min\left(2^{-2n}E_2, S_2\right)$$

and either

$$2^{-n-1}E_1 \le \text{size}\left(\langle f_1, \Phi_{P_1}\rangle_{P\in\mathbb{P}_n}\right) \le \min\left(2^{-n}E_1, S_1\right)$$

or

$$2^{-2n-2}E_2 \le \text{size}\left(\left\langle \Phi_{P_1}\chi_{\{x|N(x)\in\omega_{P_2}\}}, f_2\right\rangle_{P\in\mathbb{P}_n}\right) \le \min\left(2^{-2n}E_2, S_2\right)$$

and such that $\mathbb{P}_n$ can be decomposed as a union of trees T in $\mathbb{T}_n$ for which

$$\sum_{T\in\mathbb{T}_n} |I_T| \lesssim 2^{2n}.$$

Clearly, S_1, E_1, S_2, E_2 in the above corollary denote the corresponding sizes and energies of f_1 and f_2 with respect to the entire collection $\mathbb{P}$.

7.9. Generic estimate of the bilinear Carleson form

We are finally ready to describe the proof of Proposition 7.7.

Recall that we want to show that

$$\left|\int_{\mathbb{R}} C_{\mathbb{P}}(f_1)(x)(f_2)(x)\,dx\right| \lesssim \prod_{j=1}^{2} S_j^{\theta_j} E_j^{1-\theta_2} = E_1E_2\left(\frac{S_1}{E_1}\right)^{\theta_1}\left(\frac{S_2}{E_2}\right)^{\theta_2} \tag{7.55}$$

holds for any $0 \le \theta_1 < 1$ and $0 < \theta_2 \le 1/2$ with $\theta_1 + 2\theta_2 = 1$. The idea is to use Corollary 7.12. Observe that by a proper rescalling the indices for which $\mathbb{P}_n$ is not empty must satisfy $2^{-n} \le \max(S_1/E_1, (S_2/E_2)^{1/2})$. We need to consider several cases.

Case 1: $S_1/E_1 \le (S_2/E_2)^{1/2}$. There are two possibilities here.

Case 1_a: the power n has the property that $S_1/E_1 \le 2^{-n} \le (S_2/E_2)^{1/2}$. By using Corollary 7.12 together with the earlier tree estimate, we can bound the left-hand side of (7.55) by

$$E_1E_2\sum_n\left(\frac{S_1}{E_1}\right)2^{-2n}\left(\sum_{T\in\mathbb{T}_n}|I_T|\right)$$

$$\lesssim E_1E_2\sum_n\left(\frac{S_1}{E_1}\right)2^{-2n}2^{2n} = E_1E_2\left(\frac{S_1}{E_1}\right)\left(\sum_n 1\right). \tag{7.56}$$

Now we need to count how many indices n can have the case 1_a property. Since n is between $\log(E_2/S_2)^{1/2}$ and $\log(E_1/S_1)$ there are at most

$$\log \frac{(E_1/S_1)}{(E_2/S_2)^{1/2}}$$

such indices n. This means that (7.56) is smaller than

$$E_1 E_2 \left(\frac{S_1}{E_1}\right) \log \frac{(E_1/S_1)}{(E_2/S_2)^{1/2}} \lesssim E_1 E_2 \left(\frac{S_1}{E_1}\right) \frac{(E_1/S_1)^{\varepsilon}}{(E_2/S_2)^{\varepsilon/2}}$$
$$= E_1 E_2 \left(\frac{S_1}{E_1}\right)^{1-\varepsilon} \left(\frac{S_2}{E_2}\right)^{\varepsilon/2},$$

which proves (7.55) on choosing $\varepsilon > 0$ appropriately.

Case 1_b: the power n has the property that $2^{-n} \leq S_1/E_1 \leq (S_2/E_2)^{1/2}$. This time one bounds the left-hand side of (7.55) by

$$E_1 E_2 \sum_n 2^{-n} 2^{-2n} \sum_{T \in \mathbb{T}_n} |I_T| \lesssim E_1 E_2 \sum_n 2^{-n} 2^{-2n} 2^{2n}$$
$$= E_1 E_2 \sum_n 2^{-n} \lesssim E_1 E_2 \left(\frac{S_1}{E_1}\right).$$

We now just need to check that

$$\frac{S_1}{E_1} \lesssim \left(\frac{S_1}{E_1}\right)^{\theta_1} \left(\frac{S_2}{E_2}\right)^{\theta_2}$$

or equivalently that

$$\left(\frac{S_1}{E_1}\right)^{1-\theta_1} \lesssim \left(\frac{S_2}{E_2}\right)^{\theta_2}.$$

This is clear since we are in Case 1 and $\theta_1 + 2\theta_2 = 1$.

Case 2: $(S_2/E_2)^{\frac{1}{2}} \leq S_1/E_1$. There are again two possibilities.

Case 2_a: the power n has the property that $(S_2/E_2)^{1/2} \leq 2^{-n} \leq S_1/E_1$. In this case the left-hand side of (7.55) is smaller than

$$E_1 E_2 \sum_n 2^{-n} \left(\frac{S_2}{E_2}\right) \left(\sum_{T \in \mathbb{T}} |I_T|\right) \lesssim E_1 E_2 \left(\frac{S_2}{E_2}\right) \left(\sum_n 2^n\right)$$

and, since $2^n \leq (E_2/S_2)^{1/2}$, this is smaller than

$$E_1 E_2 \left(\frac{S_2}{E_2}\right) \left(\frac{E_2}{S_2}\right)^{1/2} = E_1 E_2 \left(\frac{S_2}{E_2}\right)^{1/2}. \tag{7.57}$$

Now we just need to observe that

$$\left(\frac{S_2}{E_2}\right)^{1/2} \leq \left(\frac{S_1}{E_1}\right)^{\theta_1} \left(\frac{S_2}{E_2}\right)^{\theta_2}, \tag{7.58}$$

which again follows easily from the fact that we are in case 2 and $\theta_1 + 2\theta_2 = 1$.

Case 2_b*:* the power n has the property that $2^{-n} \leq (S_2/E_2)^{1/2} \leq S_1/E_1$. Here, the left-hand side of (7.55) can be estimated by

$$E_1 E_2 \sum_n 2^{-n} 2^{-2n} \sum_{T \in \mathbb{T}_n} |I_T| \lesssim E_1 E_2 \sum_n 2^{-n} \lesssim E_1 E_2 \left(\frac{S_2}{E_2}\right)^{1/2},$$

which is the same as (7.57). This ends our proof.

7.10. Fefferman's counterexample

Suppose now that $f(x, y)$ is a function of two variables that is 2π-periodic in each variable separately. Assume also that $f \in L^p([0, 2\pi] \times [0, 2\pi])$ for some $1 < p < \infty$. We denote by $S_{MN} f(x, y)$ the partial sum of the *double Fourier series* of $f(x, y)$, given by

$$S_{MN} f(x, y) = \sum_{\substack{-M \leq m \leq M \\ -N \leq n \leq N}} \widehat{f}(m, n) e^{imx} e^{iny},$$

where the Fourier coefficients are defined by

$$\widehat{f}(m, n) = \frac{1}{2\pi} \frac{1}{2\pi} \int_0^{2\pi} \int_0^{2\pi} f(x, y) e^{-imx} e^{-iny} \, dx dy.$$

A very natural question is whether

$$S_{MN} f(x, y) \xrightarrow[M,N \to \infty]{} f(x, y) \tag{7.59}$$

for almost every $(x, y) \in [0, 2\pi] \times [0, 2\pi]$.

The goal of this section is to describe a counterexample of Fefferman, which provides a negative answer to the above question. As we will see, this counterexample is essentially identical to that which disproved the boundedness of the bilinear Hilbert transform on polydisk, in Chapter 6.

Let us denote by Q the square $\{(x, y) \mid 1/10 \leq x, y \leq 2\pi - 1/10\}$ and let λ be a large positive real number. If f_λ denotes the function $f_\lambda(x, y) = e^{i\lambda xy}$ we will prove that for every $(x, y) \in Q$ there exist M and N such that $|S_{MN} f_\lambda(x, y)|$ is as large as $\log \lambda$.

To understand why this is true we need to recall the relationship between the partial sums of a one-variable Fourier series, the harmonic conjugate, and the Hilbert transform that we used at the beginning of the chapter. If one takes a tensor product of those calculations then one obtains an analogous formula for $S_{MN}f$, namely

$$S_{MN}f(x, y) \sim \frac{1}{\pi^2}(T_{MN}f(x, y) - T_{-MN}f(x, y) - T_{M-N}f(x, y) + T_{-M-N}f(x, y)), \tag{7.60}$$

where, by definition,

$$T_{ab}f(x, y) = e^{iax+by}\,p.v. \int_0^{2\pi}\int_0^{2\pi} \frac{e^{-i(ax'+by')}}{(x-x')(y-y')} f(x', y')\,dx'dy'$$

for all real numbers a, b and where the $\sim$ sign means equality modulo some error terms that are uniformly bounded for $f = f_\lambda$ and $(x, y) \in Q$, as one can easily check.

Exercise 7.6 Check the formula (7.60) and the uniform boundedness of the corresponding error terms.

Now define $M = [\lambda y]$ and $N = [\lambda x]$ where in general $[z]$ denotes the greatest integer smaller than z. We claim that the main contribution in (7.60) comes from the first term $T_{MN}f_\lambda(x, y)$. To see this, let us first replace M and N by $M = \lambda y$ and $N = \lambda x$ respectively, since this alters $T_{MN}f_\lambda(x, y)$ by at most a uniformly bounded quantity. Then, one can write

$$\begin{aligned} |T_{\lambda y \lambda x} f_\lambda(x, y)| &= \left| \int_0^{2\pi}\int_0^{2\pi} \frac{e^{-i(\lambda y x' + \lambda x y')}}{(x-x')(y-y')} e^{i\lambda x' y'}\,dx'dy' \right| \\ &= \left| \int_0^{2\pi}\int_0^{2\pi} \frac{e^{-i\lambda(x-x')(y-y')}}{(x-x')(y-y')}\,dx'dy' \right| = \left| \int_{I_x}\int_{I_y} \frac{e^{-i\lambda st}}{st}\,dsdt \right| \\ &= \left| \int_{I_x^\lambda}\int_{I_y^\lambda} \frac{e^{ist}}{st}\,dsdt \right|, \end{aligned}$$

where the intervals I_x^λ and I_y^λ have the property that they both contain $[-\sqrt{\lambda}/100, \sqrt{\lambda}/100]$ as a consequence of the fact that $(x, y) \in Q$. In particular, one can see from the above that

$$|T_{[\lambda y][\lambda x]} f_\lambda(x, y)| \geq C_1 \left| \int_{-\sqrt{\lambda}/100}^{\sqrt{\lambda}/100}\int_{-\sqrt{\lambda}/100}^{\sqrt{\lambda}/100} \frac{e^{ist}}{st}\,dsdt \right| \geq C_2 \log\lambda, \tag{7.61}$$

as we showed in Chapter 6. One can treat the other terms in (7.60) similarly but they all turn out to be uniformly bounded, since the perfect combination of signs that gave rise to (7.61) does not occur this time. This proves our logarithmic claim.

Finally, to obtain the desired counterexample to the almost everywhere convergence of a double Fourier series, one just needs to take an infinite series of the type

$$f = \sum_{k=1}^{\infty} c_k f_{\lambda_k}. \tag{7.62}$$

If on the one hand the coefficients c_k decay exponentially (say $c_k = 1/2^k$) then the series is absolutely convergent and f is bounded and even continuous. If, on the other hand, $\log \lambda_k$ grows in a double exponential way (say $\log \lambda_k = 2^{2^k}$) then the above calculations guarantee that $\limsup_{M,N} |S_{MN} F(x, y)| = \infty$ for every $(x, y) \in Q$, and this completes the argument.

Exercise 7.7 Fill in the missing steps in the above argument.

Notes

The L^2 case of the main theorem of the chapter was proved by Carleson [12], answering the so-called Lusin conjecture. The L^p case for every $1 < p < \infty$ is due to Hunt [56]. Other significant papers on the subject include Billard [8], Fefferman [40], Sjölin [104], and Lacey and Thiele [72]. The presentation here follows closely Muscalu, Tao, and Thiele [95] which also contains the new result in Theorem 7.10. The bi-disk counterexample is in Fefferman [39]. More recent extensions include Li and Muscalu [75], Oberlin, Seeger, Tao *et al.* [98], and Lie [77].

Problems

Problem 7.1 Let f be a 2π-periodic function in $L^p([0, 2\pi])$ for $1 < p < \infty$. Prove directly (without using Theorem 7.1) that one has the following convergence of partial Fourier sums along a lacunary sequence:

$$\sum_{n=-2^N}^{2^N} \widehat{f}(n) e^{in\theta} \xrightarrow[N\to\infty]{} f(\theta)$$

for almost every $\theta \in [0, 2\pi]$.

Problem 7.2 Let m be a classical symbol. Prove that the maximal operator

$$f \mapsto \sup_N \left| \int_{\mathbb{R}} m(\xi - N) \widehat{f}(\xi) e^{2\pi i x \xi} \, d\xi \right|$$

is bounded from L^p into L^p for every $1 < p < \infty$.

Hint: Obtain a translation-invariant Littlewood–Paley decomposition of the symbol m, similar to that described in Section 7.2 for $m(\xi) = \chi_{(-\infty,0]}(\xi)$.

Problem 7.3 It is clear that Carleson's maximal operator $C_{\mathbb{R}}$ does not map L^∞ into L^∞ since even the Hilbert transform does not have this property as we saw in Chapter I.3. However, there are particular cases of bounded functions $f(x)$ for which $C_{\mathbb{R}} f(x)$ remains bounded. Show that $f(x) = e^{2\pi i x^2}$ is such a function.

8

Flag paraproducts

An interesting heuristical fact that we learned in the last two chapters is that the maximal Carleson operator $C_{\mathbb{R}}$ can be thought of as being an *infinite sum of maximal Hilbert transforms* and similarly the bilinear Hilbert transform BHT can be thought of as being an *infinite sum of paraproducts*. There is an alternative terminology that is commonly used when people refer to this fact. It is said that the lacunary version of the Carleson maximal operator is the maximal Hilbert transform while the lacunary version of the bilinear Hilbert transform is a paraproduct. The word lacunary comes from the lacunary dyadic frequency intervals that appear naturally in classical Littlewood–Paley decompositions.

Let us recall now the trilinear operator T_3 introduced in Chapter 5:

$$T_3(f_1, f_2, f_3)(x) = \int_{\xi_1<\xi_2<\xi_3} \widehat{f_1}(\xi_1)\widehat{f_2}(\xi_2)\widehat{f_3}(\xi_3)e^{2\pi i x(\xi_1+\xi_2+\xi_3)}\, d\xi_1 d\xi_2 d\xi_3. \tag{8.1}$$

It is the continuous analogue of an *iterated Fourier series* coming from a product of three functions.

One might ask, what is its lacunary version? A simple way to pass from the bilinear Hilbert transform to its lacunary version is to replace its symbol $\chi_{\{\xi_1<\xi_2\}}$ by a classical symbol $m(\xi_1, \xi_2)$. In the case of T_3 the symbol $\chi_{\{\xi_1<\xi_2<\xi_3\}}$ splits into $\chi_{\{\xi_1<\xi_2\}}\chi_{\{\xi_2<\xi_3\}}$ and so, to obtain its lacunary version, one would need to replace both $\chi_{\{\xi_1<\xi_2\}}$ and $\chi_{\{\xi_2<\xi_3\}}$ by classical symbols $a(\xi_1, \xi_2)$ and $b(\xi_2, \xi_3)$, thus obtaining the trilinear operator

$$\begin{aligned} &T_{ab}(f_1, f_2, f_3)(x) \\ &= \int_{\mathbb{R}^3} a(\xi_1, \xi_2)b(\xi_2, \xi_3)\widehat{f_1}(\xi_1)\widehat{f_2}(\xi_2)\widehat{f_3}(\xi_3)e^{2\pi i x(\xi_1+\xi_2+\xi_3)}\, d\xi_1 d\xi_2 d\xi_3. \end{aligned} \tag{8.2}$$

This operator is no longer a paraproduct; it is an example of a *flag paraproduct* and the goal of the present chapter is to study its boundedness properties. The previous heuristics continue to hold, since it turns out that T_3 should also be thought of as *an infinite sum of flag paraproducts*. However, we cannot describe the analysis of T_3 and its generalizations in the present volume. The interested reader is invited to consult the original papers dedicated to these interesting objects. We shall concentrate instead on the above case of T_{ab}. As we will see these flag paraproducts appear quite naturally in nonlinear PDEs.

8.1. Generic flag paraproducts

From now on let us denote by $\mathcal{M}(\mathbb{R}^n)$ the set of all classical Marcinkiewicz–Mikhlin–Hörmander symbols in $\mathbb{R}^n$ and by $\mathcal{M}_{\text{flag}}(\mathbb{R}^n)$ the set of all symbols m given by arbitrary products of the form

$$m(\xi) := \prod_{S\subseteq\{1,\ldots,n\}} m_S(\xi_S), \tag{8.3}$$

where $m_S \in \mathcal{M}(\mathbb{R}^{\text{card(S)}})$, the vector $\xi_S \in \mathbb{R}^{\text{card(S)}}$ is defined by $\xi_S := (\xi_i)_{i\in S}$, and $\xi \in \mathbb{R}^n$ is the vector $\xi := (\xi_i)_{i=1}^n$. Every such symbol $m \in \mathcal{M}_{\text{flag}}(\mathbb{R}^n)$ defines a natural n-linear operator:

$$T_m(f_1, \ldots, f_n)(x) = \int_{\mathbb{R}^n} m(\xi)\widehat{f_1}(\xi_1)\cdots\widehat{f_n}(\xi_n)e^{2\pi i x(\xi_1+\cdots+\xi_n)}\,d\xi_1\ldots d\xi_n. \tag{8.4}$$

Operators of the type (8.4) are called flag paraproducts. The word flag comes from linear algebra. A flag of subspaces of a vector spaces V is simply a sequence of subspaces satisfying $\{0\} = V_0 \subseteq V_1 \subseteq \cdots \subseteq V_k = V$. It is easy to see that a generic symbol $m \in \mathcal{M}_{\text{flag}}(\mathbb{R}^n)$ is singular along every possible flag of subspaces spanned by the coordinate system of $\mathbb{R}^n$. By comparison, recall that a classical paraproduct is defined by a single symbol $m \in \mathcal{M}(\mathbb{R}^n)$ and this is singular only at the origin.

Flag paraproducts have a natural kernel representation. For instance, our operator T_{ab} in (8.2) can also be written as

$$\begin{aligned} &T_{ab}(f_1, f_2, f_3)(x) \\ &= p.v.\int_{\mathbb{R}^4} f_1(x-t_1)f_2(x-t_2-s_1)f_3(x-s_2)K(t_1,t_2)\widetilde{K}(s_1,s_2)\,dtds, \end{aligned} \tag{8.5}$$

where K and $\widetilde{K}$ are classical Calderón–Zygmund kernels in the plane. More precisely, one has $\widehat{K} = a$ and $\widehat{\widetilde{K}} = b$.

Exercise 8.1 Check formula (8.5).

The goal of this chapter is to study in detail the trilinear operator T_{ab}, which is the simplest flag paraproduct whose complexity goes beyond that of a classical Coifman–Meyer paraproduct. More precisely, we will prove the following result.

Theorem 8.1 *The operator T_{ab} maps $L^{p_1} \times L^{p_2} \times L^{p_3}$ into L^p boundedly for every $1 < p_1, p_2, p_3 < \infty$ with $1/p_1 + 1/p_2 + 1/p_3 = 1/p$ and $0 < p < \infty$. In addition, T_{ab} also maps $L^\infty \times L^p \times L^q$ into L^r, $L^p \times L^\infty \times L^q \to L^r$, and $L^\infty \times L^s \times L^\infty \to L^s$ boundedly, for every $1 < p, q, s < \infty$ and $1/p + 1/q = 1/r$.*

The only L^∞ estimates that are not available are those of the form $L^\infty \times L^\infty \times L^s \to L^s$ or $L^s \times L^\infty \times L^\infty \to L^s$ and of course $L^\infty \times L^\infty \times L^\infty \to L^\infty$. But these are false in general, as the reader can easily check by taking f_2 to be identically equal to 1 in (8.2).

It is also believed that *all* flag paraproducts behave well and satisfy the natural generic L^p estimates as in Hölder inequalities.

8.2. Mollifying a product of two paraproducts

The original motivation for the introduction of flag paraproducts has just been mentioned. Briefly, particular examples are the lacunary versions of the T_3 operator and its generalizations, which appeared naturally in Chapter 5.

In the next three sections we will describe three other instances (two coming from nonlinear PDE theory and one from probability theory) where flag paraproducts appear naturally. The present section addresses the problem of mollifying nonlinearities of arbitrary algebraic complexity. As we will see, this problem can be essentially reduced to that of understanding the correct way of mollifying products of paraproducts.

Let us start by recalling the Leibnitz rule proved in the second chapter of this volume. This rule says that for any $1 < p_i, q_i \leq \infty$ with $1/p_i + 1/q_i = 1/p$ for $i = 1, 2$ and $1/(1+\alpha) < p < \infty$ one has

$$\|D^\alpha(f, g)\|_p \lesssim \|D^\alpha f\|_{p_1}\|g\|_{q_1} + \|f\|_{p_2}\|D^\alpha g\|_{q_2}. \tag{8.6}$$

We also know that the inequality holds for an arbitrary number of factors and that it is in general false if one of the indices p_i, q_i is strictly smaller than 1. For simplicity, we shall also assume for the rest of the discussion that the derivative order α is large enough that the inequality $1/(1+\alpha) < p < \infty$ becomes superfluous. Recall that p is always at least as large as $1/2$, as a consequence of the rest of the conditions.

Given (8.6) it is of course natural to ask whether one has similar estimates for more complex expressions, such as

$$\left\| D^{\alpha}(D^{\beta}(f_1 f_2 f_3) D^{\gamma}(f_4 f_5)) \right\|_p. \tag{8.7}$$

To estimate (8.7) one can first apply (8.6) for $f = D^{\beta}(f_1 f_2 f_3)$ and $g = D^{\gamma}(f_4 f_5)$ and bound (8.7) by

$$\|D^{\alpha+\beta}(f_1 f_2 f_3)\|_{p_1} \|D^{\gamma}(f_4 f_5)\|_{q_1} + \|D^{\beta}(f_1 f_2 f_3)\|_{p_2} \|D^{\alpha+\gamma}(f_4 f_5)\|_{q_2}. \tag{8.8}$$

Then one can apply again (8.6) to two or three factors to obtain a final upper bound.

However, it is not difficult to see that such an iterative argument has a drawback. It clearly does not work if one would like to end up with products of terms involving (for instance) only L^2 norms since then one must have $p_1 = p_2 = 2/3$ and $q_1 = q_2 = 1$ in (8.8), for which a relation corresponding to (8.6) does not hold.

Recall from Chapter 2 that the way in which we prove the Leibnitz rule (8.6) is first to write the product fg as a sum of paraproducts $\Pi(f, g)$ and then to reduce the inequality to the corresponding Coifman–Meyer theorem, after taking advantage of the fact that paraproducts behave nicely when they are acted upon by a derivative. Alternatively, one can say that paraproducts are the right objects to mollify nonlinearities of type $D^{\alpha}(fg)$.

Let us consider now the nonlinear expression $D^{\alpha}(D^{\beta}(fg)h)$, which is the simplest nonlinearity of the same complexity as that in (8.7). We say that expressions such as that in (8.6) are nonlinearities of complexity 1 expressions such as those in (8.7) are nonlinearities of complexity 2, and so on.

The above observations show that paraproducts, or even compositions of paraproducts (which one obtains in an iterative argument) are not enough to handle expressions of complexity 2. As we will argue, in order to obtain a generic estimate the correct point of view is not to write (for instance) $D^{\alpha}(D^{\beta}(fg)h)$ as a sum of a composition of paraproducts but instead as a sum of flag paraproducts. Clearly, given the above discussion, we need to understand two things, first, how to mollify generic products of the type $\Pi(f, g)h$ and then, eventually, how to mollify $D^{\alpha}(\Pi(f, g)h)$. We will first decompose $\Pi(f, g)h$ in a natural way as a sum of flag paraproducts; then we will see that flag paraproducts have the capacity of absorbing derivatives as well.

Let us take a concrete example and assume that the paraproduct $\Pi(f, g)$ is given by[1]

$$\Pi(f, g) = \sum_{k_1 \ll k_2} (f * \psi_{k_1})(g * \psi_{k_2}),$$

[1] As before, in this paraproduct context $k_a \ll k_b$ means $k_a < k_b - 100$.

where the expression on the right-hand side comes from the interaction of two Littlewood–Paley decompositions. Decompose h similarly, as

$$h = \sum_{k_3} h * \psi_{k_3},$$

again using Littlewood–Paley projections.

As a consequence, one can write[2]

$$\begin{aligned}\Pi(f, g)h &= \sum_{k_3} \sum_{k_1 \ll k_2} (f * \psi_{k_1})(g * \psi_{k_2})(h * \psi_{k_3}) \\ &= \sum_{\substack{k_1 \ll k_2 \\ k_3 \ll k_2}} (f * \psi_{k_1})(g * \psi_{k_2})(h * \psi_{k_3}) \\ &\quad + \sum_{\substack{k_1 \ll k_2 \\ k_3 \simeq k_2}} (f * \psi_{k_1})(g * \psi_{k_2})(h * \psi_{k_3}) \\ &\quad + \sum_{k_1 \ll k_2 \ll k_3} (f * \psi_{k_1})(g * \psi_{k_2})(h * \psi_{k_3}) := I + II + III.\end{aligned}$$

It is not difficult to see that both I and II are just trilinear paraproducts, while III can be written as

$$\int_{\mathbb{R}^3} m(\xi_1, \xi_2, \xi_3)\widehat{f}(\xi_1)\widehat{g}(\xi_2)\widehat{h}(\xi_3)e^{2\pi i x(\xi_1+\xi_2+\xi_3)}\, d\xi_1 d\xi_2 d\xi_3,$$

where

$$\begin{aligned}m(\xi_1, \xi_2, \xi_3) &= \sum_{k_1 \ll k_2 \ll k_3} \widehat{\psi}_{k_1}(\xi_1)\widehat{\psi}_{k_2}(\xi_2)\widehat{\psi}_{k_3}(\xi_3) \\ &=: \sum_{k_2 \ll k_3} \widehat{\phi}_{k_2}(\xi_1)\widehat{\psi}_{k_2}(\xi_2)\widehat{\psi}_{k_3}(\xi_3) \\ &= \sum_{k_2 \ll k_3} \widehat{\phi}_{k_2}(\xi_1)\widehat{\psi}_{k_2}(\xi_2)\widehat{\phi}_{k_3}(\xi_2)\widehat{\psi}_{k_3}(\xi_3) \qquad (8.9)\end{aligned}$$

for some suitable family $(\widehat{\phi}_{k_3}(\xi_2))_{k_3}$. (The reader should remember our previously used terminology. Families such as $(\psi_{k_1}(\xi_1))_{k_1}$ are said to be of ψ *type* while families such as $(\phi_{k_3}(\xi_2))_{k_3}$ are said to be of ϕ *type*. The supports of the first type are disjoint for different scales, while the supports of the latter type overlap for different scales.)

[2] As before, $k_a \simeq k_b$ means $k_a - 100 \le k_b \le k_a + 100$. Note that $k_a \ll k_b \ll k_c$ means $k_a < k_b - 100 < k_c - 200$.

Then one sees that, modulo a classical symbol that corresponds to the situation where $|k_2 - k_3|$ is bounded by a universal constant, (8.9) splits as

$$\left(\sum_{k_2} \widehat{\phi}_{k_2}(\xi_1)\widehat{\psi}_{k_2}(\xi_2)\right)\left(\sum_{k_3} \widehat{\phi}_{k_3}(\xi_2)\widehat{\psi}_{k_3}(\xi_3)\right)$$

since the only way in which $\widehat{\psi}_{k_2}(\xi_2)\widehat{\phi}_{k_3}(\xi_2) \neq 0$ is to have $k_2 \leq k_3$. This shows that (again modulo a classical symbol) the symbol of *III* can be written as $a(\xi_1, \xi_2)b(\xi_2, \xi_3)$ and, as a consequence, it is an element of $\mathcal{M}_{\mathrm{flag}}(\mathbb{R}^3)$.

In order finally to understand how to mollify the expression $D^\alpha(\Pi(f, g)h)$, given that *I* and *II* are paraproducts (and for paraproducts this is well understood from the analysis of Chapter 2), we are left to understand how to differentiate (8.9), which we now rewrite as

$$\sum_{k_2 \ll k_3} \widehat{\phi}_{k_2}(\xi_1)\widehat{\psi}_{k_2}(\xi_2)\widehat{\psi}_{k_2}(\xi_1 + \xi_2)\widehat{\phi}_{k_3}(\xi_1 + \xi_2)\widehat{\psi}_{k_3}(\xi_3)\widehat{\psi}_{k_3}(\xi_1 + \xi_2 + \xi_3) \quad (8.10)$$

where as usual $\widehat{\psi}_{k_2}(\xi_1 + \xi_2)$, $\widehat{\phi}_{k_3}(\xi_1 + \xi_2)$, and $\widehat{\psi}_{k_3}(\xi_1 + \xi_2 + \xi_3)$ have been inserted into (8.9). The advantage of (8.10) is that its corresponding trilinear operator can easily be written as

$$\sum_{k_3}\left(\left(\sum_{k_2 \ll k_3} ((f * \phi_{k_2})(g * \psi_{k_2})) * \psi_{k_2}\right) * \phi_{k_3}(h * \psi_{k_3})\right) * \psi_{k_3}. \quad (8.11)$$

Thus, as in the proof of the Leibnitz rule (2.1), one can write

$$\begin{aligned} &D^\alpha(III) \\ &= \sum_{k_3}\left(\left(\sum_{k_2 \ll k_3} ((f * \phi_{k_2})\,(g * \psi_{k_2})) * \psi_{k_2}\right) * \phi_{k_3}(h * \psi_{k_3})\right) * D^\alpha \psi_{k_3} \\ &= \qquad \cdots \\ &= \sum_{k_3}\left(\left(\sum_{k_2 \ll k_3} ((f * \phi_{k_2})\,(g * \psi_{k_2})) * \psi_{k_2}\right) * \phi_{k_3}(D^\alpha h * \widetilde{\psi}_{k_3})\right) * \widetilde{\psi}_{k_3}. \end{aligned}$$

Now, this trilinear operator has the symbol

$$\sum_{k_2 \ll k_3} \widehat{\phi}_{k_2}(\xi_1)\widehat{\psi}_{k_2}(\xi_2)\widehat{\psi}_{k_2}(\xi_1 + \xi_2)\widehat{\phi}_{k_3}(\xi_1 + \xi_2)\widehat{\widetilde{\psi}}_{k_3}(\xi_3)\widehat{\widetilde{\psi}}_{k_3}(\xi_1 + \xi_2 + \xi_3)$$

and, given the way in which $\widehat{\psi}_{k_2}(\xi_1+\xi_2)$ and $\widehat{\phi}_{k_3}(\xi_1+\xi_2)$ have been inserted earlier, this can be simplified as

$$\sum_{k_2 \ll k_3} \widehat{\phi}_{k_2}(\xi_1)\widehat{\psi}_{k_2}(\xi_2)\widehat{\widetilde{\psi}}_{k_3}(\xi_3)\widehat{\widetilde{\psi}}_{k_3}(\xi_1+\xi_2+\xi_3)$$

$$= \sum_{k_2 \ll k_3} \widehat{\phi}_{k_2}(\xi_1)\widehat{\psi}_{k_2}(\xi_2)\widehat{\phi}_{k_3}(\xi_2)\widehat{\widetilde{\psi}}_{k_3}(\xi_3)\widehat{\widetilde{\psi}}_{k_3}(\xi_1+\xi_2+\xi_3), \tag{8.12}$$

where again $\widehat{\phi}_{k_3}(\xi_2)$ has been added into the formula. Now one observes that (8.12) can be factored as

$$\left(\sum_{k_2} \widehat{\phi}_{k_2}(\xi_1)\widehat{\psi}_{k_2}(\xi_2)\right)\left(\sum_{k_3} \widehat{\phi}_{k_3}(\xi_2)\widehat{\widetilde{\psi}}_{k_3}(\xi_3)\widehat{\widetilde{\psi}}_{k_3}(\xi_1+\xi_2+\xi_3)\right)$$

$$=: c(\xi_1,\xi_2)d(\xi_1,\xi_2,\xi_3), \tag{8.13}$$

with $c \in \mathcal{M}(\mathbb{R}^2)$ and $d \in \mathcal{M}(\mathbb{R}^3)$. The reader should recall from the beginning that the derivative α has been assumed to be large enough to avoid the awkward discussion of paraproducts of the type Π^α, as in Chapter 2. To prove all the cases rigorously (when α may be small), one will have to take these paraproducts into consideration as well.

To conclude, this discussion shows that, flag paraproducts are indeed the right objects to mollify expressions of complexity 2 and, in fact, similar expressions of arbitrary complexity. We will see in Problem 8.1 at the end of this chapter that trilinear operators with symbols of the type (8.13) also satisfy the estimates in Theorem 8.1, and this fact will allow us to prove as a consequence the following *generic Leibnitz rule* for the simplest nonlinearity of complexity 2, namely

$$\begin{aligned} &\|D^\alpha(D^\beta(fg)h)\|_p \\ &\lesssim \|D^{\alpha+\beta}f\|_{p_1}\|g\|_{q_1}\|h\|_{r_1} + \|f\|_{p_2}\|D^{\alpha+\beta}g\|_{q_2}\|h\|_{r_2} \\ &\quad + \|D^\beta f\|_{p_3}\|g\|_{q_3}\|D^\alpha h\|_{r_3} + \|f\|_{p_4}\|D^\beta g\|_{q_4}\|D^\alpha h\|_{r_4} \end{aligned} \tag{8.14}$$

which holds for every $1 < p_i, q_i, r_i < \infty$ with $1/p_i + 1/q_i + 1/r_i = 1/p, i \in \{1,2,3,4\}$, as long as $\max(1/(1+\alpha), 1/(1+\beta)) < p < \infty$.

Let us end this section with the observation that *any* product of flag paraproducts can be mollified naturally as a sum of more complex flag paraproducts and that this can be done in several distinct ways. Consider for instance $T_{m_1}(f_1,\ldots,f_n)$ and $T_{m_2}(g_1,\ldots,g_m)$, where $m_1 \in \mathcal{M}_{\text{flag}}(\mathbb{R}^n)$ and $m_2 \in \mathcal{M}_{\text{flag}}(\mathbb{R}^m)$. Their product $T_{m_1}(f_1,\ldots,f_n)T_{m_2}(g_1,\ldots,g_m)$ can be seen as a $(n+m)$-linear operator with symbol $m_1(\xi_1,\ldots,\xi_n)m_2(\eta_1,\ldots,\eta_m)$. Mollyfying the product of the operators is then equivalent to decomposing this product symbol further, for example as a sum of various longer products of the type $m_1(\xi_1, \ldots, \xi_n)m_2(\eta_1, \ldots, \eta_m)m_3(\xi_1, \eta_1)$, where $m_3 \in \mathcal{M}(\mathbb{R}^2)$, a fact that can be achieved by combining various Littlewood–Paley decompositions, as before.

8.3. Flag paraproducts and quadratic NLS

The goal of this section is to describe very briefly some recent studies of Germain, Masmoudi, and Shatah on various dispersive PDEs where flag paraproducts appear naturally and play an important role.

In a few words, the main goal of these works has been to develop a general method for understanding the global existence of solutions corresponding to small initial data for various nonlinear evolution equations.

Let us consider the following case study of a *quadratic* nonlinear Schrödinger equation (NLS),

$$\partial_t u + i\Delta u = u^2, \tag{8.15}$$

$$u|_{t=t_0} = u_0, \tag{8.16}$$

for $(t, x) \in \mathbb{R} \times \mathbb{R}$. For technical reasons one prefers the initial time to be $t_0 > 0$ rather than the more common $t_0 = 0$, but this is not going to play an important role for us. More generally the above equation should be considered for $x \in \mathbb{R}^n$, but for simplicity we will work in one dimension only. Our plan here is to simply *isolate* the calculations that naturally gave rise to flag paraproducts in the particular one-dimensional quadratic case (8.15).

It will be convenient during this section to work with the following definitions of the Fourier transform $\mathcal{F}$ and the inverse Fourier transform $\mathcal{F}^{-1}$:

$$\mathcal{F}f(\xi) = \widehat{f}(\xi) := \frac{1}{2\pi}\int_{\mathbb{R}} f(x)e^{-ix\xi}\,dx$$

and

$$\mathcal{F}^{-1}g(x) := \frac{1}{2\pi}\int_{\mathbb{R}} g(\xi)e^{ix\xi}\,d\xi.$$

Coming back to equation (8.15), the Duhamel formula associated with it and written in terms of the Fourier transform is

$$\widehat{u}(t, \xi) = \widehat{u_0}(\xi)e^{it|\xi|^2} + \int_{t_0}^{t} e^{-i(s-t)|\xi|^2}\widehat{u^2}(s, \xi)\,ds. \tag{8.17}$$

The reader may remember that in Chapter 1, when we discussed the generalized KdV equation, we derived at some point the Duhamel formula associated with the latter. Equation (8.17) can be deduced in precisely the same way.

Exercise 8.2 Check formula (8.17).

It is natural to look for a solution of type $u = e^{-it\Delta} f$ and as a consequence the corresponding formula for f becomes

$$\widehat{f}(t,\xi) = \widehat{u_0}(\xi) + \int_{t_0}^{t} \int_{\mathbb{R}} e^{is(-|\xi|^2+|\eta|^2+|\xi-\eta|^2)} \widehat{f}(s,\eta)\widehat{f}(s,\xi-\eta)\, ds d\eta. \quad (8.18)$$

Exercise 8.3 Check formula (8.18).

An important part of the argument depends on how well one can estimate the integral expression in (8.18). The general idea is to take advantage of the oscillation of the term $e^{is\phi}$, where

$$\phi := -|\xi|^2 + |\eta|^2 + |\xi - \eta|^2.$$

However, ϕ is too degenerate in the sense that $\phi = 0$ along the lines $\xi = \eta$ and $\eta = 0$ and, because of this, a simple integration by parts with respect to the parameter s will not by itself be very helpful.

To fix this problem, one would need a better integration-by-parts argument, not only for the s variable but also for the η variable, with the hope that in this way one might be able to overcome the singularity of ϕ. To see how this works, let us write

$$P := -\eta + \tfrac{1}{2}\xi$$

and

$$Q := \phi + P(\partial_\eta \phi).$$

Then we observe by a direct calculation that

$$Q = -(|\eta|^2 + |\xi - \eta|^2),$$

which is much less degenerate than the above expression for ϕ, since it is equal to zero only when $\xi = \eta = 0$.

Using these, one has

$$\frac{1}{iQ}\left(\partial_s + \frac{P}{s}\partial_\eta\right) e^{is\phi} = e^{is\phi}$$

and, as a consequence, the inverse Fourier transform of the integral term in (8.18) can be written as

$$\mathcal{F}^{-1} \int_{t_0}^{t} \int_{\mathbb{R}} \frac{1}{iQ}\left(\partial_s + \frac{P}{s}\partial_\eta\right) e^{is\phi} \widehat{f}(s,\xi-\eta)\widehat{f}(s,\eta)\, ds d\eta := I + II.$$

Using the that both $|\xi - \eta|^2/iQ$ and $|\eta|^2/iQ$ are classical symbols in $\mathcal{M}(\mathbb{R}^2)$, the Coifman–Meyer theorem, Theorem 2.15, proves that terms I and II are *smoothing expressions*, which is clearly very good news from a PDE perspective. To summarize, expressions of the type

$$\mathcal{F}^{-1}\int_{t_0}^{t}\int_{\mathbb{R}} m(\xi,\eta)\widehat{g}(s,\eta)\widehat{h}(s,\xi-\eta)\,ds\,d\eta \tag{8.19}$$

for some $m \in \mathcal{M}(\mathbb{R}^2)$ appear in this way, and it is not difficult to see that if one keeps the parameter s fixed then the rest of the formula is just a bilinear paraproduct.

Exercise 8.4 Show that formula (8.19) (for a fixed s) can indeed be seen as a bilinear paraproduct of type T_m for some symbol $m \in \mathcal{M}(\mathbb{R}^2)$.

Coming back to the first term, I, an expression related to it (given the discussion of (8.19), after differentiation with respect to s) is of the form

$$\mathcal{F}^{-1}\int_{t_0}^{t}\int_{\mathbb{R}} e^{is\phi}m(\xi,\eta)\partial_s\widehat{f}(s,\eta)\widehat{F}(s,\xi-\eta)\,ds\,d\eta \tag{8.20}$$

for a certain new function F. Given that u has been assumed to be of the form $u = e^{-is\Delta}f$, it follows that $f = e^{is\Delta}u$; thus, using the fact that u solves the equation (8.15), we deduce that $\partial_s f = e^{is\Delta}u^2$ or, equivalently,

$$\begin{aligned}\widehat{\partial_s f}(s,\eta) &= e^{-is|\eta|^2}\widehat{u^2}(\eta)\\ &= e^{-is|\eta|^2}\int_{\eta_1+\eta_2=\eta}\widehat{u}(\eta_1)\widehat{u}(\eta_2)\,d\eta_1 d\eta_2\\ &= e^{-is|\eta|^2}\int_{\eta_1+\eta_2=\eta} e^{is|\eta_1|^2}\widehat{f}(s,\eta_1)e^{is|\eta_2|^2}\widehat{f}(s,\eta_2)\,d\eta_1 d\eta_2\\ &= e^{-is|\eta|^2}\int_{\mathbb{R}} e^{is|\tau|^2}\widehat{f}(s,\tau)e^{is|\eta-\tau|^2}\widehat{f}(s,\eta-\tau)\,d\tau.\end{aligned}$$

Using this formula in (8.20) we obtain

$$\mathcal{F}^{-1}\int_{t_0}^{t}\int_{\mathbb{R}^2} e^{is\widetilde{\phi}}m(\xi,\eta)\widehat{f}(s,\tau)\widehat{f}(s,\eta-\tau)\widehat{F}(s,\xi-\eta)\,ds\,d\eta\,d\tau,$$

where $\widetilde{\phi} := -|\xi|^2 + |\xi-\eta|^2 + |\tau|^2 + |\eta-\tau|^2$.

The idea now would be to continue in precisely the same way as we did before, more specifically to use a similar *integration by parts argument* but in all three variables s, η, τ this time. We will ignore the details but the outcome of this is that one obtains expressions of type

$$\mathcal{F}^{-1}\int_{t_0}^{t}\int_{\mathbb{R}^2} e^{is\widetilde{\phi}}m(\xi,\eta)m(\xi,\eta,\tau)\widehat{g}(s,\tau)\widehat{f}(s,\eta-\tau)\widehat{h}(s,\xi-\eta)\,ds\,d\eta\,d\tau.$$

If we fix the variable s, the integral becomes

$$\mathcal{F}^{-1} \int_{\mathbb{R}^2} m(\xi, \eta) m(\xi, \eta, \tau) \widehat{g}(\tau) \widehat{f}(\eta - \tau) \widehat{h}(\xi - \eta)\, d\eta d\tau$$

and we claim that it is related in a natural way to the flag paraproduct operators described earlier. To see this, let us first rewrite the above expression as

$$\int_{\mathbb{R}^3} m(\xi, \eta) m(\xi, \eta, \gamma) \widehat{f}(\eta - \gamma) \widehat{g}(\gamma) \widehat{h}(\xi - \eta) e^{ix\xi}\, d\xi d\eta d\gamma. \tag{8.21}$$

Here we have ignored the factor $1/2\pi$ that appears in the definition of the inverse Fourier transform $\mathcal{F}^{-1}$. It is natural now to change variables, setting $\xi - \eta =: \xi_3$, $\eta - \gamma =: \xi_1$, and $\gamma =: \xi_2$ and to rewrite (8.21) as

$$\begin{aligned}
&\int_{\mathbb{R}^3} m(\xi_1 + \xi_2 + \xi_3, \xi_1 + \xi_2) m(\xi_1 + \xi_2 + \xi_2, \xi_1 + \xi_2, \xi_2) \\
&\quad \times \widehat{f}(\xi_1) \widehat{g}(\xi_2) \widehat{h}(\xi_3) e^{ix(\xi_1 + \xi_2 + \xi_3)}\, d\xi \\
&=: \int_{\mathbb{R}^3} \widetilde{m}(\xi_1 + \xi_2, \xi_3) \widetilde{\widetilde{m}}(\xi_1, \xi_2, \xi_3) \widehat{f}(\xi_1) \widehat{g}(\xi_2) \widehat{h}(\xi_3) e^{ix(\xi_1 + \xi_2 + \xi_3)}\, d\xi,
\end{aligned} \tag{8.22}$$

for $\widetilde{m} \in \mathcal{M}(\mathbb{R}^2)$ and $\widetilde{\widetilde{m}} \in \mathcal{M}(\mathbb{R}^3)$ and where $d\xi$ represents $d\xi_1 d\xi_2 d\xi_3$.

Clearly, as it stands (8.22) is not a flag paraproduct but, as we will see in the next few lines, it can be expressed as the sum of a paraproduct and an adjoint of a flag paraproduct.

Let us assume for simplicity that

$$\widetilde{m}(\xi_1 + \xi_2, \xi_3) = \sum_{k_1} \widehat{\phi}_{k_1}(\xi_1 + \xi_2) \widehat{\phi}_{k_1}(\xi_3) \tag{8.23}$$

and

$$\widetilde{\widetilde{m}}(\xi_1, \xi_2, \xi_3) = \sum_{k_2} \widehat{\phi}_{k_2}(\xi_1) \widehat{\phi}_{k_2}(\xi_2) \widehat{\phi}_{k_2}(\xi_3). \tag{8.24}$$

As we saw in Chapter 2, modulo some minor technical details, one can always assume that (8.23) and (8.24) hold. Recall also the important fact that at least one family of functions appearing in (8.23) and (8.24) must be of the ψ *type*.

In particular,

$$\widetilde{m}(\xi_1 + \xi_2, \xi_3) \widetilde{\widetilde{m}}(\xi_1, \xi_2, \xi_3) = \sum_{k_1, k_2} \widehat{\phi}_{k_1}(\xi_1 + \xi_2) \widehat{\phi}_{k_1}(\xi_3) \widehat{\phi}_{k_2}(\xi_1) \widehat{\phi}_{k_2}(\xi_2) \widehat{\phi}_{k_2}(\xi_3). \tag{8.25}$$

Given that when $k_1 \simeq k_2$ (i.e., when $k_1 - 100 \leq k_2 \leq k_1 + 100$) equation (8.25) becomes a classical paraproduct symbol we have two cases to consider.

Case 1: $k_2 \ll k_1$. In this case, the only option is that $(\widehat{\phi}_{k_1}(\xi_3))_{k_1}$ is ϕ type, which means that $(\widehat{\phi}_{k_1}(\xi_1+\xi_2))_{k_1}$ has to be ψ type. Since (8.25) can also be written as

$$\sum_{k_2 \ll k_1} \widehat{\phi}_{k_1}(\xi_1+\xi_2)\widehat{\phi}_{k_1}(\xi_3)\widehat{\phi}_{k_2}(\xi_1)\widehat{\phi}_{k_2}(\xi_2)\widehat{\widehat{\phi}}_{k_2}(\xi_1+\xi_2)\widehat{\phi}_{k_2}(\xi_3)$$

for some other suitably chosen family $(\widehat{\widehat{\phi}}_{k_2}(\xi_1+\xi_2))_{k_2}$, we see that the only way in which we can have $\widehat{\phi}_{k_1}(\xi_1+\xi_2)\widehat{\widehat{\phi}}_{k_2}(\xi_1+\xi_2) \neq 0$ is if $|k_1-k_2| < 10$. But this contradicts our assumption.

Case 2: $k_2 \gg k_1$. This case is more interesting. First, observe that the only possibility is that $(\widehat{\phi}_{k_2}(\xi_3))_{k_2}$ is ϕ type. Then as usual we can rewrite the expression in (8.25) as

$$\sum_{k_2 \gg k_1} \widehat{\phi}_{k_2}(\xi_1)\widehat{\phi}_{k_2}(\xi_2)\widehat{\phi}_{k_2}(\xi_3)\widehat{\widehat{\phi}}_{k_2}(\xi_1+\xi_2)\widehat{\phi}_{k_1}(\xi_1+\xi_2)\widehat{\phi}_{k_1}(\xi_3)\widehat{\widehat{\phi}}_{k_1}(\xi_1+\xi_2+\xi_3). \tag{8.26}$$

If one considers the family $(\widehat{\widehat{\phi}}_{k_2}(\xi_1+\xi_2))_{k_2}$ one sees that we have two options. Either it is ψ type, in which case the only nonzero terms would correspond to $|k_1-k_2| < 10$ (and that is impossible by our assumption) or it is ϕ type, but this can only happen when both families $(\widehat{\phi}_{k_2}(\xi_1))_{k_2}$ and $(\widehat{\phi}_{k_2}(\xi_2))_{k_2}$ are ψ type (and their oscillations cancel each other out). Given that we also know that either $(\widehat{\phi}_{k_1}(\xi_1+\xi_2))_{k_1}$ or $(\widehat{\phi}_{k_1}(\xi_3))_{k_1}$ has to be ψ type, we deduce that (8.26) has to factor as

$$\left(\sum_{k_2} \widehat{\phi}_{k_2}(\xi_1)\widehat{\phi}_{k_2}(\xi_2)\widehat{\phi}_{k_2}(\xi_3)\widehat{\widehat{\phi}}_{k_2}(\xi_1+\xi_2)\right)$$
$$\times\left(\sum_{k_1} \widehat{\phi}_{k_1}(\xi_1+\xi_2)\widehat{\phi}_{k_1}(\xi_3)\widehat{\widehat{\phi}}_{k_1}(\xi_1+\xi_2+\xi_3)\right).$$

Then, if one denotes by $T(f,g,h)$ the trilinear operator defined by the above symbol, one can write its 4-linear form as

$$\Lambda_T(f,g,h,k) := \int_{\mathbb{R}} T(f,g,h)(x)k(x)\,dx$$
$$= \int_{\mathbb{R}^3}\left(\sum_{k_2} \widehat{\phi}_{k_2}(\xi_1)\widehat{\phi}_{k_2}(\xi_2)\widehat{\phi}_{k_2}(\xi_3)\widehat{\widehat{\phi}}_{k_2}(\xi_1+\xi_2)\right)$$

$$\times \left(\sum_{k_1} \widehat{\phi}_{k_1}(\xi_1 + \xi_2)\widehat{\phi}_{k_1}(\xi_3)\widehat{\widehat{\phi}}_{k_1}(\xi_1 + \xi_2 + \xi_3) \right)$$

$$\times \widehat{f}(\xi_1)\widehat{g}(\xi_2)\widehat{h}(\xi_3)\widehat{k}(-\xi_1 - \xi_2 - \xi_3)\, d\xi_1\, d\xi_2 d\xi_3$$

$$= \int_{\mathbb{R}^3} \left(\sum_{k_2} \widehat{\widehat{\phi}}_{k_2}(-\xi_1)\widehat{\phi}_{k_2}(\xi_2)\widehat{\phi}_{k_2}(\xi_3)\widehat{\widehat{\widehat{\phi}}}_{k_2}(-\xi_1 - \xi_2) \right)$$

$$\times \left(\sum_{k_1} \widehat{\widehat{\phi}}_{k_1}(-\xi_1 - \xi_2)\widehat{\phi}_{k_1}(\xi_3)\widehat{\widehat{\widehat{\phi}}}_{k_1}(-\xi_1 - \xi_2 - \xi_3) \right)$$

$$\times \widehat{f}(\xi_1)\widehat{g}(\xi_2)\widehat{h}(\xi_3)\widehat{k}(-\xi_1 - \xi_2 - \xi_3)\, d\xi_1 d\xi_2 d\xi_3.$$

If we now write $\lambda := -\xi_1 - \xi_2 - \xi_3$, the above expression becomes

$$\int_{\mathbb{R}^3} \left(\sum_{k_2} \widehat{\widehat{\phi}}_{k_2}(\lambda + \xi_2 + \xi_3)\widehat{\phi}_{k_2}(\xi_2)\widehat{\phi}_{k_2}(\xi_3)\widehat{\widehat{\widehat{\phi}}}_{k_2}(\lambda + \xi_3) \right)$$

$$\times \left(\sum_{k_1} \widehat{\widehat{\phi}}_{k_1}(\lambda + \xi_3)\widehat{\phi}_{k_1}(\xi_3)\widehat{\widehat{\widehat{\phi}}}_{k_1}(\lambda) \right)$$

$$\times \widehat{f}(-\lambda - \xi_2 - \xi_3)\widehat{g}(\xi_2)\widehat{h}(\xi_3)\widehat{k}(\lambda)\, d\xi_2 d\xi_3 d\lambda$$

$$=: \int_{\mathbb{R}^3} m(\xi_2, \xi_3, \lambda)m(\xi_3, \lambda)\widehat{g}(\xi_2)\widehat{h}(\xi_3)\widehat{k}(\lambda)\widehat{f}(-\lambda - \xi_2 - \xi_3)\, d\xi_2 d\xi_3 d\lambda$$

$$=: \int_{\mathbb{R}} \Pi_{\text{flag}}(g, h, k)(x) f(x)\, dx.$$

To conclude, there does indeed exist a flag paraproduct Π_{flag} with the property that

$$\Lambda_T(f, g, h, k) = \int_{\mathbb{R}} \Pi_{\text{flag}}(g, h, k)(x) f(x)\, dx,$$

which proves that T is an adjoint of a flag paraproduct, as claimed earlier. All the above calculations have natural higher-dimensional analogues.

In this way paraproducts appear in the study of the three-dimensional quadratic Schrödinger equation, while flag paraproducts are in addition necessary to deal with the more difficult two-dimensional case and also with the three-dimensional water-wave equation.

8.4. Flag paraproducts and U-statistics

Before starting the proof of the main theorem, we remark that *probabilistic analogues* of some flag paraproducts have been considered in the first volume of the book; see Chapter I.5.

More precisely, suppose that $(X_n)_{n=1}^{\infty}$ is a sequence of independent random variables of mean zero. Then, it is well known that the sequence of partial sums $(S_n)_n$ defined by

$$S_n = \sum_{1\le i\le n} X_i$$

forms a martingale. However, it is also known that the *iterated sequence* $(S_n^k)_n$ given by

$$S_n^k = \sum_{1\le i_1<\cdots<i_k\le n} X_{i_1}\cdots X_{i_k}, \tag{8.27}$$

forms a martingale as well, even though it does not have structure of a sum of independent random variables. Such expressions play an important role in unbiased statistics (U-statistics).

The *harmonic analysis* analogue of (8.27) is clearly the expression

$$\sum_{1\le i_1<\cdots<i_k\le n} (f * \psi_{i_1})\cdots(f * \psi_{i_k}),$$

where $(\psi_\ell)_\ell$ is the usual Littlewood–Paley sequence, and it is not difficult to see that when viewed as a k-linear operator the above formula defines a flag paraproduct.

Exercise 8.5 Prove the above assertion.

8.5. Discrete operators and interpolation

With this section we begin the proof of the main theorem in this chapter, Theorem 8.1. As in the case of classical paraproducts, we now describe some discrete well-localized operators and claim that Theorem 8.1 can be reduced to the study of them. Even more, we will see that it is enough to prove some *restricted weak-type* estimates for these model operators in order to obtain the desired *strong estimates* in Theorem 8.1.

We will use the language of lacunary and nonlacunary families introduced in Chapter 2. Consider $\mathcal{I}_1$ and $\mathcal{J}_1$, two finite families of dyadic intervals. Let $(\phi_I^j)_{I\in\mathcal{I}_1}$ for $j = 1, 2, 3$ be three families of L^2-normalized bump functions such that the family $(\phi_I^2)_{I\in\mathcal{I}_1}$ is nonlacunary while the families $(\phi_I^j)_{I\in\mathcal{I}_1}$ for $j \neq 2$ are both lacunary. Also consider $(\phi_J^j)_{J\in\mathcal{J}_1}$ for $j = 1, 2, 3$: three other families of L^2-normalized bump functions such that at least two are lacunary. Using them, we define the discrete model operator T_1 by

$$T_1(f_1, f_2, f_3)(x) := \sum_{I\in\mathcal{I}_1} \frac{1}{|I|^{1/2}} \langle f_1, \phi_I^1\rangle \langle B_I^1(f_2, f_3), \phi_I^2\rangle \phi_I^3, \tag{8.28}$$

where we have implicitly defined

$$B_I^1(f_2, f_3)(x) := \sum_{\substack{J \in \mathcal{J}_1 \\ |\omega_J^3| \leq |\omega_I^2| \\ \omega_J^3 \cap \omega_I^2 \neq \emptyset}} \frac{1}{|J|^{1/2}} \langle f_2, \phi_J^1 \rangle \langle f_3, \phi_J^2 \rangle \phi_J^3. \tag{8.29}$$

Then, if ℓ_0 is a positive integer, we define also the trilinear operator T_{1,ℓ_0}, by

$$T_{1,\ell_0}(f_1, f_2, f_3)(x) := \sum_{I \in \mathcal{I}_1} \frac{1}{|I|^{1/2}} \langle f_1, \phi_I^1 \rangle \langle B_{I,\ell_0}^1(f_2, f_3), \phi_I^2 \rangle \phi_I^3, \tag{8.30}$$

where this time

$$B_{I,\ell_0}^1(f_2, f_3)(x) := \sum_{\substack{J \in \mathcal{J}_1 \\ 2^{\ell_0} |\omega_J^3| \simeq |\omega_I^2| \\ \omega_J^3 \cap \omega_I^2 \neq \emptyset}} \frac{1}{|J|^{1/2}} \langle f_2, \phi_J^1 \rangle \langle f_3, \phi_J^2 \rangle \phi_J^3. \tag{8.31}$$

Similarly, now consider two other finite families $\mathcal{I}_2$ and $\mathcal{J}_2$ of dyadic intervals. As before consider families $(\phi_I^j)_{I \in \mathcal{I}_2}$ and $(\phi_J^j)_{J \in \mathcal{J}_2}$ for $j = 1, 2, 3$ of L^2-normalized bump functions, where we assume that the family $(\phi_I^1)_{I \in \mathcal{I}_2}$ is nonlacunary while the families $(\phi_I^j)_{I \in \mathcal{I}_2}$ for $j \neq 1$ are both lacunary and at least two of the families $(\phi_J^j)_{J \in \mathcal{J}_2}$ are lacunary. Then we define the trilinear operator T_2 by the formula

$$T_2(f_1, f_2, f_3)(x) := \sum_{I \in \mathcal{I}_2} \frac{1}{|I|^{1/2}} \langle B_I{}^2(f_1, f_2), \phi_I^1 \rangle \langle f_3, \phi_I^2 \rangle \phi_I^3 \tag{8.32}$$

where

$$B_I^2(f_1, f_2)(x) := \sum_{\substack{J \in \mathcal{J}_2 \\ |\omega_J^3| \leq |\omega_I^1| \\ \omega_J^3 \cap \omega_I^1 \neq \emptyset}} \frac{1}{|J|^{1/2}} \langle f_1, \phi_J^1 \rangle \langle f_2, \phi_J^2 \rangle \phi_J^3. \tag{8.33}$$

Finally, also as before, for any positive integer ℓ_0 define the trilinear operator T_{2,ℓ_0} by the formula

$$T_{2,\ell_0}(f_1, f_2, f_3)(x) := \sum_{I \in \mathcal{I}_2} \frac{1}{|I|^{1/2}} \langle B_{I,\ell_0}^2(f_1, f_2), \phi_I^1 \rangle \langle f_3, \phi_I^2 \rangle \phi_I^3, \tag{8.34}$$

where

$$B_{I,\ell_0}^2(f_1, f_2)(x) := \sum_{\substack{J \in \mathcal{J}_2 \\ 2^{\ell_0} |\omega_J^3| \simeq |\omega_I^1| \\ \omega_J^3 \cap \omega_I^1 \neq \emptyset}} \frac{1}{|J|^{1/2}} \langle f_1, \phi_J^1 \rangle \langle f_2, \phi_J^2 \rangle \phi_J^3. \tag{8.35}$$

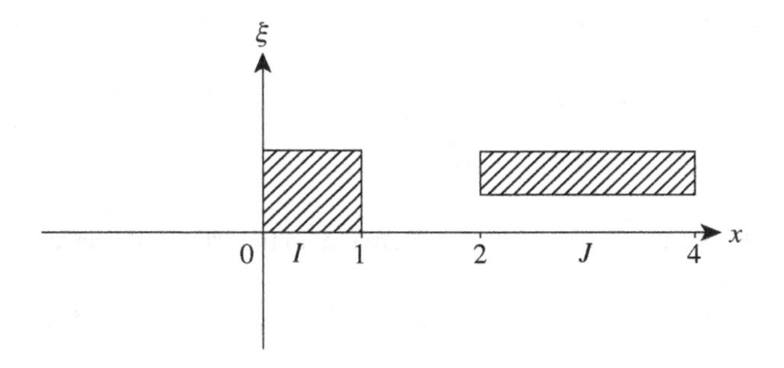

Figure 8.1. Heisenberg boxes of ϕ_I^1 and ϕ_J^3.

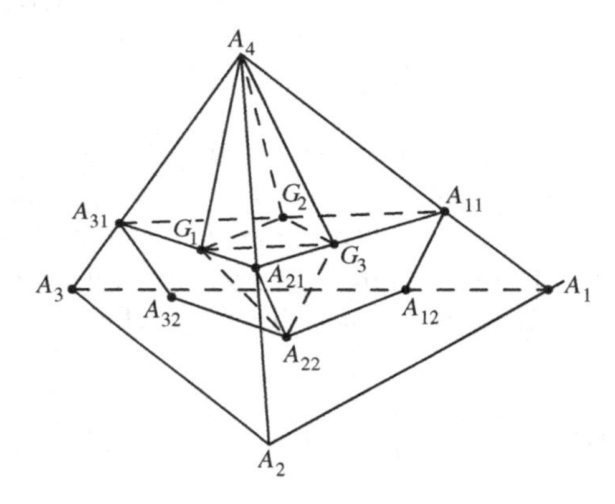

Figure 8.2

Figure 8.1 shows the Heisenberg boxes for generic functions ϕ_I^1 and ϕ_J^2.

To be able to describe the restricted weak-type estimates that we mentioned earlier, let us consider the three-dimensional hyperspace $\mathcal{H}$ defined by

$$\mathcal{H} := \{(\beta_1, \beta_2, \beta_3, \beta_4) \in \mathbb{R}^4 | \beta_1 + \beta_2 + \beta_3 + \beta_4 = 1\}.$$

Then, denote by $\mathbb{B}$ the open interior of the convex hull of the points $A_{11}, A_{12}, A_{21}, A_{22}, A_{31}, A_{32}$, and A_4 in Figure 8.2. All these points belong to $\mathcal{H}$ and have coordinates $A_{11}(-1, 1, 1, 0)$, $A_{12}(-1, 1, 0, 1)$, $A_{21}(1, -1, 1, 0)$, $A_{22}(0, 0, 0, 1)$, $A_{31}(1, 1, -1, 0)$, $A_{32}(0, 1, -1, 1)$, and $A_4(1, 1, 1, -2)$. We will denote by $\widetilde{\mathbb{B}}$ the open interior of the convex hull of the points A_{22}, G_1, G_2, G_3, and A_4, where G_1, G_2, G_3 have the coordinates $(1, 0, 0, 0)$, $(0, 1, 0, 0)$ and $(0, 0, 1, 0)$ respectively. The other three points that appear in Figure 8.2 are $A_1(-2, 1, 1, 1)$, $A_2(1, -2, 1, 1)$, and $A_3(1, 1, -2, 1)$.

The following theorem holds.

Theorem 8.2 *For simplicity, denote by T any of the previous discrete trilinear operators $T_1, T_2, T_{1,\ell_0}, T_{2,\ell_0}$. Then*

(i) *There exist points*

$$\left(\frac{1}{p_1}, \frac{1}{p_2}, \frac{1}{p_3}, \frac{1}{p_4'}\right) \in \mathbb{B}$$

arbitrarily close to A_4 such that T is of the restricted weak type (p_1, p_2, p_3, p_4).

(ii) *Similarly, there exist points*

$$\left(\frac{1}{p_1^{ij}}, \frac{1}{p_2^{ij}}, \frac{1}{p_3^{ij}}, \frac{1}{p_4'^{ij}}\right) \in \mathbb{B},$$

arbitrarily close to A_{ij}, such that the adjoint operator T^{*i} is of the restricted weak type $\left(p_1^{ij}, p_2^{ij}, p_3^{ij}, p_4^{ij}\right)$ for $i = 1, 2, 3$ and $j = 1, 2$.

Moreover, all the implicit bounds are independent of ℓ_0 and of the cardinalities of $\mathcal{I}_1, \mathcal{I}_2, \mathcal{J}_1, \mathcal{J}_2$. Likewise, the subsets $\left(E_j'\right)_{j=1}^4$ that appear implicitly in the restricted weak-type estimates can also be chosen independently of the L^2-normalized families considered in the definition of the operators.

Strictly speaking, so far we have defined the *restricted weak-type* property for bilinear operators only (see Definition 6.4) but it clearly extends naturally to arbitrary multilinear operators. More details of this can be found in the appendix. In the next section we will discretize our original operator T_{ab}, and this will indeed allow us to understand why Theorem 8.2 implies Theorem 8.1.

8.6. Reduction to the model operators

Recall that the symbol of our operator is $a(\xi_1, \xi_2)b(\xi_2, \xi_3)$. As we have already seen in Section 2.13, modulo some technical issues by now standard, one can assume that a and b are given by

$$a(\xi_1, \xi_2) = \sum_{k_1} \widehat{\phi_{k_1}}(\xi_1)\widehat{\phi_{k_1}}(\xi_2),$$

where at least one of the families $(\widehat{\phi_{k_1}}(\xi_1))_{k_1}$ and $(\widehat{\phi_{k_1}}(\xi_2))_{k_1}$ is ψ type and

$$b(\xi_2, \xi_3) = \sum_{k_2} \widehat{\phi_{k_2}}(\xi_2)\widehat{\phi_{k_2}}(\xi_3),$$

where again at least one of the families $(\widehat{\phi_{k_2}}(\xi_2))_{k_2}$ and $(\widehat{\phi_{k_2}}(\xi_3))_{k_2}$ is ψ type.

In particular, this means that

$$a(\xi_1, \xi_2)b(\xi_2, \xi_3) = \sum_{k_1,k_2} \widehat{\phi_{k_1}}(\xi_1)\widehat{\phi_{k_1}}(\xi_2)\widehat{\phi_{k_2}}(\xi_2)\widehat{\phi_{k_2}}(\xi_3).$$

As usual, there are three cases; $k_1 \simeq k_2$, $k_1 \ll k_2$, and $k_2 \ll k_1$. The first is simple since it generates paraproducts and can be treated by the Coifman–Meyer theorem of Chapter 2.

Let us study then the case $k_1 \ll k_2$; the case $k_2 \ll k_1$ is similar. We therefore have to understand the symbol

$$\sum_{k_1 \ll k_2} \widehat{\phi_{k_1}}(\xi_1)\widehat{\phi_{k_1}}(\xi_2)\widehat{\phi_{k_2}}(\xi_2)\widehat{\phi_{k_2}}(\xi_3). \tag{8.36}$$

Observe that in this case $\widehat{\phi_{k_2}}(\xi_2)$ must be ϕ type. In fact, instead of (8.36) we would have preferred to deal with an expression of the type

$$\sum_{k_1 \ll k_2} \widehat{\phi_{k_1}}(\xi_1)\widehat{\phi_{k_1}}(\xi_2)\widehat{\phi_{k_2}}(\xi_1 + \xi_2)\widehat{\phi_{k_2}}(\xi_3). \tag{8.37}$$

Indeed, we could have "completed" (8.37) as

$$\sum_{k_1 \ll k_2} \widehat{\phi_{k_1}}(\xi_1)\widehat{\phi_{k_1}}(\xi_2)\widehat{\phi_{k_1}}(\xi_1 + \xi_2)\widehat{\phi_{k_2}}(\xi_1 + \xi_2)\widehat{\phi_{k_2}}(\xi_3)\widehat{\phi_{k_2}}(\xi_1 + \xi_2 + \xi_3)$$

and then the trilinear operator defined by this symbol could have been conveniently rewritten as

$$\sum_{k_2} \left(\left(\sum_{k_1 \ll k_2} ((f * \phi_{k_1})(g * \phi_{k_1})) * \phi_{k_1} \right) * \phi_{k_2}(h * \phi_{k_2}) \right) * \phi_{k_2}. \tag{8.38}$$

Such operators can easily be reduced to discrete operators of the type T_2, using the standard discretization procedure described in context of paraproducts in Chapter 2.

Exercise 8.6 Prove that the trilinear operator in (8.38) can be discretized and written as an average of operators of the type T_2.

Next we will explain how to turn an expression of the type (8.36) into one much closer to (8.37). The idea behind it is that when $k_1 \ll k_2$ then $\widehat{\phi_{k_2}}(\xi_2)$ becomes close to $\widehat{\phi_{k_2}}(\xi_1 + \xi_2)$ and this suggests the use of a Taylor expension. More precisely, one can write

$$\widehat{\phi_{k_2}}(\xi_2) = \widehat{\phi_{k_2}}(\xi_1 + \xi_2) + \frac{\widehat{\phi}'_{k_2}(\xi_1 + \xi_2)}{1!}(-\xi_1)$$
$$+ \frac{\widehat{\phi}''_{k_2}(\xi_1 + \xi_2)}{2!}(-\xi_1)^2 + \cdots + \frac{\widehat{\phi}^M_{k_2}(\xi_1 + \xi_2)}{M!}(-\xi_1)^M + R_M(\xi_1, \xi_2).$$

Clearly, the zeroth term will generate a model operator of the type (8.38), as desired, but we have to understand what is happening with the operators corresponding to the rest of the terms.

Fix $0 < \ell \leq M$ and consider the intermediate term

$$\frac{\widehat{\Phi}^{\ell}_{k_2}(\xi_1 + \xi_2)}{\ell!}(-\xi_1)^{\ell}.$$

If we insert this in the place of $\widehat{\phi}_{k_2}(\xi_2)$ in (8.36), the multiplier becomes

$$\sum_{k_1 \ll k_2} \widehat{\phi}_{k_1}(\xi_1)\widehat{\phi}_{k_1}(\xi_2)\frac{\widehat{\phi}^{\ell}_{k_2}(\xi_1 + \xi_2)}{\ell!}(-\xi_1)^{\ell}\widehat{\phi}_{k_2}(\xi_3)$$

$$= \sum_{\mu=101}^{\infty} \sum_{k_2 = k_1 + \mu} \widehat{\phi}_{k_1}(\xi_1)\widehat{\phi}_{k_1}(\xi_2)\frac{\widehat{\phi}^{\ell}_{k_2}(\xi_1 + \xi_2)}{\ell!}(-\xi_1)^{\ell}\widehat{\phi}_{k_2}(\xi_3)$$

$$= \sum_{\mu=101}^{\infty} \sum_{k_2 = k_1 + \mu} \widehat{\phi}_{k_1}(\xi_1)(-\xi_1)^{\ell}\widehat{\phi}_{k_1}(\xi_2)\frac{\widehat{\phi}^{\ell}_{k_2}(\xi_1 + \xi_2)}{\ell!}\widehat{\phi}_{k_2}(\xi_3)$$

$$=: \sum_{\mu=101}^{\infty} \sum_{k_2 = k_1 + \mu} \frac{2^{k_1 \ell}}{2^{k_2 \ell}}\widehat{\phi}_{k_1,\ell}(\xi_1)\widehat{\phi}_{k_1}(\xi_2)\widehat{\phi}_{k_2,\ell}(\xi_1 + \xi_2)\widehat{\phi}_{k_2}(\xi_3)$$

$$= \sum_{\mu=101}^{\infty} 2^{-\mu\ell} \sum_{k_2 = k_1 + \mu} \widehat{\phi}_{k_1,\ell}(\xi_1)\widehat{\phi}_{k_1}(\xi_2)\widehat{\phi}_{k_2,\ell}(\xi_1 + \xi_2)\widehat{\phi}_{k_2}(\xi_3). \quad (8.39)$$

For any fixed μ, the terms in (8.39) are of a *good type*, and so (8.39) can be rewritten in the form of (8.38), the only difference being that now we have $k_2 = k_1 + \mu$. Then the usual discretization procedure reduces the analysis of the corresponding operator to the analysis of $T_{2,\mu}$, defined in (8.34). Of course, one can add all these contributions together, thanks to the decay factor $2^{-\mu\ell}$ in (8.39). Finally, it is easy to see that the *rest operator* (the operator corresponding to $R_M(\xi_1, \xi_2)$ in the Taylor formula) can be also written in the form

$$\sum_{\mu=101}^{\infty} 2^{-\mu M} T_{\mu},$$

where T_{μ} is a Coifman–Meyer operator whose symbol m_{μ} satisfies estimates

$$\left|\partial^{\alpha} m_{\mu}(\xi)\right| \lesssim 2^{\mu|\alpha|}\frac{1}{|\xi|^{|\alpha|}}$$

for sufficiently many multi-indices α. In particular, the Coifman–Meyer theorem implies that T_{μ} is bounded, with a bound that is of the type $O(2^{100\mu})$, say. This is acceptable if we recall the large decay factor $2^{-M\mu}$.

Now, if we assume Theorem 8.2, then Theorem 8.1 follows as usual from standard approximation arguments and the multilinear interpolation explained in the appendix, using the observation that every point of the form $(1/p_1, 1/p_2, 1/p_3, 1/p_4')$ coming from the 4-tuple (p_1, p_2, p_3, p_4) for which $1 < p_1, p_2, p_3 < \infty$, $1/p_1 + 1/p_2 + 1/p_3 = 1/p_4$, and $0 < p_4 < \infty$ lies in the interior of $\mathbb{B}$ (in fact, it lies in $\widetilde{\mathbb{B}}$). The L^∞ estimates follow similarly, since we observe that points of the form $(0, \alpha, \beta, \gamma)$, $(\alpha, 0, \beta, \gamma)$, $(\alpha, \beta, 0, \gamma)$, with $\alpha, \beta > 0$ and $\alpha + \beta + \gamma = 1$ and $(0, \widetilde{\alpha}, 0, \widetilde{\beta})$ with $\widetilde{\alpha}, \widetilde{\beta} > 0$ and $\widetilde{\alpha} + \widetilde{\beta} = 1$ belong to $\mathbb{B}$ as well.

It is therefore enough to prove Theorem 8.2 as claimed.

8.7. Rewriting the 4-linear forms

In order to prove Theorem 8.2, it is clearly enough to consider the operators T_1 and T_{1,ℓ_0}, since T_2 and T_{2,ℓ_0} can be treated similarly.

For technical reasons that will become clearer later (briefly, the I intervals are smaller than the corresponding J intervals), we will switch the I and J summations and rewrite the form Λ_1 associated with T_1 as

$$\Lambda_1(f_1, f_2, f_3, f_4) = \sum_{J \in \mathcal{J}_1} \frac{1}{|J|^{1/2}} a_J^{(1)} a_J^{(2)} a_J^{(3)}, \tag{8.40}$$

where

$$a_J^{(1)} = \langle f_2, \phi_J^1 \rangle,$$
$$a_J^{(2)} = \langle f_3, \phi_J^2 \rangle,$$

and

$$a_J^{(3)} := \left\langle \sum_{\substack{I \in \mathcal{I}_1 \\ \omega_J^3 \cap \omega_I^2 \neq \emptyset; |\omega_J^3| \leq |\omega_I^2|}} \frac{1}{|I|^{1/2}} \langle f_1, \phi_I^1 \rangle \langle f_4, \phi_I^3 \rangle \phi_I^2, \phi_J^3 \right\rangle.$$

Similarly, we will rewrite the form Λ_{1,ℓ_0} associated with T_{1,ℓ_0} as

$$\Lambda_{1,\ell_0}(f_1, f_2, f_3, f_4) = \sum_{J \in \mathcal{J}_1} \frac{1}{|J|^{1/2}} a_J^{(1)} a_J^{(2)} a_{J,\ell_0}^{(3)} \tag{8.41}$$

where

$$a_{J,\ell_0}^{(3)} = \left\langle \sum_{\substack{I \in \mathcal{I}_1 \\ \omega_J^3 \cap \omega_I^2 \neq \emptyset; 2^{\ell_0} |\omega_J^3| \simeq |\omega_I^2|}} \frac{1}{|I|^{1/2}} \langle f_1, \phi_I^1 \rangle \langle f_4, \phi_I^3 \rangle \phi_I^2, \phi_J^3 \right\rangle.$$

Let us also recall that the family $(\phi_I^2)_I$ may be nonlacunary, while the families $(\phi_I^i)_I$ for $i \neq 2$ are both lacunary.

However, we also know that there exists a unique $j = 1, 2$, or 3, which we fix from now on in such a way that the corresponding family $(\phi_J^j)_J$ is nonlacunary, while the families $(\phi_J^i)_J$ for $i \neq j$ are both lacunary.

8.8. The new size and energy estimates

Clearly, as in the case of paraproducts, the proof of Theorem 8.2 is related to the way in which one is able to estimate the 4-linear forms in (8.40) and (8.41). As before, the generic way to do this will be by using Proposition 2.12, which says that one can estimate (8.40) by

$$\prod_{j=1}^{3} \left(\text{size}_{\mathcal{J}_1}^{(j)}\left(\left(a_J^j\right)_J\right)\right)^{1-\theta_j} \left(\text{energy}_{\mathcal{J}_1}^{(j)}\left(\left(a_J^j\right)_J\right)\right)^{\theta_j}, \tag{8.42}$$

for any $0 \leq \theta_1, \theta_2, \theta_3 < 1$ with $\theta_1 + \theta_2 + \theta_3 = 1$, and (8.41) by

$$\begin{aligned} &\left(\text{size}_{\mathcal{J}_1}^{(1)}((a_J^1)_J)\right)^{1-\theta_1} \left(\text{size}_{\mathcal{J}_1}^{(2)}((a_J^2)_J)\right)^{1-\theta_2} \left(\text{size}_{\mathcal{J}_1}^{(3)}\left((a_{J,\ell_0}^3)_J\right)\right)^{1-\theta_3} \\ &\quad \times \left(\text{energy}_{\mathcal{J}_1}^{(1)}((a_J^1)_J)\right)^{\theta_1} \left(\text{energy}_{\mathcal{J}_1}^{(2)}((a_J^2)_J)\right)^{\theta_2} \left(\text{energy}_{\mathcal{J}_1}^{(3)}\left((a_{J,\ell_0}^3)_J\right)\right)^{\theta_3}, \end{aligned} \tag{8.43}$$

again for $\theta_1, \theta_2, \theta_3$ as before.

The reader should recall the definitions of all these quantities carefully, since they will play an important role from now on. One should also observe that only the third size factor and the third energy factor are new; the others are similar to those appearing in the study of classical paraproducts in Chapter 2. The following two lemmas describe the way in which one can estimate them.

Lemma 8.3 *Let $E_1, E_4 \subseteq \mathbb{R}$ be sets of finite measure and let $|f_1| \leq \chi_{E_1}$ and $|f_4| \leq \chi_{E_4}$. Then one has*

$$\begin{aligned} &\text{size}_{\mathcal{J}_1}^{(3)}((a_J^3)_J),\ \text{size}_{\mathcal{J}_1}^{(3)}\left((a_{J,\ell_0}^3)_J\right) \\ &\quad \lesssim \left(\sup_{J\in\mathcal{J}_1} \frac{1}{|J|}\int_{E_1} \widetilde{\chi}_J^N dx\right)^{1-\theta} \left(\sup_{J\in\mathcal{J}_1} \frac{1}{|J|}\int_{E_4} \widetilde{\chi}_J^N\, dx\right)^{\theta} \end{aligned}$$

for any $0 < \theta < 1$ and any positive integer N, where the implicit constants depend on θ and N.

Lemma 8.4 *With the same notation as before, one has*

$$\text{energy}^{(3)}_{\mathcal{J}_1}\left((a^3_J)_J\right), \text{energy}^{(3)}_{\mathcal{J}_1}\left((a^3_{J,\ell_0})_J\right)$$

$$\lesssim \left(\sup_{I\in\mathcal{I}_1}\frac{1}{|I|}\int_{E_1}\widetilde{\chi}_I^N\,dx\right)^{1-\theta_1}\left(\sup_{I\in\mathcal{I}_1}\frac{1}{|I|}\int_{E_4}\widetilde{\chi}_I^N\,dx\right)^{1-\theta_2}|E_1|^{\theta_1}|E_4|^{\theta_2}$$

for any $0 \le \theta_1, \theta_2 < 1$ with $\theta_1 + \theta_2 = 1$ and for any integer N, with the implicit constants depending on θ_1, θ_2, and N.

We will prove both these lemmas in detail later. For the next two sections will assume that they hold true and use them to develop the proof of Theorem 8.2.

8.9. Estimates for T_1 and T_{1,ℓ_0} near A_4

Let us start by considering a 4-tuple (p_1, p_2, p_3, p_4) having the property that $(1/p_1, 1/p_2, 1/p_3, 1/p_4') \in \mathbb{B}$ and which is arbitrarily close to A_4. Recall that A_4 has coordinates $(1, 1, 1, -2)$. Also consider $E_1, E_2, E_3, E_4 \subseteq \mathbb{R}$ of finite measure with $|E_4| = 1$. As we know already, by scaling invariance this can be assumed without loss of generality. The goal is to construct $E_4' \subseteq E_4$ with the property $|E_4'| \simeq 1$ and such that

$$|\Lambda_1(f_1, f_2, f_3, f_4)|, |\Lambda_{1,\ell_0}(f_1, f_2, f_3, f_4)| \lesssim |E_1|^{1/p_1}|E_2|^{1/p_2}|E_3|^{1/p_3} \quad (8.44)$$

for any $|f_i| \le \chi_{E_i}$, $i \in \{1, 2, 3\}$, and $|f_4| \le \chi_{E_4'}$.

Define first an exceptional set Ω by

$$\Omega = \bigcup_{j=1}^{3}\left(M\left(\frac{\chi_{E_j}}{|E_j|}\right) > C\right)$$

and notice as always that $|\Omega| \ll 1$ if C is large enough. Then define $E_4' := E_4 \backslash \Omega$, which has the desired property $|E_4'| \simeq 1$. After that, we decompose $\mathcal{J}_1$ and $\mathcal{I}_1$ as

$$\mathcal{J}_1 = \bigcup_{d\ge 0}\mathcal{J}_1^d \quad \text{and} \quad \mathcal{I}_1 := \bigcup_{d'\ge 0}\mathcal{I}_1^{d'},$$

where $\mathcal{J}_1^d$ is the collection of all the intervals $J \in \mathcal{J}_1$ having the property that

$$\left(1 + \frac{\text{dist}(J, \Omega^c)}{|J|}\right) \simeq 2^d$$

and $\mathcal{I}_1^{d'}$ is the collection of all intervals $I \in \mathcal{I}_1$ having the property that

$$\left(1 + \frac{\operatorname{dist}(I, \Omega^c)}{|I|}\right) \simeq 2^{d'}.$$

Now using the definition of the exceptional set Ω, we know that

$$\frac{1}{|J|} \int_{E_j} \widetilde{\chi}_J \, dx \lesssim 2^d |E_j|$$

for $j = 1, 2, 3$ and, since clearly the left-hand side is also smaller than 1, we deduce that

$$\frac{1}{|J|} \int_{E_j} \widetilde{\chi}_J \, dx \lesssim 2^{\alpha d} |E_j|^\alpha$$

for any $0 \leq \alpha \leq 1$ and $j = 1, 2, 3$.

In a completely similar way, we deduce that

$$\frac{1}{|I|} \int_{E_j} \widetilde{\chi}_I \, dx \lesssim 2^{\beta d'} |E_j|^\beta$$

for every $0 \leq \beta \leq 1$ and $j = 1, 2, 3$.

However, given that $E_4' \subseteq \Omega^c$ we also know that

$$\frac{1}{|J|} \int_{E_4'} \widetilde{\chi}_J^N \, dx \lesssim 2^{-Nd} \quad \text{and} \quad \frac{1}{|I|} \int_{E_4'} \widetilde{\chi}_I^N \, dx \lesssim 2^{-Nd'}$$

for an arbitrary integer $N > 0$. Using all these observations together with Lemmas 2.13 and 2.14 estimating the size and energy and also Lemmas 8.3 and 8.4 above, one concludes that

$$\operatorname{size}^{(1)}_{\mathcal{J}_1^d}\left((a_J^1)_J\right) \lesssim 2^{d\alpha_2} |E_2|^{\alpha_2}, \quad \operatorname{size}^{(2)}_{\mathcal{J}_1^d}\left((a_J^2)_J\right) \lesssim 2^{d\alpha_3} |E_3|^{\alpha_3},$$

$$\operatorname{size}^{(3)}_{\mathcal{J}_1^d}\left((a_J^3)_J\right), \ \operatorname{size}^{(3)}_{\mathcal{J}_1^d}\left(\left(a_{J,\ell_0}^3\right)_J\right) \lesssim (2^{d\alpha_1} |E_1|^{\alpha_1})^{1-\theta} (2^{-Nd})^\theta$$

and also that

$$\operatorname{energy}^{(1)}_{\mathcal{J}_1^d}\left((a_J^1)_J\right) \lesssim |E_2|, \quad \operatorname{energy}^{(2)}_{\mathcal{J}_1^d}\left((a_J^2)_J\right) \lesssim |E_3|,$$

$$\operatorname{energy}^{(3)}_{\mathcal{J}_1^d}\left((a_J^3)_J\right), \ \operatorname{energy}^{(3)}_{\mathcal{J}_1^d}\left(\left(a_{J,\ell_0}^3\right)_J\right) \lesssim (2^{d'\beta_1} |E_1|^{\beta_1})^{1-\theta_1'} (2^{-Nd'})^{1-\theta_2'} |E_1|^{\theta_1'}.$$

These hold whenever $0 \leq \alpha_1, \alpha_2, \alpha_3, \beta_1 \leq 1$, $0 < \theta < 1$, and $0 \leq \theta_1', \theta_2' < 1$ with $\theta_1' + \theta_2' = 1$. Now using (8.42) and (8.43) we can estimate the left-hand side of (8.44) by

$$(2^{d\alpha_2}|E_2|^{\alpha_2})^{1-\theta_1}(2^{d\alpha_3}|E_3|^{\alpha_3})^{1-\theta_2}\left((2^d|E_1|)^{1-\theta}(2^{-Nd})^{\theta}\right)^{1-\theta_3}$$
$$\times |E_2|^{\theta_1}|E_3|^{\theta_2}\left((2^{d'}|E_1|)^{1-\theta_1'}(2^{-Nd'})^{1-\theta_2'}|E_1|^{\theta_1'}\right)^{\theta_3}$$
$$= |E_1|^{(1-\theta)(1-\theta_3)+\theta_3}|E_2|^{\alpha_2(1-\theta_1)+\theta_1}|E_3|^{\alpha_3(1-\theta_2)+\theta_2}2^{-\delta_1 d}2^{-\delta_2 d'},$$

where both δ_1 and δ_2 are positive real numbers depending on all the previous parameters and also on N. Then, if one takes $\theta_1, \theta_2, \theta_3$ such that θ_1 is close to 0 and both α_2 and α_3 are close to 1, one can then define $1/p_1 := (1-\theta)(1-\theta_3)+\theta_3$, $1/p_2 := \alpha_2(1-\theta_1)+\theta_1$, and $1/p_3 := \alpha_3(1-\theta_2)+\theta_2$ and notice that they can be chosen as close as we want to the coordinate point $(1, 1, 1)$.

Finally, one can clearly sum over $d, d' \geq 0$ if N is chosen large enough.

It is not difficult to observe that a similar argument proves the desired estimates for T_1^{*1} and T_{1,ℓ_0}^{*1} near to the points A_{11} and A_{12}. In fact one could even prove the estimates near to the point A_1, but these would not be useful.

8.10. Estimates for T_1^{*3} and T_{1,ℓ_0}^{*3} near A_{31} and A_{32}

We will use ideas similar to the previous argument. The details are as follows. Pick (p_1, p_2, p_3, p_4) such that $(1/p_1, 1/p_2, 1/p_3', 1/p_4) \in \mathbb{B}$ and is arbitrarily close to either A_{31} or A_{32}. Consider as before measurable sets $E_1, E_2, E_3, E_4 \subseteq \mathbb{R}$ of finite measure and assume that $|E_3| = 1$. As usual we want to construct $E_3' \subseteq E_3$ with $|E_3'| \simeq 1$ such that

$$|\Lambda_1(f_1, f_2, f_3, f_4)|,\ |\Lambda_{1,\ell_0}(f_1, f_2, f_3, f_4)| \lesssim |E_1|^{1/p_1}|E_2|^{1/p_2}|E_4|^{1/p_4} \quad (8.45)$$

for every $|f_i| \leq \chi_{E_i}$ $i = 1, 2, 4$, and $|f_3| \leq \chi_{E_3'}$.

To be able to do this, define first the exceptional set

$$\Omega = \bigcup_{j=1,2,4}\left\{M\left(\frac{\chi_{E_j}}{|E_j|}\right) > C\right\}$$

and define $E_3' := E_3 \backslash \Omega$ for a sufficiently large constant $C > 0$. Arguing as in the previous section one obtains the estimates (this time, as we shall see, there is no need to decompose the family $\mathcal{I}_1$ as we did there)

$$\text{size}^{(1)}_{\mathcal{J}_1^d}\left((a_J^1)_J\right) \lesssim 2^{d\alpha_2}|E_2|^{\alpha_2}, \quad \text{size}^{(2)}_{\mathcal{J}_1^d}\left((a_J^2)_J\right) \lesssim 2^{-Nd},$$
$$\text{size}^{(3)}_{\mathcal{J}_1^d}\left((a_J^3)_J\right),\ \text{size}^{(3)}_{\mathcal{J}_1^d}\left(\left(a_{J,\ell_0}^3\right)_J\right) \lesssim (2^{d\alpha_1}|E_1|^{\alpha_1})^{1-\theta}(2^{d\alpha_4}|E_4|^{\alpha_4})^{\theta}$$

and also

$$\text{energy}^{(1)}_{\mathcal{J}_1^d}\left((a_J^1)_J\right) \lesssim |E_2|, \quad \text{energy}^{(2)}_{\mathcal{J}_1^d}\left((a_J^2)_J\right) \lesssim 1,$$
$$\text{energy}^{(3)}_{\mathcal{J}_1^d}\left((a_J^3)_J\right), \ \text{energy}^{(3)}_{\mathcal{J}_1^d}\left((a_{J,\ell_0}^3)_J\right) \lesssim |E_1|^{\widetilde{\theta_1}}|E_4|^{\widetilde{\theta_2}},$$

which hold as long as $0 \leq \alpha_1, \alpha_2, \alpha_4 \leq 1,\ 0 < \theta < 1$, and $0 \leq \widetilde{\theta_1}, \widetilde{\theta_2} < 1$ such that $\widetilde{\theta_1} + \widetilde{\theta_2} = 1$.

Then, using all these inequalities one can as before estimate the left-hand side of (8.45) by

$$(2^{d\alpha_2}|E_2|^{\alpha_2})^{1-\theta_1}(2^{-Nd})^{1-\theta_2}\left((2^d|E_1|)^{1-\theta}(2^d|E_4|)^{\theta}\right)^{1-\theta_3}|E_2|^{\theta_1}\left(|E_1|^{\widetilde{\theta_1}}|E_4|^{\widetilde{\theta_2}}\right)^{\theta_3}$$
$$= |E_1|^{(1-\theta)(1-\theta_3)+\widetilde{\theta_1}\theta_3}|E_2|^{\alpha_2(1-\theta_1)+\theta_1}|E_4|^{\theta(1-\theta_3)+\widetilde{\theta_2}\theta_3}2^{-\delta d},$$

where δ is a positive real number depending on all the parameters including N. Now we define $1/p_1 := (1-\theta)(1-\theta_3) + \widetilde{\theta_1}\theta_3$, $1/p_2 := \alpha_2(1-\theta_1)+\theta_1$, and $1/p_4 := \theta(1-\theta_3)+\widetilde{\theta_2}\theta_3$ and, given that $(1-\theta)(1-\theta_3)+\widetilde{\theta_1}\theta_3+\theta(1-\theta_3)+\widetilde{\theta_2}\theta_3 = 1$, we can easily see that p_2 can be indeed chosen very close to 1 by choosing α_2 close to 1 and likewise the pair $\left(1/p_1, 1/p_4\right)$ can be chosen as close as we wish to either $(0,1)$ or $(1,0)$, as desired. Then once again we can sum over $d \geq 0$ for large enough N.

A similar argument proves the desired estimates for the operators T_1^{*2} and T_{1,ℓ_0}^{*2} near the points A_{21} and A_{22}. This completes the proof of Theorem 8.2 since, as we mentioned, by symmetry the analyses of T_2 and T_{2,ℓ_0} are identical to those for the above operators respectively. It therefore remains to prove Lemmas 8.3 and 8.4.

8.11. Upper bounds for flag sizes

In this section we will prove Lemma 8.3. First let us observe that we can assume from the beginning, for simplicity, that both collections $\mathcal{I}_1$ and $\mathcal{J}_1$ are *sparse* in the sense of the definition given in Chapter 6.

Case 1: Estimates for $\text{size}^{(3)}_{\mathcal{J}_1}((a_J^3)_J)$

Recall that the index j, which has been fixed earlier, denotes which position is nonlacunary in the families $(\phi_J^i)_J \ \ i = 1, 2, 3$.

Case I_1***:*** $j \neq 3$. Since $j \neq 3$ we have to work with the *lacunary* definition of sizes in Chapter 2. Fix an arbitrary dyadic interval $J_0 \in \mathcal{J}_1$. It is easy to see that in order to prove the estimate for $\text{size}^{(3)}_{\mathcal{J}_1}((a_J^3)_J)$ it is enough to show that

$$\left\| \left(\sum_{J \subseteq J_0} \frac{|a_J^3|^2}{|J|} \chi_J \right)^{1/2} \right\|_{1,\infty} \lesssim \| f_1 \widetilde{\chi}_{J_0}^N \|_p \| f_4 \widetilde{\chi}_{J_0}^N \|_q$$

for every $1 < p, q < \infty$ with $1/p + 1/q = 1$.

Recall now that a_J^3 is defined as follows:

$$a_J^3 = \left\langle \sum_{\substack{I \in \mathcal{I}_1 \\ \omega_J^3 \cap \omega_I^2 \neq \emptyset; |\omega_J^3| \leq |\omega_I^2|}} \frac{1}{|I|^{1/2}} \langle f_1, \phi_I^1 \rangle \langle f_4, \phi_I^3 \rangle \phi_I^2, \phi_J^3 \right\rangle. \tag{8.46}$$

Consider a new family of dyadic intervals $\widetilde{\mathcal{I}}$, the set of all $I \in \mathcal{I}_1$ for which there exists $J \in \mathcal{J}_1$ such that $\omega_J^3 \cap \omega_I^2 \neq \emptyset$ and $|\omega_J^3| \leq |\omega_I^2|$. We claim now that in fact, for every J, one has

$$a_J^3 = \langle B(f_1, f_4), \phi_J^3 \rangle, \tag{8.47}$$

where $B(f_1, f_4)$ is given by

$$B(f_1, f_4) = \sum_{I \in \widetilde{\mathcal{I}}} \frac{1}{|I|^{1/2}} \langle f_1, \phi_I^1 \rangle \langle f_4, \phi_I^3 \rangle \phi_I^2.$$

To prove this claim let us first observe that, for each I, in order for $\langle \phi_I^2, \phi_J^3 \rangle$ to be nonzero one must have $\omega_J^3 \cap \omega_I^2 \neq \emptyset$. Then, observe that in (8.47) we have, implicitly, many terms. Some correspond to the situation when $|\omega_J^3| \leq |\omega_I^2|$, which are clearly acceptable in (8.46), but then there is also the case when $|\omega_I^2| < |\omega_J^3|$ and $\langle \phi_I^2, \phi_J^3 \rangle \neq \emptyset$. The observation is that this case cannot appear. Indeed, if there were an $I \in \widetilde{\mathcal{I}}$ such that $|\omega_I^2| < |\omega_J^3|$ and $\langle \phi_I^2, \phi_J^3 \rangle \neq \emptyset$ then, since $I \in \widetilde{\mathcal{I}}$, there would exist another $J' \in \mathcal{J}_1$ such that $\omega_I^2 \cap \omega_{J'}^3 \neq \emptyset$ and such that $|\omega_{J'}^3| \leq |\omega_I^2|$. But this would mean that $\omega_I^2 \cap \omega_J^3 \neq \emptyset$, $\omega_I^2 \cap \omega_{J'}^3 \neq \emptyset$, and $|\omega_{J'}^3| \leq |\omega_I^2| < |\omega_J^3|$, and the last inequality implies $|\omega_{J'}^3| \ll |\omega_J^3|$ by the sparseness assumption. But then this would imply that (for instance) $2\omega_J^3 \cap \omega_{J'}^3 \neq \emptyset$, which would contradict the lacunarity of the third position (recall that $j \neq 3$), i.e., of the family $(\phi_J^3)_J$.

Using (8.47) and also the size estimate in Lemma 2.13, we see that our desired inequality can be reduced to

$$\left\| B(f_1, f_4)\widetilde{\chi}_{J_0}^{N'} \right\|_1 \lesssim \| f_1 \widetilde{\chi}_{J_0}^{N} \|_p \| f_4 \widetilde{\chi}_{J_0}^{N} \|_q . \tag{8.48}$$

By scaling invariance, one can of course assume in the above that $|J_0| = 1$. Our goal now is to prove a slightly weaker variant of (8.48), namely

$$\| B(f_1, f_4)\chi_J \|_1 \lesssim \| f_1 \widetilde{\chi}_J^{N} \|_p \| f_4 \widetilde{\chi}_J^{N} \|_q , \tag{8.49}$$

which we claim to hold true for every dyadic interval $J \subseteq \mathbb{R}$ of length 1. We now show that (8.49) easily implies (8.48). To see this, consider a partition $(J_n)_{n\in\mathbb{Z}^*}$ of the real line with disjoint intervals of length 1:

$$\left(\bigcup_{n\in\mathbb{Z}^*} J_n \right) \cup J_0 = \mathbb{R}.$$

Then, using (8.49), (8.48) can be estimated by

$$\begin{aligned} \left\| B(f_1, f_4)\widetilde{\chi}_{J_0}^{N'} \right\|_1 &\lesssim \left\| B(f_1, f_4)\chi_{J_0} \right\|_1 + \sum_{n\in\mathbb{Z}^*} \left\| B(f_1, f_4)\widetilde{\chi}_{J_0}^{N'} \chi_{J_n} \right\|_1 \\ &\lesssim \left\| B(f_1, f_4)\chi_{J_0} \right\|_1 + \sum_{n\in\mathbb{Z}^*} \frac{1}{|n|^{N'}} \left\| B(f_1, f_4)\chi_{J_n} \right\|_1 . \end{aligned}$$

From (8.49), the first term satisfies the desired estimate in (8.48). The rest can be further estimated by

$$\begin{aligned} &\sum_{n\in\mathbb{Z}^*} \frac{1}{|n|^{N'}} \left\| f_1 \widetilde{\chi}_{J_n}^{N''} \right\|_p \left\| f_4 \widetilde{\chi}_{J_n}^{N''} \right\|_q \\ &\lesssim \left(\sum_{n\in\mathbb{Z}^*} \frac{1}{|n|^{N'}} \left\| f_1 \widetilde{\chi}_{J_n}^{N''} \right\|_p^p \right)^{1/p} \left(\sum_{n\in\mathbb{Z}^*} \frac{1}{|n|^{N'}} \left\| f_4 \widetilde{\chi}_{J_n}^{N''} \right\|_q^q \right)^{1/q} \\ &\lesssim \left(\int_{\mathbb{R}} |f_1|^p \left(\sum_{n\in\mathbb{Z}^*} \frac{1}{|n|^{N'}} \widetilde{\chi}_{J_n}^{pN''} \right) dx \right)^{1/p} \\ &\quad \times \left(\int_{\mathbb{R}} |f_4|^q \left(\sum_{n\in\mathbb{Z}^*} \frac{1}{|n|^{N'}} \widetilde{\chi}_{J_n}^{qN''} \right) dx \right)^{1/q} \\ &\lesssim \| f_1 \widetilde{\chi}_{J_0}^{N} \|_p \| f_1 \widetilde{\chi}_{J_0}^{N} \|_q , \end{aligned}$$

if N' and N'' are large enough in comparison with N.

Hence we are left with proving (8.49).

Case I_{1a}: $\mathrm{supp}\, f_1, \mathrm{supp} f_4 \subseteq 5J$. In this case our desired inequality (8.49) follows clearly from the known inequalities for discretized paraproducts in Chapter 2.

Case I_{1b}: Either $\mathrm{supp}\, f_1 \subseteq (5J)^c$ or $\mathrm{supp} f_4 \subseteq (5J)^c$. Let us assume that for example

$$\mathrm{supp}\, f_1 \subseteq (5J)^c. \tag{8.50}$$

Decompose $B(f_1, f_4)$ as follows:

$$B(f_1, f_4) = B'(f_1, f_4) + B''(f_1, f_4)$$

where

$$B'(f_1, f_4) = \sum_{\substack{I \in \widetilde{\mathcal{I}} \\ I \cap 3J \neq \emptyset}} \frac{1}{|I|^{1/2}} \langle f_1, \phi_I^1 \rangle \langle f_4, \phi_I^3 \rangle \phi_I^2$$

and

$$B''(f_1, f_4) = \sum_{\substack{I \in \widetilde{\mathcal{I}} \\ I \cap 3J = \emptyset}} \frac{1}{|I|^{1/2}} \langle f_1, \phi_I^1 \rangle \langle f_4, \phi_I^3 \rangle \phi_I^2.$$

Given that $|J_0| = 1$ we know now that the lengths of our intervals in $\widetilde{\mathcal{I}}$ are smaller than or equal to 1.

If now $h \in L^\infty$ with $\|h\|_\infty \leq 1$, one has

$$\left| \int_{\mathbb{R}} B'(f_1, f_4)(x) h(x) \chi_J(x)\, dx \right|$$

$$\lesssim \sum_{\substack{I \in \widetilde{\mathcal{I}} \\ I \cap 3J \neq \emptyset}} \frac{1}{|I|^{1/2}} |\langle f, \phi_I^1 \rangle| \, |\langle f_4, \phi_I^3 \rangle| \, |\langle h\chi_J, \phi_I^2 \rangle|$$

$$= \sum_{k=0}^{\infty} \sum_{\substack{I \in \widetilde{\mathcal{I}} \\ I \cap 3J \neq \emptyset \\ |I| = 2^{-k}}} 2^{2k} \, |\langle f_1, 2^{-k/2} \phi_I^1 \rangle| \, |\langle f_4, 2^{-k/2} \phi_I^3 \rangle| \, |\langle h\chi_J, 2^{-k/2} \phi_I^2 \rangle| \, .$$

Notice that the smooth bump functions $2^{-k/2}\phi_I^1$, $2^{-k/2}\phi_I^3$, and $2^{-k/2}\phi_I^2$ are all L^∞-normalized. In particular, we can further estimate the above expression by (taking also advantage of (8.50))

$$\sum_{k=0}^{\infty} 2^{2k} \left(\sup_{\substack{I \cap 3J \neq \emptyset \\ |I| = 2^{-k}}} |\langle f_1, 2^{-k/2} \phi_I^1 \rangle| \right) \left(\sup_{\substack{I \cap 3J \neq \emptyset \\ |I| = 2^{-k}}} |\langle f_4, 2^{-k/2} \phi_I^3 \rangle| \right)$$

$$\times \left\langle h\chi_J, \sum_{I \cap 3J \neq \emptyset : |I| = 2^{-k}} \widetilde{\chi}_I^N \right\rangle$$

$$\lesssim \sum_{k=0}^{\infty} 2^{2k} 2^{-100k} \| f_1 \widetilde{\chi}_J^N \|_1 \| f_4 \widetilde{\chi}_J^N \|_1 \lesssim \| f_1 \widetilde{\chi}_J^N \|_p \| f_4 \widetilde{\chi}_J^N \|_q .$$

Similarly, one has

$$\left| \int_{\mathbb{R}} B''(f_1, f_4)(x) h(x) \chi_J(x)\, dx \right|$$

$$\lesssim \sum_{k=0}^{\infty} \sum_{\substack{I \in \widetilde{\mathcal{I}}: \\ I \cap 3J = \emptyset \\ |I| = 2^{-k}}} 2^{2k} |\langle f_1, 2^{-k/2} \phi_I^1 \rangle| \, |\langle f_4, 2^{-k/2} \phi_I^3 \rangle| \, |\langle h\chi_J, 2^{-k/2} \phi_I^2 \rangle|$$

$$\lesssim \sum_{k=0}^{\infty} 2^{2k} \sum_{\substack{I \in \widetilde{\mathcal{I}}: \\ I \cap 3J = \emptyset \\ |I| = 2^{-k}}} \operatorname{dist}(I, J)^{2N} \| f_1 \widetilde{\chi}_J^N \|_1 \| f_4 \widetilde{\chi}_J^N \|_1 \left(\frac{\operatorname{dist}(I, J)}{|I|} \right)^{-N'}$$

$$\lesssim \| f_1 \widetilde{\chi}_J^N \|_p \| f_4 \widetilde{\chi}_J^N \|_q \sum_{k=0}^{\infty} 2^{-(N'-2)k} \sum_{\substack{I \in \widetilde{\mathcal{I}} \\ I \cap 3J = \emptyset \\ |I| = 2^{-k}}} (\operatorname{dist}(I, J))^{-(N'-2N)}.$$

Finally, taking N' much bigger than $2N$ we deduce that the inner sum above is smaller than

$$\sum_{n=1}^{\infty} \frac{1}{\left(n 2^{-k}\right)^{N'-2N}} \leq 2^{k(N'-2N)},$$

which makes the geometric sum in k convergent.

The case I_2, when $j = 3$, clearly follows by the same arguments and is left to the reader.

Exercise 8.7 Recall carefully the definition of the *nonlacunary* sizes in Chapter 2 and complete Case I_2 above.

Case II: Estimates for $\text{size}_J^{(3)}((a^3_{J,\ell_0})_J)$

Case II$_1$***:*** $j \neq 3$. Exactly as before, fix an interval $J_0 \in \mathcal{J}_1$ and assume, also as before, without loss of generality, that $|J_0| = 1$ by scale invariance. To prove our estimate, it is enough to show that

$$\left\| \left(\sum_{J \subseteq J_0} \frac{\left|a^3_{J,\ell_0}\right|^2}{|J|} \chi_J \right)^{1/2} \right\|_{1,\infty} \lesssim \| f_1 \widetilde{\chi}^N_{J_0} \|_p \| f_4 \widetilde{\chi}^N_{J_0} \|_q . \tag{8.51}$$

Recall that a^3_{J,ℓ_0} was defined as follows:

$$a^3_{J,\ell_0} = \left\langle \sum_{\substack{I \in \mathcal{I}_1 : \omega^3_J \cap \omega^2_I \neq \emptyset \\ 2^{\ell_0} |\omega^3_J| \sim |\omega^2_I|}} \frac{1}{|I|^{1/2}} \langle f, \phi^1_I \rangle \langle f_4, \phi^3_I \rangle \phi^2_I, \phi^3_J \right\rangle.$$

Given that the frequency intervals ω^2_I and ω^3_J depend only on the scales $|I|$ and $|J|$, we can assume that given $|J|$ there exists only one $|I|$ such that $\omega^2_{|J|} \cap \omega^3_{|J|} \neq \emptyset$ and $2^{\ell_0} \left|\omega^3_{|J|}\right| \simeq \left|\omega^2_{|I|}\right|$. Fix a pair of such intervals I and J. Using Plancherel's theorem, we obtain

$$\langle \phi^2_I, \phi^3_J \rangle = \langle \widehat{\phi}^2_I, \widehat{\phi}^3_J \rangle . \tag{8.52}$$

Since $|J| \sim 2^{\ell_0} |I|$, consider a Schwartz function $\psi_{|I|,\ell_0}$ with the property that $\text{supp}\, \widehat{\psi}_{|I|,\ell_0} \subseteq 2\omega^3_{|J|}$ and $\widehat{\psi}_{|I|,\ell_0} \equiv 1$ on $\omega^3_{|J|}$. Then (8.52) can also be rewritten as

$$\begin{aligned}\left\langle \widehat{\phi^2_I}, \widehat{\phi^3_J} \cdot \widehat{\psi}_{|I|,\ell_0} \right\rangle &= \left\langle \widehat{\phi^2_I * \psi_{|I|,\ell_0}}, \widehat{\phi^3_J} \right\rangle = 2^{-\ell_0/2} \left\langle \widehat{2^{\ell_0/2} \phi^2_I * \psi}_{|I|,\ell_0}, \widehat{\phi^3_J} \right\rangle \\ &= 2^{-\ell_0/2} \left\langle \widehat{\widetilde{\phi^2_I}}, \widehat{\phi^3_J} \right\rangle\end{aligned}$$

where

$$\widetilde{\phi^2_I} := 2^{\ell_0/2} \phi^2_I * \psi_{|I|,\ell_0};$$

we also observe the important fact that $\widetilde{\phi^2_I}$ is now an L^2-normalized bump function adapted to $\widetilde{I}$, where $\widetilde{I}$ is the unique dyadic interval of length $2^{\ell_0}|I|$ containing I. We also note that for different scales, the supports of the Fourier transforms $\widehat{\widetilde{\phi^2_I}}$ are all disjoint. Using all this we deduce that

$$a^3_{J},_{\ell_0} = \langle \widetilde{B}_{\ell_0}(f_1, f_4), \phi^3_J \rangle$$

with

$$\widetilde{B}_{\ell_0}(f_1, f_4) := 2^{-\ell_0/2} \sum_I \frac{1}{|I|^{1/2}} \langle f_1, \phi^1_I \rangle \langle f_4, \phi^3_I \rangle \widetilde{\phi}^2_I .$$

Thus the inequality (8.51) will follow exactly as in case I once we can prove that

$$\|\widetilde{B}_{\ell_0}(f_1, f_4)\chi_J\|_1 \lesssim \|f_1\widetilde{\chi}_J^N\|_p \, \|f_4\widetilde{\chi}_J^N\|_q \tag{8.53}$$

for every dyadic interval $J \subseteq \mathbb{R}$ of length 1. Fix such an interval J.

Case II_{1a}: supp f_1, supp $f_4 \subseteq 5J$. To understand this case, let us consider again $h \in L^\infty$ with $\|h\|_\infty \leq 1$. Then

$$\left| \int_{\mathbb{R}} \widetilde{B}_{\ell_0}(f_1, f_2)(x)h(x)\chi_J(x)\,dx \right|$$
$$\lesssim 2^{-\ell_0/2} \sum_I \frac{1}{|I|^{1/2}} |\langle f_1, \phi_I^1\rangle| \, |\langle f_4, \phi_I^3\rangle| \, |\langle h\chi_J, \widetilde{\phi}_I^2\rangle|$$
$$= \sum_I |\langle f_1, \phi_I^1\rangle| \, |\langle f_4, \phi_I^3\rangle| \, \frac{|\langle h\chi_J, \widetilde{\phi}_I^2\rangle|}{2^{\ell_0/2}|I|^{1/2}}$$

and, since the function $\widetilde{\phi}_I^2/(2^{\ell_0/2}|I|^{1/2})$ is now L^1-normalized, we can estimate the above expression by

$$\sum_I |\langle f_1, \phi_I^1\rangle| \, |\langle f_4, \phi_I^3\rangle| = \sum_I \frac{|\langle f_1, \phi_I^1\rangle|}{|I|^{1/2}} \frac{|\langle f_4, \phi_I^3\rangle|}{|I|^{1/2}} |I|$$
$$= \int_{\mathbb{R}} \sum_I \frac{|\langle f_1, \phi_I^1\rangle|}{|I|^{1/2}} \frac{|\langle f_4, \phi_I^3\rangle|}{|I|^{1/2}} \chi_I(x)\,dx$$
$$\lesssim \int_{\mathbb{R}} \left(\sum_I \frac{|\langle f_1, \phi_I^1\rangle|^2}{|I|} \chi_I \right)^{1/2} \left(\sum_I \frac{|\langle f_4, \phi_I^3\rangle|^2}{|I|} \chi_I \right)^{1/2} dx$$
$$\lesssim \int_{\mathbb{R}} S(f_1)(x)S(f_4)(x)\,dx \lesssim \|S(f_1)\|_p \, \|S(f_4)\|_q$$
$$\lesssim \|f_1\|_p \, \|f_4\|_q \lesssim \|f_4\widetilde{\chi}_J^N\|_p \, \|f_4\widetilde{\chi}_J^N\|_q,$$

since we have assumed that both supp f_1 and supp f_4 are within $5J$. We have also used the fact that the families $(\phi_I^1)_I$ and $(\phi_I^3)_I$ are both lacunary to estimate the corresponding square functions.

Case II_{1b}: either supp$f_1 \subseteq (5J)^c$ or supp$f_4 \subseteq (5J)^c$. Assume for instance that supp $f_1 \subseteq (5J)^c$. Then, as before, decompose $\widetilde{B}_{\ell_0}(f_1, f_4)$ as follows:

$$\widetilde{B}_{\ell_0}(f_1, f_4) = \widetilde{B}'_{\ell_0}(f_1, f_4) + \widetilde{B}''_{\ell_0}(f_1, f_4),$$

where

$$\widetilde{B}'_{\ell_0}(f_1, f_3) = 2^{-\ell_0/2} \sum_{I:\widetilde{I}\cap 3J \neq \emptyset} \frac{1}{|I|^{1/2}} \langle f_1, \phi_I^1 \rangle \langle f_4, \phi_I^3 \rangle \widetilde{\phi}_I^2$$

and

$$\widetilde{B}''_{\ell_0}(f_1, f_4) = 2^{-\ell_0/2} \sum_{I:\widetilde{I}\cap 3J = \emptyset} \frac{1}{|I|^{1/2}} \langle f_1, \phi_I^1 \rangle \langle f_4, \phi_I^3 \rangle \widetilde{\phi}_I^2.$$

Recall that, for every I, $\widetilde{I}$ denotes the unique dyadic interval 2^{ℓ_0} times larger than and containing I. Observe also that, by construction, all the lenghts $|\widetilde{I}|$ are smaller than 1.

To estimate the first sum, for h as before, one can write

$$\left| \int_{\mathbb{R}} \widetilde{B}'_{\ell_0}(f_1, f_4)(x) h(x) \chi_J(x)\, dx \right|$$

$$\lesssim 2^{-\ell_0} \sum_{k=0}^{\infty} \sum_{\substack{I:\widetilde{I}\cap 3J \neq \emptyset \\ |I|=2^{-k}}} 2^{2k} \left|\langle f_1, 2^{-k/2}\phi_I^1 \rangle\right| \; \left|\langle f_4, 2^{-k/2}\phi_I^3 \rangle\right|$$

$$\times \left|\langle h\chi_J, 2^{\ell_0/2} 2^{-k/2} \widetilde{\phi}_I^2 \rangle\right|$$

and observe that the functions $2^{-k/2}\phi_I^1$, $2^{-k/2}\phi_I^3$, and $2^{\ell_0/2}2^{-k/2}\widetilde{\phi}_I^{\,2}$ are all L^∞-normalized. Then one can further estimate the previous expression by

$$2^{-\ell_0} \sum_{k=0}^{\infty} 2^{2k} \left(\sup_{\substack{\widetilde{I}\cap 3J \neq \phi \\ |I|=2^{-k}}} \left|\langle f_1, 2^{-k/2}\phi_I^1 \rangle\right| \right) \left(\sup_{\substack{\widetilde{I}\cap 3J \neq \emptyset \\ |I|=2^{-k}}} \left|\langle f_4, 2^{-k/2}\phi^3{}_I \rangle\right| \right)$$

$$\times \left| \left\langle |h|\chi_J, \sum_{\substack{\widetilde{I}\cap 3J \neq \emptyset \\ |I|=2^{-k}}} \widetilde{\chi}_{\widetilde{I}}^N \right\rangle \right|.$$

Given also that

$$\left| \left\langle |h|\chi_J, \sum_{\substack{\widetilde{I}\cap 3J \neq \emptyset; \\ |I|=2^{-k}}} \widetilde{\chi}_{\widetilde{I}}^N \right\rangle \right| \lesssim 2^{\ell_0},$$

the desired estimate follows by an argument simlar to that used in case I_{1b}.

Then, to estimate the contribution of $\widetilde{B}''_{\ell_0}$, we write

$$
\begin{aligned}
&\left| \int_{\mathbb{R}} \widetilde{B}''_{\ell_0}(f_1, f_4)(x)h(x)\chi_J(x)dx \right| \\
&\lesssim 2^{-\ell_0/2} \sum_{\widetilde{I}\cap 3J=\emptyset} \frac{1}{|I|^{1/2}} |\langle f_1, \phi_I^1\rangle| \, |\langle f_4, \phi_I^3\rangle| \, |\langle h\chi_J, \widetilde{\phi}_I^2\rangle| \\
&= \sum_{\widetilde{I}\cap 3J=\emptyset} |\langle f_1, \phi_I^1\rangle| \, |\langle f_4, \phi_I^3\rangle| \, \frac{|\langle h\chi_J, \widetilde{\phi}_I^2\rangle|}{2^{\ell_0/2}|I|^{1/2}} \\
&\leq \sum_{L:L\cap 3J=\emptyset} \sum_{I:\widetilde{I}=L} |\langle f_1, \phi_I^1\rangle| \, |\langle f_4, \phi_I^3\rangle| \, \frac{|\langle h\chi_J, \widetilde{\chi}_L^{N'}\rangle|}{|L|} \\
&\lesssim \sum_{L:L\cap 3J=\emptyset} \left(\frac{\mathrm{dist}(L, J)}{|L|}\right)^{-N'} \sum_{I:\widetilde{I}=L} |\langle f_1, \phi_I^1\rangle| \, |\langle f_4, \phi_I^3\rangle| \\
&\leq \sum_{L:L\cap 3J=\emptyset} \left(\frac{\mathrm{dist}(L, J)}{|L|}\right)^{-N'} \\
&\quad \times \int_{\mathbb{R}} \left(\sum_{I:\widetilde{I}=L} \frac{|\langle f_1, \phi_I^1\rangle|^2}{|I|} \chi_I \right)^{1/2} \left(\sum_{I:\widetilde{I}=L} \frac{|\langle f_4, \phi_I^3\rangle|^2}{|I|} \chi_I \right)^{1/2} dx \\
&\lesssim \sum_{L:L\cap 3J=\emptyset} \left(\frac{\mathrm{dist}(L, J)}{|L|}\right)^{-N'} \|f_1\widetilde{\chi}_L^N\|_p \, \|f_4\widetilde{\chi}_L^N\|_q,
\end{aligned}
$$

by the usual argument based on Lemma 2.13. This can be further estimated by

$$
\sum_{L:L\cap 3J=\emptyset} \left(\frac{\mathrm{dist}(L, J)}{|L|}\right)^{-N'} (\mathrm{dist}(L, J))^{2N} \|f_1\widetilde{\chi}_J^N\|_p \, \|f_4\widetilde{\chi}_J^N\|_q.
$$

Finally, as before, one observes that the sum

$$
\sum_{L:L\cap 3J=\emptyset} \left(\frac{\mathrm{dist}(L, J)}{|L|}\right)^{-N'} (\mathrm{dist}(L, J))^{2N}
$$

is uniformly bounded if N' is much larger than $2N$, and this completes the estimate.

Case II_2: $j = 3$. This case follows the same ideas and is left to the reader. This ends the proof of Lemma 8.3.

Exercise 8.8 Complete the details of case II_2 above.

8.12. Upper bounds for flag energies

In this final section we will prove Lemma 8.4. As before we distinguish several cases.

Case I: estimates for $\text{energy}^{(3)}_{\mathcal{J}_1}((a^3_J)_J)$

Case I_1: $j \neq 3$. Consider an integer $n \in \mathbb{Z}$ and a set $\mathbb{D}$ such that the supremum in the definition of the energy (see Definition 2.9) is attained. Then we have

$$\text{energy}^{(3)}_{\mathcal{J}_1}\left((a^3_J)_J\right) \simeq 2^n \left(\sum_{J\in\mathbb{D}} |J|\right) = 2^n \left\|\sum_{J\in\mathbb{D}} \chi_J\right\|_1 = \left\|\sum_{J\in\mathbb{D}} 2^n \chi_J\right\|_{1,\infty}$$

$$\lesssim \left\|\sum_{J\in\mathbb{D}} \frac{1}{|J|}\left\|\left(\sum_{J'\subseteq J} \frac{|a^3_{J'}|^2}{|J'|}\chi_{J'}\right)^{1/2}\right\|_{1,\infty} \chi_J\right\|_{1,\infty}.$$

Using the notation in the proof of Lemma 8.3 we can observe, exactly as there, that

$$a^3_{J'} = \langle B(f_1, f_4), \phi^3_{J'}\rangle \tag{8.54}$$

and this fact, together with some other by now standard arguments, allows us to estimate our expression by

$$\left\|\sum_{J\in\mathbb{D}}\left(\frac{1}{|J|}\int_{\mathbb{R}} |B(f_1, f_4)|\widetilde{\chi}^N_J dx\right)\chi_J\right\|_{1,\infty} \lesssim \|M(B(f_1, f_4))\|_{1,\infty}$$

$$\lesssim \|B(f_1, f_4)\|_1$$

$$\lesssim \sum_{I\in\widetilde{\mathcal{I}}} \frac{1}{|I|^{1/2}} |\langle f_1, \phi^1_I\rangle|\, |\langle f_4, \phi^3_I\rangle|\, |\langle h, \phi^2_I\rangle|$$

for a certain $h \in L^\infty$ with $\|h\|_\infty \leq 1$, where M is the usual Hardy–Littlewood maximal operator. Since $\phi^2_I/|I|^{1/2}$ is L^1-normalized, we can further estimate our expression as

$$\sum_{I\in\widetilde{\mathcal{I}}} |\langle f_1, \phi^1_I\rangle|\, |\langle f_4, \phi^3_I\rangle|. \tag{8.55}$$

Given that both families $(\phi_I^1)_I$ and $(\phi_I^2)_I$ are lacunary, one can estimate (8.55) as in Chapter 2 by

$$\left(\text{size}_{\widetilde{\mathcal{I}}}^{(1)}(((f_1, \phi_I^1))_I)\right)^{1-\theta_1}\left(\text{size}_{\widetilde{\mathcal{I}}}^{(2)}(((f_4, \phi_I^3))_I)\right)^{1-\theta_2}$$
$$\times\left(\text{energy}_{\widetilde{\mathcal{I}}}^{(1)}(((f_1, \phi_I^1))_I)\right)^{\theta_1}\left(\text{energy}_{\widetilde{\mathcal{I}}}^{(2)}(((f_4, \phi_I^3))_I)\right)^{\theta_2},$$

for any $0 \le \theta_1, \theta_2 < 1$, where these sizes and energies are defined naturally as before. Using now the well-known upper bounds for sizes and energies, we can finally estimate the previous expression as

$$\left(\sup_I \frac{1}{|I|}\int_{E_1} \widetilde{\chi}_I^N\,dx\right)^{1-\theta_1}\left(\sup_I \frac{1}{|I|}\int_{E_4} \widetilde{\chi}_I^N\,dx\right)^{1-\theta_2}|E_1|^{\theta_1}|E_4|^{\theta_2},$$

and this ends the proof of the first subcase.

The second subcase, where $j = 3$, is in fact easier (as before) and is left to the reader. Finally, to obtain similar estimates for $\text{energy}_{\mathcal{J}}^{(3)}\left(\left(a_{J,\ell_0}^3\right)_J\right)$ one argues in a similar manner. The only difference is that instead of (8.54) one uses the same argument as in the second case of Lemma 8.3 together with the fact that, for any interval I, the function $\widetilde{\phi}_I^2/\left(2^{\ell_0/2}|I|^{1/2}\right)$ is L^1-normalized, and this completes our proof.

Exercise 8.9 Fill in the details in the above proof.

Notes

Flag paraproducts were introduced in Muscalu [84]. The material of this chapter follows closely this paper and also Muscalu [85]. For more about the connection between the water-wave equation, the two-dimensional Schrödinger equation, and flag paraproducts see Germain, Masmoudi, and Shatah [45, 46]. A good reference on U statistics is Chow and Teicher [14].

Problems

Problem 8.1 Consider the symbol $a(\xi_1, \xi_2)b(\xi_1, \xi_2, \xi_3)$ with $a \in \mathcal{M}(\mathbb{R}^2)$ and $b \in \mathcal{M}(\mathbb{R}^3)$ and denote by T_{ab} the trilinear multiplier defined by it. Prove that T_{ab} maps $L^{p_1} \times L^{p_2} \times L^{p_3}$ into L^p for every $1 < p_1, p_2, p_3 < \infty$ with $1/p_1 + 1/p_2 + 1/p_3 = 1/p$ and $0 < p < \infty$. Can you prove some L^∞ estimates as well?

Hint: Show that the *model operators* associated with T_{ab} are the same as those considered earlier.

Problem 8.2 Prove the general Leibnitz rule in (8.14).

Problem 8.3 Consider the symbol $a(\xi_1, \xi_2)b(\xi_1 + \xi_2, \xi_3)c(\xi_1, \xi_2, \xi_3)$ with $a, b \in \mathcal{M}(\mathbb{R}^2)$ and $c \in \mathcal{M}(\mathbb{R}^3)$. Prove that the trilinear operator associated with it can be factored out naturally, as a sum of paraproducts, flag paraproducts, and adjoints of flag paraproducts.

(Note: These are the one-dimensional variants of the operators that appear in the water-wave problem mentioned earlier.)

Problem 8.4 Consider the following k-linear operator for $k \geq 3$:

$$(f_1, \ldots, f_k) \mapsto \sum_{1 \leq i_1 < \ldots < i_k \leq n} (f_1 * \psi_{i_1}) \cdots (f_k * \psi_{i_k}).$$

Prove that it maps $L^{p_1} \times \cdots \times L^{p_k}$ into L^p for every $1 < p_1, \ldots, p_k < \infty$ with $1/p_1 + \cdots + 1/p_k = 1/p$ and $0 < p < \infty$. Can you decide which L^∞ estimates are available as well?

9

Appendix: Multilinear interpolation

Our main task here is to describe the details of the *multilinear interpolation* used many times throughout this volume. Let n be an integer larger than 3 and Λ a generic n-linear form. The reader should regard Λ as one of the many multilinear forms encountered in the previous chapters. Let T be the $(n-1)$-linear operator whose form is Λ. As we have seen, one usually wants to prove *Hölder estimates* for T of the type

$$T : L^{p_1} \times \ldots \times L^{p_{n-1}} \to L^{p_n} \tag{9.1}$$

where the indices $(p_j)_{j=1}^n$ satisfy $1 < p_1, \ldots, p_n \le \infty$, $1/p_1 + \cdots + 1/p_{n-1} = 1/p_n$, and $0 < p_n < \infty$. It will be more convenient to write $\beta_j = 1/p_j$ for $j = 1, \ldots, n-1$ and $\beta_n = 1/p_n'$. Then, we observe that

$$\sum_{j=1}^{n} \beta_j = 1. \tag{9.2}$$

Notice that while $0 \le \beta_j < 1$ for $j = 1, \ldots, n-1$, β_n can be negative if the corresponding index p_n is strictly smaller than 1. The following definitions will be useful.

Definition 9.1 An n-tuple $\beta = (\beta_1, \ldots, \beta_n)$ is said to be *admissible* if and only if $\beta_j < 1$ for every $1 \le j \le n$, (9.2) holds, and there is at most one index j for which $\beta_j < 0$. An index j is called *good* if $\beta_j \ge 0$ and *bad* if $\beta_j < 0$. A *good tuple* is an admissible tuple that contains only good indices; a *bad tuple* is an admissible tuple that contains precisely one bad index.

Clearly, good tuples correspond to Banach space estimates while bad tuples correspond to quasi-Banach space estimates.

Definition 9.2 Let β be an admissible tuple. If β is good we say that the n-linear form Λ is of *restricted weak type* β if and only if, for every measurable set $E_1, \dots, E_n$ of positive and finite measure and every measurable function $|f_1| \leq \chi_{E_1}, \dots, |f_n| \leq \chi_{E_n}$, one has

$$|\Lambda(f_1, \dots, f_n)| \lesssim |E_1|^{\beta_1} \cdots |E_n|^{\beta_n}. \tag{9.3}$$

If β is bad, with bad index j_0, we say similarly that the n-linear form Λ is of *restricted weak type* β if and only if, for every measurable set $E_1, \dots, E_n$ of positive and finite measure, there exists a subset $E'_{j_0} \subseteq E_{j_0}$ with $|E'_{j_0}| > |E_{j_0}|/2$ such that

$$|\Lambda(f_1, \dots, f_n)| \lesssim |E_1|^{\beta_1} \cdots |E_n|^{\beta_n} \tag{9.4}$$

for every $|f_j| \leq \chi_{E_j}$, for $j \neq j_0$ and $|f_{j_0}| \leq \chi_{E'_{j_0}}$. If it happens that one can choose the subset $E'_{j_0} \subseteq E_{j_0}$ in a way that depends only on the measurable sets $E_1, \dots, E_n$ and not on β, we say that Λ is of *uniformly restricted weak type.*

Definition 9.3 Let α, β be two fixed n-tuples. If both are good we say that β is a *majorant* of α if and only if there exists j_0 such that

$$\alpha_j < \beta_j \quad \text{for all } j \neq j_0. \tag{9.5}$$

If α is good and β is bad with bad index j_0, we say similarly that β is a majorant of α if and only if (9.5) holds true. Finally, if both α and β are bad with bad index j_0, we say as before that β is a majorant of α if and only if (9.5) is available.

Now that we have set up the terminology, we can prove the interpolation results. The first is quite simple but very helpful.

Lemma 9.4 *Let $\alpha^{(1)}, \dots, \alpha^{(n)}$ be good tuples. Assume that the n-linear form Λ is of restricted weak type $\alpha^{(j)}$ for every $j = 1, \dots, n$. Then Λ is also of restricted weak type α for every α of the form*

$$\alpha = \sum_{j=1}^{n} \theta_j \alpha^{(j)},$$

where $0 < \theta_j < 1$ are such that $\sum_{j=1}^{n} \theta_j = 1$.

Proof The proof is, as we said, straightforward. One just has to raise the inequality (9.3) corresponding to $\alpha^{(j)}$ to the power θ_j and then to multiply all these new inequalities together. □

The above result can be extended slightly to include bad tuples.

Lemma 9.5 *Fix an index* $1 \leq j_0 \leq n$. *Let* $\alpha^{(1)}, \ldots, \alpha^{(n)}$ *be such that each is either a good tuple or a bad tuple with bad index* j_0. *Assume that the* n*-linear form* Λ *is of uniformly restricted weak type* $\alpha^{(j)}$ *for every* $j = 1, \ldots, n$. *Then* Λ *is also of uniformly restricted weak type* α *for bad tuples* α, *with bad index* j_0, *that are of the form*

$$\alpha = \sum_{j=1}^{n} \theta_j \alpha^{(j)}$$

where $0 < \theta_j < 1$ *are such that* $\sum_{j=1}^{n} \theta_j = 1$.

Proof The proof is similar to that for Lemma 9.4. Using the fact that the subsets E'_{j_0} can be chosen independently of $\alpha^{(1)}, \ldots, \alpha^{(n)}$, one just has to raise the corresponding jth inequality (9.4) to the power θ_j and then again to multiply all these new inequalities together. □

It is also important to notice that the *boundedness constant* corresponding to α is smaller than $\max(C_1, \ldots, C_n)$, where C_j is the boundedness constant corresponding to the tuple α_j for $j = 1, \ldots, n$. In fact it is even smaller than the *convex product* $C_1^{\theta_1} \cdots C_n^{\theta_n}$. Finally, we will also prove the next result which is the trickiest of all.

Lemma 9.6 *Consider* $\alpha^{(1)}, \ldots, \alpha^{(n)}$ *with the property that, for every* $1 \leq j \leq n$, *the tuple* $\alpha^{(j)}$ *is either good or bad with bad index* j. *Let* α *be admissible and, as before, of the form*

$$\alpha = \sum_{j=1}^{n} \theta_j \alpha^{(j)}$$

where $0 < \theta_j < 1$ *are such that* $\sum_{j=1}^{n} \theta_j = 1$.

If α *happens to be good, there must exist at least one index* j_0 *such that* $\alpha_{j_0} > 0$. *Assume that for one of these indices we have that* $\alpha^{(j_0)}$ *is a majorant of* α. *If, however,* α *happens to be bad with bad index* j_0 *then we assume that* $\alpha^{(j_0)}$ *is a majorant of* α. *Suppose also that the* n*-linear form* Λ *is of restricted weak type* $\alpha^{(j)}$ *for every* $1 \leq j \leq n$. *Then* Λ *is of restricted weak type* α.

Proof If β is an n-tuple and $E_1, \ldots, E_n$ are measurable sets, we will write from now on $|E|^\beta := |E_1|^{\beta_1} \cdots |E_n|^{\beta_n}$. Assume also for simplicity that $j_0 = n$ since our argument will be independent of this choice. The majorant conditions may look somewhat unintuitive but the reader will see that they arise naturally in the proof. In practice they are not difficult to check.

We have two cases. Suppose first that α is a good tuple and denote by $A(\infty)$ the smallest constant such that the inequality

$$|\Lambda(f_1, \dots, f_n)| \leq A(\infty)|E|^\alpha \tag{9.6}$$

holds true for all measurable sets $E_1, \dots, E_n$ of finite measure and measurable functions $|f_j| \leq \chi_{E_j}$ for $1 \leq j \leq n$. Given the nature of the form Λ, we can assume that $A(\infty)$ is finite and so the goal is to prove that $A(\infty)$ is bounded by a constant depending only on the parameters in the hypothesis.

Recall that $\alpha_n > 0$. Using the fact that Λ is of restricted weak type $\alpha^{(n)}$, there exists $E_n' \subseteq E_n$ with $|E_n'| > (1/2)|E_n|$ such that we can write

$$\begin{aligned} |\Lambda(f_1, \dots, f_n)| &\leq |\Lambda(f_1, \dots, f_n\chi_{E_n'})| + |\Lambda(f_1, \dots, f_n\chi_{E_n \setminus E_n'})| \\ &\leq C_n|E|^{\alpha^{(n)}} + \frac{1}{2^{\alpha_n}} A(\infty)|E|^\alpha \\ &= C_n|E|^\alpha \left(\frac{|E|^{\alpha^{(n)}}}{|E|^\alpha} \right) + \frac{1}{2^{\alpha_n}} A(\infty)|E|^\alpha. \end{aligned}$$

Clearly, one would like to have $|E|^{\alpha^{(n)}}/|E|^\alpha$ bounded if possible, and this suggests that we should define $A(\lambda)$ to be the smallest constant such that the inequality

$$|\Lambda(f_1, \dots, f_n)| \leq A(\lambda)|E|^\alpha \tag{9.7}$$

holds true under the assumption that

$$\frac{|E|^{\alpha^{(n)}}}{|E|^\alpha} \leq \lambda. \tag{9.8}$$

The above estimates show that

$$A(\lambda) \leq C_n\lambda + \frac{1}{2^{\alpha_n}} A(\infty). \tag{9.9}$$

Observe that $A(\lambda)$ is increasing and that $A(\lambda)$ tends to $A(\infty)$ as $\lambda \to \infty$. To show that $A(\lambda)$ is uniformly bounded, it is natural first to compare it with $A(\lambda/2)$. Given (9.8), one either has

$$\frac{|E|^{\alpha^{(n)}}}{|E|^\alpha} \leq \frac{\lambda}{2}$$

or

$$\frac{\lambda}{2} \leq \frac{|E|^{\alpha^{(n)}}}{|E|^\alpha} \leq \lambda. \tag{9.10}$$

In particular, we have

$$A(\lambda) \leq \max\left(A\left(\frac{\lambda}{2}\right), B(\lambda) \right), \tag{9.11}$$

where $B(\lambda)$ is similarly defined to be the best constant under the assumption (9.10). It is not difficult to see, using convexity, that there must exist an index $j \neq n$ with the property that

$$|E|^{\alpha^{(j)}} \leq C\lambda^{-\theta_n/(1-\theta_n)}|E|^{\alpha} \tag{9.12}$$

for a certain constant C depending on θ_n. Then one can make the split

$$|\Lambda(f_1, \ldots, f_n)| \leq |\Lambda(f_1, \ldots, f_j\chi_{E'j}, \ldots, f_n)| \\ + |\Lambda(f_1, \ldots, f_j\chi_{E_j\setminus E'j}, \ldots, f_n)|$$

naturally and, after using the restricted weak type property for α_j, one concludes exactly as before that

$$B(\lambda) \leq CC_j\lambda^{-\theta_n/(1-\theta_n)} + A\left(\frac{\lambda}{2^{\alpha_j^{(n)}-\alpha_j}}\right).$$

Together with (9.11) this implies that

$$A(\lambda) \leq \max\left(A\left(\frac{\lambda}{2}\right), CC_j\lambda^{-\theta_n/(1-\theta_n)} + A\left(\frac{\lambda}{2^{\alpha_j^{(n)}-\alpha_j}}\right)\right).$$

Since $\alpha_j^{(n)} - \alpha_j$ is strictly positive (recall that $\alpha^{(n)}$ is a majorant of α) one can iterate the above inequality and obtain finally $A(\infty) \leq \widetilde{C} + A(1000)$, which proves the boundedness of $A(\infty)$.

The other case when α is a bad tuple is similar. First, one defines $A(\lambda)$ to be the smallest constant such that, for all measurable sets $E_1, \ldots, E_n$ satisfying (9.8), there exists $E'_n \subseteq E_n$ with $|E'_n| > (1/2)|E_n|$ and such that

$$|\Lambda(f_1, \ldots, f_n)| \leq A(\lambda)|E|^{\alpha}$$

for every $|f_j| \leq \chi_{E_j}$ with $j \neq n$ and $|f_n| \leq \chi_{E'_n}$. Then, by using the fact that Λ is of restricted weak type $\alpha^{(n)}$, one obtains this time the simpler inequality

$$A(\lambda) \leq C\lambda.$$

After that, the argument is identical to that used above. □

Exercise 9.1 Complete the details of the case when α is a bad triple.

As before, the proof shows that the boundedness constant corresponding to α is no greater than $C(\theta_1, \ldots, \theta_n)\max(C_1, \ldots, C_n)$.

These three lemmas, combined with various results of linear interpolation such as the Riesz–Thorin theorem or Marcinkiewicz's theorem, by fixing all but a single function, allow one to pass from *restricted weak-type estimates* to the *strong-type estimates* that one usually needs. We will exemplify the method with two case studies.

First case study

Let us first treat the case of the bilinear Hilbert transform. We have proved directly (in Theorem 6.5) that its form Λ is of *uniformly restricted weak type* β for β arbitrarily close to the 3-tuples $(1, 1/2, -1/2)$, $(1/2, 1, -1/2)$, $(-1/2, 1, 1/2)$, $(-1/2, 1/2, 1)$, $(1, -1/2, 1/2)$, and $(1/2, -1/2, 1)$. Recall also that if β_j is close to 1 for some $1 \leq j \leq 3$ then β_j is strictly smaller than 1, and if it is close to $1/2$ then it is strictly larger than $1/2$.

Consider now a tuple arbitrarily close to $(1, 0, 0)$ of the form $(1 - \epsilon_1, \epsilon_2, \epsilon_3)$ for some small numbers $\epsilon_1, \epsilon_2, \epsilon_3 > 0$. It is not dificult to see that one can apply Lemma 9.6 and conclude that Λ is of the restricted weak type $(1 - \epsilon_1, \epsilon_2, \epsilon_3)$. Similarly Λ is of restricted weak type for tuples arbitrarily close to $(0, 1, 0)$ and $(0, 0, 1)$. Then, however, we can interpolate between these good tuples (using Lemma 9.4) and obtain restricted weak-type estimates inside the convex hull of the three coordinate points $(1, 0, 0)$, $(0, 1, 0)$, and $(0, 0, 1)$. After that, we can interpolate one more time (now using Lemma 9.5) to conclude that we have restricted weak-type estimates for tuples inside the convex hull of all the six *extremal* points listed above. Finally, one can *freeze* all but one function and interpolate linearly to obtain the desired strong estimates (see Figure 9.1). This

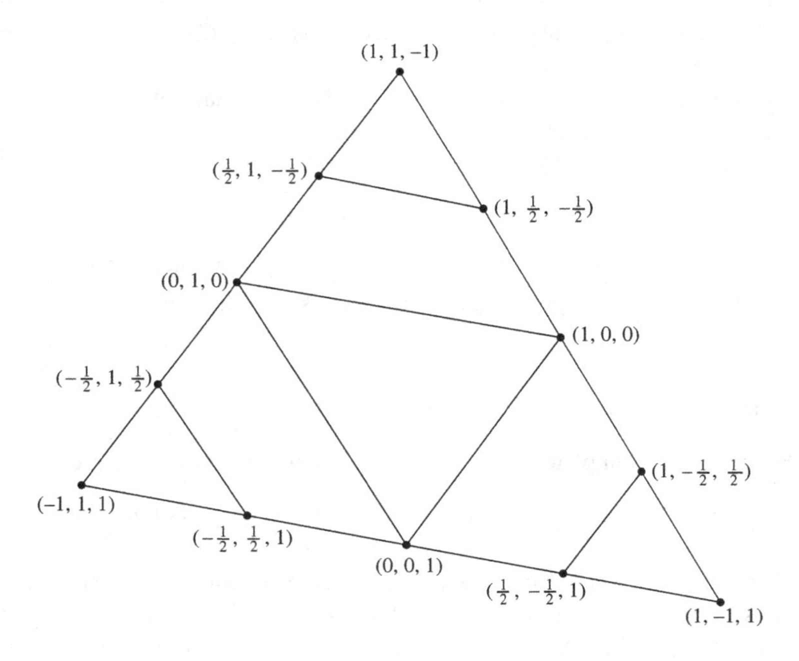

Figure 9.1. The interpolation triangle for the bilinear Hilbert transform.

is clearly possible as long as one is looking for estimates of the type $L^p \times L^q$ into L^r when $1 < p, q < \infty$. To obtain $L^p \times L^\infty$ into L^p one uses *duality* and proves instead that the *second adjoint* is bounded from $L^p \times L^{p'}$ into L^1, which is a case of the previous kind.

Second case study

We observe from the first case that when one space is L^∞ we have to be a little more careful. We will treat here the situation when two of the desired spaces are equal to L^∞. The same argument will work in general, when we consider an arbitrary number of L^∞ spaces. Suppose that Π is a discretized paraproduct and we want to show that Π maps $L^2 \times L^\infty \times L^\infty$ into L^2. Alternatively, we want to show that its 4-linear form Λ is of the strong type $(1/2, 0, 0, 1/2)$. Assume also that we have proved already that Λ is of the uniformly restricted weak type β, for β arbitrarily close to $(1, 1, 1, -2)$, $(-2, 1, 1, 1)$, $(1, -2, 1, 1)$, and $(1, 1, -2, 1)$ as is the case for paraproducts. See Figure 8.2, where the tetrahedron corresponding to these points is shown. Using the previous three lemmas as before, we first observe that Λ is of restricted weak type for *any* admissible tuple in the interior of the convex hull of the above four extremal points. In particular, Λ is of restricted weak type for good tuples of the form $(a, 0, 0, b)$ for a, b arbitrarily close to $1/2$. Since the form is 4-linear, recall that four functions are implicitly used here and satisfy $|f_j| \le \chi_{E_j}$ for some measurable sets of finite measure. We freeze f_2 and f_3 and apply linear interpolation to the linear operator defined by the corresponding bilinear form. We obtain as a consequence the *strong inequality*

$$|\Lambda(f_1, f_2, f_3, f_4)| \lesssim \|f_1\|_2 \, \|f_2\|_\infty \, \|f_3\|_\infty \, \|f_4\|_2 \tag{9.13}$$

for every $f_1, f_4 \in L^2$, every $|f_2| \le \chi_{E_2}$, and $|f_3| \le \chi_{E_3}$. To obtain the above inequality for *every* $f_2, f_3 \in L^\infty$, a simple limiting argument based on the particular formulae of discretized paraproducts is used.

Notes

The material here is from Muscalu, Tao, and Thiele [94]. The first paper on the interpolation of multilinear operators seems to be Strichartz [108]. Related results can be found in Janson [57] and Grafakos and Tao [53].

References

[1] Ablowitz, M. and Segur, H. 1981. *Solitons and the Inverse Scattering Transform*. SIAM, Philadelphia, PA.

[2] Auscher, P., Hofmann, S., Lacey, M., McIntosh, A., and Tchamitchian, Ph. 2002a. The solution of the Kato square root problem for second order elliptic operators in $\mathbb{R}^n$. *Ann. Math.* **156**, 633–654.

[3] Auscher, P., Hofmann, S., Muscalu, C., Tao, T., and Thiele, C. 2002b. Carleson measures, trees, extrapolation and *T(b)* theorems. *Publ. Mat.* **46**, 257–325.

[4] Bahouri, H., Chemin, J.-Y., and Danchin, R. 2001. *Fourier Analysis and Nonlinear Partial Differential Equations*. Grundlehren der Mathematischen Wissenschaften, vol. 343, Springer, Heidelberg.

[5] Bényi, A., Demeter, C., Nahmod, A., Thiele, C., Torres, R., and Villaroya, P. 2009. Modular invariant bilinear $T(1)$ theorem. *J. Anal. Math.* **109**, 279–352.

[6] Bernicot, F. 2008. Local estimates and global continuities in Lebesgue spaces for bilinear operators. *Anal. PDE* **1**, 1–27.

[7] Bernicot, F. 2011. L^p estimates for non-smooth bilinear Littlewood–Paley square functions on $\mathbb{R}$. *Math. Ann.* **351**, 1–49.

[8] Billard, P. 1966/1967. Sur la convergence presque partout des séries de Fourier–Walsh de fonctions de l'espace $L^2(0, 1)$. *Studia Math.* **28**, 363–388.

[9] Bony, J. M. 1981. Calcul symbolique et equations non lineaires. *Ann. Sci. Ecole Norm. Sup.* **14**, 209–246.

[10] Calderón, A. 1965. Commutators of singular integral operators. *Proc. Nat. Acad. Sci. USA* **53**, 1092–1099.

[11] Calderón, A. 1978. Commutators, singular integrals on Lipschitz curves and applications. In *Proc. Int. Congress of Math.* (Helsinki), pp. 85–96.

[12] Carleson, L. 1966. On convergence and growth of partial sums of Fourier series. *Acta Math.* **116**, 135–157.

[13] Chang, S.-Y. A. and Fefferman, R. 1985. Some recent developments in Fourier analysis and H^p-theory on product domains. *Bull. Amer. Math. Soc.* **12**, 1–43.

[14] Chow, Y. and Teicher, H. 1997. *Probability Theory. Independency, Interchangeability and Martingales*. Third edition. Springer Texts in Statistics, vol. 22, Springer, New York.

[15] Christ, M. 1990. *Lectures on Singular Integral Operators*. CBMS Regional Conference Series in Mathematics, vol. 77, published for the Conference Board of the Mathematical Sciences, Washington, DC, by the American Mathematical Society, Providence, RI.

[16] Christ, M. 2001. On certain elementary trilinear operators. *Math. Res. Lett.* **8**, 43–56.

[17] Christ, M. and Journé, J. 1987. Polynomial growth estimates for multilinear singular integrals. *Acta Math.* **159**, 51–80.
[18] Christ, M. and Kiselev, A. 2001a. Maximal functions associated to filtrations. *J. Funct. Anal.* **179**, 409–425.
[19] Christ, M. and Kiselev, A. 2001b. WKB asymptotic behaviour of almost all generalized eigenfunctions of one-dimensional Schrödinger operators. *J. Funct. Anal.* **179**, 426–447.
[20] Christ, M. and Weinstein, M. 1991. Dispersion of small amplitude solutions of the generalized KdV equation. *J. Funct. Anal.* **100**, 87–109.
[21] Christ, M., Li, X., Tao, T., and Thiele, C. 2005. On multilinear oscillatory integrals, nonsingular and singular. *Duke Math. J.* **130**, 321–351.
[22] Cohen, J. and Gosselin, J. 1986. A BMO estimate for multi-linear singular integrals. *Illinois J. Math.* **30**, 445–465.
[23] Coifman, R. and Grafakos, L. 1992. Hardy space estimates for multilinear operators I. *Revista Mat. Iberoamericana* **8**, 45–67.
[24] Coifman, R. and Meyer, Y. 1975. On commutators of singular integrals and bilinear singular integrals. *Trans. Amer. Math. Soc.* **212**, 315–331.
[25] Coifman, R. and Meyer, Y. 1978. Au delá des opérateurs pseudo-différentieles. *Astérisque* **57**.
[26] Coifman, R. and Meyer, Y. 1997. *Wavelets, Calderón Zygmund Operators and Multilinear Operators.* Translated from the 1990 and 1991 French originals by David Salinger. Cambridge Studies in Advanced Mathematics, vol. 48, Cambridge University Press, Cambridge.
[27] Coifman, R. and Rochberg, R. 1980. Another characterization of BMO. *Proc. Amer. Math. Soc.* **79**, 249–254.
[28] Coifman, R., Rochberg, R., and Weiss, G. 1976. Factorization theorems for Hardy spaces in several variables. *Ann. Math.* **103**, 611–635.
[29] Coifman, R., McIntosh, A., and Meyer, Y. 1982. L'integral de Cauchy définit un opérateur bourné sur le courbes Lipschitziennes. *Ann. Math.* **116**, 361–387.
[30] Coifman, R., Jones, P., and Semmes, S. 1989a. Two elementary proofs of the L^2-boundedness of the Cauchy integral on Lipschitz curves. *J. Amer. Math. Soc.* **2**, 553–564.
[31] Coifman, R., Lions, P., Meyer, Y., and Semmes, S. 1989b. Compacité par compensation et espaces de Hardy. *C.R. Acad. Sci. Paris* **309**, 945–949.
[32] Dahlberg, B. 1983. Real analysis and potential theory. In *Proc. Int. Congress Mathematics* (Warszawa), 953–959.
[33] David, G. and Journé, J.-L. 1984. A boundedness criterion for generalized Calderón–Zygmund operators. *Ann. Math.* (2)**120**, 371–397.
[34] Demeter, C. 2008. Divergence of some combinatorial averages and the unboundedness of the trilinear Hilbert transform. *Ergodic Theory Dynam. Systems* **28**, 1453–1464.
[35] Demeter, C. and Thiele, C. 2010. On the two-dimensional bilinear Hilbert transform. *Amer. J. Math.* **132**, 201–256.
[36] Demeter, C., Pramanik, M., and Thiele, C. 2010. Multilinear singular operators with fractional rank. *Pacific J. Math.* **246**, 293–324.
[37] Do, Y., Muscalu, C., and Thiele, C. 2012. Variational estimates on paraproducts. *Revista Math. Iberoamericana* **28**, 857–878.
[38] Egorov, Yu. V. 1975. Subelliptic operators. *Uspekhi Mat. Nauk.* **30**, 57–104.
[39] Fefferman, C. 1971. On the divergence of multiple Fourier series. *Bull. Amer. Math. Soc.* **77**, 191–195.
[40] Fefferman, C. 1973. Pointwise convergence of Fourier series. *Ann Math.* **98**, 551–571.
[41] Fefferman, C. 1983. The uncertainty principle. *Bull. Amer. Math. Soc.* **9**, 129–206.
[42] Fefferman, R. and Stein, E. 1982. Singular integrals on product spaces. *Adv. Math.* **45**, 117–143.

[43] Ferguson, S. and Lacey, M. 2002. A characterization of product BMO by commutators. *Acta Math.* **189**, 143–160.
[44] Frazier, M. and Jawert, B. 1990. A discrete transform and decompositions of distribution spaces. *J. Funct. Anal.* **190**, 34–170.
[45] Germain, P., Masmoudi, N., and Shatah, J. 2012. Global solutions for the gravity water wave equation in dimension 3. *Ann. Math.* **175**, 691–754.
[46] Germain, P., Masmoudi, N., and Shatah, J. 2010b. Global solutions for $2D$ quadratic Schrödinger equations. To appear in *J. Analyse Math.*
[47] Gilbert, J. and Nahmod, A. 2001. Bilinear operators with nonsmooth symbols I. *J. Fourier Anal. Appl.* **7**, 435–467.
[48] Grafakos, L. 1992. Hardy space estimates for multilinear operators II. *Revista Mat. Iberoamericana* **8**, 69–92.
[49] Grafakos, L. and Honzík, P. 2006. Maximal transference and summability of multilinear Fourier series. *J. Aust. Math. Soc.* **80**, 65–80.
[50] Grafakos, L. and Kalton, N. 2001. The Marcinkiewicz multiplier condition for bilinear operators. *Studia Math.* **146**, 115–156.
[51] Grafakos, L. and Li, X. 2004. Uniform estimates for the bilinear Hilbert transform I. *Ann. Math.* **159**, 889–933.
[52] Grafakos, L. and Li, X. 2006. The disc as a bilinear multiplier. *Amer. J. Math.* **128**, 91–119.
[53] Grafakos, L. and Tao, T. 2003. Multilinear interpolation between adjoint operators. *J. Funct. Anal.* **199**, 379–385.
[54] Grafakos, L. and Torres, R. 2002a. Discrete decompositions for bilinear operators and almost diagonal conditions. *Trans. Amer. Math. Soc.* **354**, 1153–1176.
[55] Grafakos, L. and Torres, R. 2002b. Multilinear Calderón Zygmund theory. *Adv. Math.* **165**, 124–164.
[56] Hunt, R. 1968. On the convergence of Fourier series. *Orthogonal Expansions and their Continuous Analogues. Proc. Conf. Edwardsville, IL, 1967.* Southern Illinois University Press, Carbondale, IL, pp. 235–255.
[57] Janson, S. 1988. On interpolation of multilinear operators. *Lecture Notes in Mathematics*, vol. 1302, Springer, pp. 290–302.
[58] Jerison, D. and Kenig, C. 1981. The Dirichlet problem in non-smooth domains. *Ann. Math.* **113**, 367–382.
[59] Journé, J. L. 1985. Calderón–Zygmund operators on product spaces. *Revista Mat. Iberoamericana* **1**, 55–91.
[60] Journé, J. L. 1986. A covering lemma for product spaces. *Ann. Inst. Fourier* **96**, 593–598.
[61] Journé, J. L. 1988. Two problems of Calderón–Zygmund theory on product-spaces. *Ann. Inst. Fourier* **38**, 111–132.
[62] Kato, T. and Ponce, G. 1988. Commutator estimates and the Euler and Navier–Stokes equations. *Comm. Pure Appl. Math.* **41**, 891–907.
[63] Kenig, C. 1994. *Harmonic Analysis Techniques for Second Order Elliptic Boundary Value Problems.* CBMS Conference Series in Mathematics, vol. 83, American Mathematical Society, Providence, RI.
[64] Kenig, C. 2004. On the local and global well-posedness theory for the KP-I equation. *Ann. Inst. H. Poincaré Anal. Non Linéaire* **21**, 827–838.
[65] Kenig, C. and Stein, E. M. 1999. Multilinear estimates and fractional integration. *Math. Res. Lett.* **6**, 1–15.
[66] Kenig, C., Ponce, G., and Vega, L. 1989. On the (generalized) KdV equations. *Duke Math. J.* **59**, 585–610.
[67] Kovać, V. 2010. Boundedness of the twisted paraproduct. arXiv:1011:6140, 21 pp.
[68] Lacey, M. 1998. On the bilinear Hilbert transform. In *Proc. ICM Berlin*, pp. 647–656.

[69] Lacey, M. 2000. The bilinear maximal functions map into L^p for $2/3 < p \leq 1$. *Ann. Math.* **155**, 35–57.

[70] Lacey, M. and Thiele, C. 1997. L^p estimates for the bilinear Hilbert transform. *Ann. Math.* **146**, 693–724.

[71] Lacey, M. and Thiele, C. 1999. On Calderón's conjecture. *Ann. Math.* **150**, 475–496.

[72] Lacey, M. and Thiele, C. 2000. A proof of boundedness of the Carleson operator. *Math. Res. Lett.* **7**, 361–370.

[73] Lacey, M., Petermichl, S., Pipher, J., and Wick, B. 2009. Multiparameter Riesz commutators. *Amer. J. Math.* **131**, 731–769.

[74] Li, X. 2006. Uniform estimates for the bilinear Hilbert transform II. *Revista Mat. Iberoamericana* **22**, 1069–1126.

[75] Li, X. and Muscalu, C. 2007. Generalizations of the Carleson–Hunt theorem I. The classical singularity case. *Amer. J. Math.* **129**, 983–1019.

[76] Lie, V. 2009. The (weak) L^2 boundedness of the quadratic Carleson operator. *Geom. Funct. Anal.* **19**, 457–497.

[77] Lie, V. 2011. The polynomial Carleson operator. arXiv:1105.4504, 37 pp.

[78] Linares, F. and Ponce, G. 2009. *Introduction to Nonlinear Dispersive Equations*. Springer, New York.

[79] Mitrea, M. 1994. Clifford wavelets, singular integrals and Hardy spaces. *Lecture Notes in Mathematics,* vol. 1575, Springer, New York.

[80] Murai, T. 1984. *A Real Variable Method for the Cauchy Transform and Analytic Capacity*. Lecture Notes in Mathematics, vol. 1307, Springer, New York.

[81] Muscalu, C. 2000. Ph. D. thesis. Brown University.

[82] Muscalu, C. 2003a. Unpublished notes I. IAS Princeton.

[83] Muscalu, C. 2003b. Unpublished notes II. IAS Princeton.

[84] Muscalu, C. 2007. Paraproducts with flag singularities I. A case study. *Revista Mat. Iberoamericana* **23**, 705–742.

[85] Muscalu, C. 2010. Flag paraproducts. *Contemp. Math.* **505**, 131–151.

[86] Muscalu, C. 2012a. Calderón commutators and the Cauchy integral on Lipschitz curves revisited I. First commutator and generalizations. arXiv:1201.3845, 23 pp.

[87] Muscalu, C. 2012b. Calderón commutators and the Cauchy integral on Lipschitz curves revisited II. Cauchy integral and generalizations. arXiv:1201.3850, 29 pp.

[88] Muscalu, C. 2012c. Calderón commutators and the Cauchy integral on Lipschitz curves revisited III. Polydisc extensions. arXiv:1201.3855, 25 pp.

[89] Muscalu, C., Tao, T., and Thiele, C. 2002a. A discrete model for the Bi-Carleson operator. *Geom. Funct. Anal.* **12**, 1324–1364.

[90] Muscalu, C., Tao, T., and Thiele, C. 2002b. Multilinear multipliers given by singular symbols. *J. Amer. Math. Soc.* **15**, 469–496.

[91] Muscalu, C., Tao, T., and Thiele, C. 2003a. A Carleson theorem for a Cantor group model of the scattering transform. *Nonlinearity* **16**, 219–246.

[92] Muscalu, C., Tao, T., and Thiele, C. 2003b. A counterexample to a multilinear endpoint question of Christ and Kiselev. *Math. Res. Lett.* **10**, 237–246.

[93] Muscalu, C., Pipher, J., Tao, T., and Thiele, C. 2004a. Bi-parameter paraproducts. *Acta Math.* **193**, 269–296.

[94] Muscalu, C., Tao, T., and Thiele, C. 2004b. L^p estimates for the biest II. The Fourier case. *Math. Ann.* **329**, 427–461.

[95] Muscalu, C., Tao, T., and Thiele, C. 2004c. The bi-Carleson operator. *Geom. Funct. Anal.* **16**, 230–277.

[96] Muscalu, C., Pipher, J., Tao, T., and Thiele, C. 2006. Multi-parameter paraproducts. *Revista Mat. Iberoamericana* **22**, 963–976.

[97] Muscalu, C., Tao, T., and Thiele, C. 2007. Multi-linear multipliers associated to simplexes of arbitrary length. arXiv:0712.2420, 52 pp.

[98] Oberlin, R., Seeger, A., Tao, T., Thiele, C., and Wright, J. 2009. A variation norm Carleson operator. arXiv:0910.1555, 41 pp.

[99] Palsson, E. 2011. Ph. D. thesis. Cornell University.

[100] Pipher, J. 1986. Journé's covering lemma and its extension to higher dimensions. *Duke Math. J.* **53**, 683–690.

[101] Pipher, J. 2001. Personal communication. Mount Holyoke.

[102] Ponce, G. and Vega, L. 1988. Nonlinear small data for the gKdV equation. *J. Funct. Anal.* **90**, 445–457.

[103] Rochberg, R. and Weiss, G. 1983. Derivatives of analytic families of Banach spaces. *Ann. Math.* **118**, 315–347.

[104] Sjölin, P. 1971. Convergence almost everywhere of certain singular integrals and multiple Fourier series. *Ark. Mat.* **9**, 65–90.

[105] Stein, E. 1993. *Harmonic Analysis: Real-Variable Methods, Orthogonality and Oscillatory Integrals*. Princeton University Press, Princeton, NJ.

[106] Stein, E. and Wainger, S. 2001. Oscillatory integrals related to Carleson theorem. *Math. Res. Lett.* **8**, 789–800.

[107] Strauss, W. 1974. Dispersion of low energy waves for two conservative equations. *Arch. Rational Mech. Anal.* **55**, 110–133.

[108] Strichartz, R. 1969. A multilinear version of Marcinkiewicz interpolation theorem. *Proc. Amer. Math. Soc.* **21**, 441–444.

[109] Tao, T. 2002. Lectures in harmonic analysis. Notes from UCLA.

[110] Tao, T. and Thiele, C. 2003. Nonlinear Fourier analysis. *Lectures at Park City*.

[111] Taylor, M. 1998. *Tools for PDEs. Pseudodifferential Operators, Paradifferential Operators and Layer Potentials*. Mathematical Surveys Monographs, vol. 10, American Mathematical Society, Providence, RI.

[112] Thiele, C. 1995. Ph. D. thesis. Yale University.

[113] Thiele, C. 2000. The quartile operator and pointwise convergence of Walsh series. *Trans. Amer. Math. Soc.* **351**, 5745–5766.

[114] Thiele, C. 2002a. A uniform estimate. *Ann. Math.* **156**, 519–563.

[115] Thiele, C. 2002b. Singular integrals meet modulation invariance. In *Proc. ICM* (Beijing), pp. 721–732.

[116] Thiele, C. 2006. *Wave Packet Analysis*. CBMS Conference Series in Mathematics, vol. 105, American Mathematical Society, Providence, RI.

[117] Verchota, G. 1984. Layer potentials and regularity for the Dirichlet problem for Laplace's equation in Lipschitz domains. *J. Funct. Anal.* **59**, 572–611.

[118] Verdera, J. 2001. L^2 boundedness of the Cauchy integral and Menger curvature. *Contemp. Math.* **277**, 139–158.

[119] Workman, J. 2008. Ph. D. thesis. Cornell University.

[120] Wu, S. 2009. Almost global wellposedness of the 2D full water wave problem. *Invent. Math.* **177**, 45–135.

[121] Wu, S. 2011. Global wellposedness of the 3D full water wave problem. *Invent. Math.* **184**, 125–220.

Index